ADVANCES IN ELECTRONICS AND ELECTRON PHYSICS

VOLUME 58

Contributors to This Volume

R. L. Champion
J. N. Churchill
T. W. Collins
D. K. Ferry
F. E. Holmstrom
U. Kaufmann
Heiner Ryssel
J. Schneider
John E. Walsh

Advances in Electronics and Electron Physics

EDITED BY
CLAIRE MARTON
Smithsonian Institution
Washington, D. C.

VOLUME 58

1982

ACADEMIC PRESS
A Subsidiary of Harcourt Brace Jovanovich, Publishers
New York   London
Paris   San Diego   San Francisco   São Paulo   Sydney   Tokyo   Toronto

COPYRIGHT © 1982, BY ACADEMIC PRESS, INC.
ALL RIGHTS RESERVED.
NO PART OF THIS PUBLICATION MAY BE REPRODUCED OR
TRANSMITTED IN ANY FORM OR BY ANY MEANS, ELECTRONIC
OR MECHANICAL, INCLUDING PHOTOCOPY, RECORDING, OR ANY
INFORMATION STORAGE AND RETRIEVAL SYSTEM, WITHOUT
PERMISSION IN WRITING FROM THE PUBLISHER.

ACADEMIC PRESS, INC.
111 Fifth Avenue, New York, New York 10003

United Kingdom Edition published by
ACADEMIC PRESS, INC. (LONDON) LTD.
24/28 Oval Road, London NW1 7DX

LIBRARY OF CONGRESS CATALOG CARD NUMBER: 49–7504

ISBN 0–12–014658–4

PRINTED IN THE UNITED STATES OF AMERICA

82 83 84 85   9 8 7 6 5 4 3 2 1

CONTENTS

CONTRIBUTORS TO VOLUME 58 vii
PUBLISHER'S FOREWORD ix
FOREWORD xi

Modeling of Irradiation-Induced Changes in the Electrical Properties of Metal–Oxide–Semiconductor Structures
J. N. Churchill, F. E. Holmstrom, and T. W. Collins

I. Brief History and Introduction 1
II. Modeling Considerations 12
III. General Relationships between V_{fb} and $V_{gr}(D)$ 22
IV. Simple Descriptive Models 41
V. Complete Computer Simulation 53
VI. Discussion 72
    Nomenclature 75
    References 76

Point Defects in GaP, GaAs, and InP
U. Kaufmann and J. Schneider

I. Introduction 81
II. Experimental Techniques for Point Defect Assessment 84
III. Donors and Acceptors 86
IV. Intrinsic Defects 96
V. 3d Transition Metals 109
    References 135

Collisional Detachment of Negative Ions
R. L. Champion

I. Introduction 143
II. Nomenclature and Experimental Techniques 146
III. Atomic Reactants 148
IV. Molecular Reactants 177
V. Summary 186
    References 187

Ion Implantation for Very Large Scale Integration
Heiner Ryssel

1. Introduction	191
2. Range Distributions of Implanted Ions	193
3. Annealing of Implanted Layers	210
4. Nondoping and Other New Applications of Implantation	231
5. Application of Implantation to Devices	251
6. Conclusions	265
References	266

Stimulated Čerenkov Radiation
John E. Walsh

I. Introduction	271
II. Theory	275
III. Experiment	301
IV. Conclusion	309
References	310

Materials Considerations for Advances in Submicron Very Large Scale Integration
D. K. Ferry

I. Introduction	312
II. Submicron MOSFETs	314
III. Submicron MESFETs	332
IV. Switching of High-Speed Logic	340
V. Band Structure Considerations	361
VI. Comparisons and Limitations of Logic for Ultra-Large-Scale Integration	375
References	387

Author Index	391
Subject Index	405

CONTRIBUTORS TO VOLUME 58

Numbers in parentheses indicate the pages on which the authors' contributions begin.

R. L. CHAMPION, Department of Physics, The College of William and Mary, Williamsburg, Virginia 23185 (143)

J. N. CHURCHILL,* General Products Division, International Business Machines Corporation, San Jose, California 95193 (1)

T. W. COLLINS,† General Products Division, International Business Machines Corporation, San Jose, California 95193 (1)

D. K. FERRY, Department of Electrical Engineering, Colorado State University, Fort Collins, Colorado 80523 (311)

F. E. HOLMSTROM,‡ General Products Division, International Business Machines Corporation, San Jose, California 95193 (1)

U. KAUFMANN, Fraunhofer-Institut für Angewandte Festkörperphysik, D-7800 Freiburg, West Germany (81)

HEINER RYSSEL, Fraunhofer-Institut für Festkörpertechnologie, 8000 Munich 60, West Germany (191)

J. SCHNEIDER, Fraunhofer-Institut für Angewandte Festkörperphysik, D-7800 Freiburg, West Germany (81)

JOHN E. WALSH, Department of Physics and Astronomy, Dartmouth College, Hanover, New Hampshire 03755 (271)

---

* Present address: Department of Electrical and Computer Engineering, University of California, Davis, California 95616.

† Present address: Tandem Computers, Cupertino, California 95014.

‡ Present address: Department of Physics, San Jose State University, San Jose, California 95192.

PUBLISHER'S FOREWORD

It is with sadness that we inform readers of *Advances in Electronics and Electron Physics* of the passing of Dr. Claire Marton. Both Claire and her late husband, Bill, were associated with Academic Press almost from its founding. Their passing is both a personal and professional loss to us.

Future volumes will be edited by Dr. Peter Hawkes, Laboratoire d'Optique Electronique du CNRS, 29 rue Jeanne-Marvig, 31055 Toulouse-Cedex, France. He will process those articles listed in the Foreword and develop future volumes. We are pleased to have a person of his high caliber assume this position.

FOREWORD

Materials considerations dominate this volume. Two of the articles deal with semiconductor topics related to device needs, one with materials problems related to very large scale integration (VLSI) and the other with processing by ion beams for VLSI. The trend of modern electronic devices is clearly documented in these articles. The remaining contribution on collisional detachment of negative ions allows us to keep at least part of one foot in classical electron physics.

As is our custom, we now list articles scheduled for publication in future volumes of *Advances in Electronics and Electron Physics*:

Critical Reviews:

Atomic Frequency Standards	C. Audouin
Electron Scattering and Nuclear Structure	G. A. Peterson
Large Molecules in Space	M. and G. Winnewisser
The Impact of Integrated Electronics in Medicine	J. D. Meindl
Electron Storage Rings	D. Trines
Radiation Damage in Semiconductors	N. D. Wilsey and J. W. Corbett
Visualization of Single Heavy Atoms with the Electron Microscope	J. S. Wall
Light Valve Technology	J. Grinberg
Electrical Structure of the Middle Atmosphere	L. C. Hale
Microwave Superconducting Electronics	R. Adde
Diagnosis and Therapy Using Microwaves	M. Gautherie and A. Priou
Computer Microscopy Image Analysis of Biological Tissues Seen in the Light Microscope	E. M. Glaser
Low-Energy Atomic Beam Spectroscopy	E. M. Hörl and E. Semerad
History of Photoemission	W. E. Spicer
Power Switching Transistors	P. L. Hower
Radiation Technology	L. S. Birks
Diffraction of Neutral Atoms and Molecules from Crystalline Surfaces	G. Boato and P. Cantini
Auger Spectroscopy	M. Cailler, J. P. Hanachaud, and D. Roptin
High Field Effects in Semiconductor Devices	K. Hess
Digital Image Processing and Analysis	B. R. Hunt
Infrared Detector Arrays	D. Long and W. Scott
Energy Levels in Gallium Arsenide	A. G. Milnes
Polarized Electrons in Solid-State Physics	H. C. Siegmann, M. Erbudak, M. Landolt, and F. Meier
The Technical Development of the Shortwave Radio	E. Sivowitch
Chemical Trends of Deep Traps in Semiconductors	P. Vogl
Potential Calculation in Hall Plates	G. DeMey

Gamma-Ray Internal Conversion	O. Dragoun
CW Beam Annealing Process and Application for Superconducting Alloy Fabrication	J. F. Gibbons
Polarized Ion Sources	H. F. Glavish
Ultrasensitive Detection	K. H. Purser
The Interactions of Measurement Principles, Interfaces, and Microcomputers in Intelligent Instruments	W. G. Wolber
Fine-Line Pattern Definition and Etching for VLSI	Roy A. Colclaser
Recent Trends in Photomultipliers for Nuclear Physics	J. P. Boutot, J. Nussli, and D. Vallat
Waveguide and Coaxial Probes for Nondestructive Testing of Materials	F. E. Gardiol
Holography in Electron Microscopy	K. J. Hanssen
The Measurement of Core Electron Energy Levels	R. N. Lee and C. Anderson
Millimeter Radar	Robert D. Hayes
Recent Advances in the Theory of Surface Electronic Structure	Henry Krakauer
Rydberg States	R. F. Stebbings
Long-Life High-Current-Density Cathodes	Robert T. Longo
Dynamic Radiation Model for Microstrip Structures	F. E. Gardiol
Microwaves in Semiconductor Electronics	J. L. Allen
Applications of Quadrupole Mass Spectrometers	I. Berecz, S. Bohatka, and G. Langer
Advances in Materials for Thick-Film Hybrid Microcircuits	J. Sergent
Guided-Wave Circuit Technology	M. K. Barnoski
Fast-Wave Tube Devices	J. M. Baird
Spin Effects in Electron–Atom Collision Processes	H. Keinpoppen
Recent Advances in and Basic Studies of Photoemitters	H. Timan
Thermal and Electrothermal Instabilities in Semiconductors	M. P. Shaw and Y. Yildirim
High-Resolution Spectroscopy of Interstellar Molecules	G. Winnewisser
Solid State Imaging Devices	E. H. Snow
Structure of Intermetallic and Interstitial Compounds	A. C. Switendick
Smart Sensors	W. G. Wolber

Supplementary Volumes:

Microwave Field-Effect Transistors	J. Frey
Magnetic Reconnection	P. J. Baum and A. Bratenahl

Our sincere thanks to all of the authors for such splendid and valuable reviews.

C. MARTON

ADVANCES IN ELECTRONICS AND ELECTRON PHYSICS

VOLUME 58

Modeling of Irradiation-Induced Changes in the Electrical Properties of Metal–Oxide–Semiconductor Structures

J. N. CHURCHILL,* F. E. HOLMSTROM,†
AND T. W. COLLINS‡

General Products Division
International Business Machines Corporation
San Jose, California

I. Brief History and Introduction	1
II. Modeling Considerations	12
A. Low-Dose Range	15
B. High-Dose Range	15
C. Medium-Dose Range	15
D. Other Factors	16
III. General Relationships between V_{fb} and $V_{gr}(D)$	22
A. Type 1 Irradiation	25
B. Type 2 Irradiation	26
C. Applications	28
IV. Simple Descriptive Models	41
A. Captured during Transit (CDT) Models	42
B. Dynamic Equilibrium (DE) Models	45
V. Complete Computer Simulation	53
A. Describing Equations	54
B. Boundary Conditions	55
C. Comments	57
D. Results of the Computer Simulation	58
E. Discussion of the Simulation	67
VI. Discussion	72
Nomenclature	75
References	76

I. Brief History and Introduction

Qualitative attempts to describe the nature of matter in the solid state can be traced back in time to the beginning of written records (*1*). Although many useful techniques for modifying solids had been empirically discovered, little progress was made at characterizing the parameters of different solids until the introduction of the atomic theory. With the emergence of an atomic

---

* Present address: Department of Electrical and Computer Engineering, University of California, Davis, California 95616.

† Present address: Department of Physics, San Jose State University, San Jose, California 95192.

‡ Present address: Tandem Computers, Cupertino, California 95014.

view of matter, it was possible to model solid structures and thereby account for some physical properties.

With the discovery of ionizing radiations at the turn of this century, it was found that radiation can not only pass through various solids, but can also trigger physical processes and modify bonds during exposure. Acceptance of the Rutherford model of the atom in 1911, followed two years later by the Bohr structure of the atom, paved the way for more correct views of the interaction of energetic radiation with solids. With a classical, mechanistic approach to solids it became possible to grasp the significance of lattice sites where physical displacements can occur with the absorption of high-energy radiation; however, this classical picture of solids did not account for electrical conduction.

It was not until F. Bloch applied the concepts of Schrödinger waves to solids in 1928 that the scientific community was able to extend the classical atomic view of solids to include the quantized band structure to account for electrical properties (2). In particular, Bloch's contribution described the mechanisms for electrical conduction in a metallic lattice and found that electrons can readily flow through it if the lattice is perfect. Thermal motion of the impurities causes the conductivity to be finite, he noted. This view was an improvement over an approach taken by Sommerfeld (3) where he assumed that valence electrons comprise an electron gas that obeys Fermi–Dirac statistics. Such an assumption was able to explain the specific heat of metals, but it was unable to give the temperature dependence of electrical resistance.

Shortly after Bloch published his analysis for metals, A.H. Wilson (4) combined the successful quantum mechanical views with the concepts of energy bands to develop a satisfactory explanation of electron conduction in semiconductors. With his approach he was able to account for conduction through lattice-site impurities. He differentiated between intrinsic conductivity and impurity conduction in semiconducting materials. He also set the stage for the concept of donors and acceptors relative to their locations in the band-gap region. Wilson was one of the first investigators who could explain the electrical conduction properties of insulators and metals with energy band-gap differences.

In the history of events leading to an understanding of the interaction of energetic radiation with solids, Prandtl (5) formulated a model in 1928 that allowed the introduction of defects into a crystal lattice. He studied the influence of these defects on the mechanical hysteresis effects and experimentally noted that defects were rather numerous after deformation of the material. His work was fundamental to the many extensive studies of crystal imperfections that have been carried out since his original work. Dehlinger (6) verified the observations of Prandtl and noted that the imperfections in crystals can be removed by heating and recrystallization (the forerunner of present-day thermal annealing).

Shortly before Prandtl did his work, Frenkel (7) reported in 1926 that defects can occur in a solid when lattice-site atoms are displaced through the absorption of incident radiation. The combination of the vacancy and the displaced interstitial atom, which is produced by the absorption of sufficient recoil energy, has been termed a "Frenkel pair." Whenever low-energy encounters occur, the transfer effects are primarily elastic. For more energetic encounters, the production of Frenkel defects becomes significant and accompanying ionization and excitation processes must be considered.

It was not until 1949 that Seitz (8) was able to estimate the energy required to displace atoms in solids. In his work, he estimated that the displacement energy would be approximately four times the sublimation energy of the material. This suggested a displacement energy of approximately 25 eV for solids of interest in this field. For displacement-type damage, the type of incident radiation and the magnitude of the primary energy determines whether displacement, ionization, or excitation processes will take place. Many of the radiation damage studies of the past 40 years have been devoted to these three processes, which occur in various ratios depending on the nature of the experiment. For example, ^{60}Co gamma rays have enough energy to produce recoil electrons with a maximum energy of 0.95 MeV (9). There is sufficient energy in such recoil electrons to produce lattice displacements. Thus, the use of ^{60}Co as a radiation source can produce physical damage to the lattice as well as generating free charge carriers. More will be said of the use of ^{60}Co to irradiate solid-state devices later.

Although a considerable theoretical foundation for understanding conduction and radiation damage mechanisms in crystalline solids was well underway by the early 1930s, several factors emerged that undermined further rapid growth in this area. With regard to the political climate in Germany at this time, Freeman Dyson succinctly phrased it as follows: "In 1933 the era of poets and amateurs was over and the era of the professionals had begun" (10). Other European countries began bracing for the political revolutions that were soon to engulf them. Creative science would then be intertwined with military objectives.

Another factor that delayed the experimental developments in understanding conduction mechanisms in semiconductors and insulators and radiation damage mechanisms was the unavailability of solid-state devices to allow appropriate measurement of certain bulk and surface parameters. Although some work had already been done toward fabricating rectifying diodes using copper oxide and selenium materials, the fabrication of such devices was more of an art than a science. Even here, it was not until 1939 that Schottky and Mott (11, 12) published the first space-charge dipole theory that partially accounted for the rectification mechanisms for such materials. Simple galena crystal detectors had been constructed using the famous "cat whisker" structure, but no satisfactory theory to account for the parameters involved was available (*13*). By 1940 there were good explanations by Bardeen (*14*) for electrical conduction in metals.

Just prior to World War II, various silicon detectors were developed that soon appeared in microwave components of radar. These applications motivated the development of highly purified silicon, which has found so many uses today. It was Scaff and Theurer (*13*) who first reported the growth of silicon p–n junction structures showing unusual rectification properties.

Although Kronig and Penney (*15*) had published a one-dimensional analysis of the motion of an electron in a periodic potential field, full utilization of their model did not come about until after World War II. Their model was basic in that it was formulated using a structure that approximated the potential in the lattice by a series of equidistant rectangular barriers of small dimension, with the barrier height corresponding to the field potential.

Through the Kronig–Penney model it was found that the spectrum of permissible energy values consisted of continuous regions separated by finite intervals. By varying the product of the barrier width and the potential height, they were able to establish the cases of free and bound electrons by using the wave equation and its solutions. The values of the allowed and forbidden energies could thus be determined. The transition from stationary states to other states with the absorption or emission of radiation was finally justified through this model. The way was opened much later to a consistent explanation of mechanisms and the characterization of the behavior of electronic conduction in irradiated semiconductors.

During the years of the World War II programs to utilize nuclear energy, much information on the behavior of solids had been accumulated. In particular, the effects of absorbed neutrons on metals had been studied extensively. With peaceful applications of nuclear energy under way, there were renewed efforts to understand electrical conduction phenomena in insulators and semiconductors and how physical parameters might be affected by energetic radiation. (For an excellent review of the history of radiation damage during this time period, see reference *16*.)

With the reporting of transistor operation by Bardeen and Brattain (*17*) in 1948, the way was paved for device measurement of numerous physical parameters. The announcement of the transistor in 1948 was largely responsible for the direction that solid-state physics has taken in the past 30 years.

In 1949, Shockley (*18*) published a comprehensive theory of p–n junctions in semiconductors and p–n junction transistors. His comprehensive treatment of the behavior of semiconductor–semiconductor junctions helped motivate many programs for designing and fabricating various transistor configurations. In 1950, Hall and Dunlap (*19*) published a description for fabricating junctions by a diffusion method which is the most commonly used method in the industry today. The stage was now set for a plethora of configurations and combinations to appear on the electronic market, as well as for futher extension of the theories to describe electronic conduction mechanisms.

In 1952, Hall (*20*) and Shockley and Read (*21*) published descriptions on the statististics of the recombination of holes and electrons. By employing Fermi–Dirac statistics it was found that carrier recombination is dependent on several factors: (1) the recombination center density, (2) the free-carrier density, (3) the recombination probability at the recombination sites, and (4) the density of normally filled trap sites. Consideration of the emission, capture, and recombination rates led to an expression for what is referred to as a lifetime for Hall–Shockley–Read (HSR) trap sites. More will be said of this phenomenon later.

With the development of junction-type transistors, it became possible to measure various electronic parameters. When junction transistors were subjected to energetic radiation, various changes in these parameters were found (*22, 23*). The description of a unipolar (or field-effect) transistor was published by Shockley in 1952 (*24*) with the configuration consisting of p- and n-type materials forming a channel between a source and a drain that could be "pinched off" by electrically biasing the gate electrodes. This design employed semiconducting material throughout the device with no oxide.

It was not until 1963 that Hofstein and Heiman (*25*) published a description of an insulated-gate field-effect transistor (FET) that used a thin metal-covered oxide layer located on top of the semiconducting material in a metal–oxide–semiconductor (MOS) control structure. This structure, which had been utilized by Shockley and Pearson (*26*) in 1948 to study conductance by surface-charge effects, was hailed as a "device possessing a significantly increased versatility" (*26*). This configuration could be used to enhance as well as deplete the charge near the surface of the semiconductor between the source and the drain. However, it was soon to be observed that the presence of the oxide film might cause modified, or even irreversible, operating characteristics of the FET device if the oxide contained trapped charge. Such trapped charge could unknowingly be introduced at the time of the oxide growth and processing or by exposure to penetrating ionizing radiation (*27*). Thus, versatility was soon to be replaced by operational and fabrication restrictions for MOS structures. In parallel with the development of semiconducting devices, attempts to obtain a more exact understanding of the interaction mechanisms between ionizing radiation and various crystalline solids (*28, 29*) were made.

In 1960, Nelson and Crawford (*30*) reported on optical absorption in the 1850–26,000-Å range for irradiated (neutron and ^{60}Co gammas) crystalline quartz and fused silica. For gamma-ray exposure, they observed a monotonically increasing extinction coefficient with increasing dose for the various SiO_2 samples they used (up to 5×10^7 rads). The coloration was bleached by subsequent exposure to ultraviolet (UV) radiation and by thermal annealing at temperatures between 350 and 400°C. The exact mechanisms causing this behavior remained obscure.

In 1961, Compton and Arnold (*31*) studied radiation effects in fused silica. They measured the efficiency of defect production using optical absorption techniques. They found monotonically increasing coefficients for absorbed doses of up to nearly 10^8 rads in commercially available fused silica materials.

In 1963, Peck *et al.* (*32*) reported on the surface radiation effects in several types of transistors. They observed that it was the total radiation dose rather than the dose rate that was the important factor in producing the various changes in the operation of the devices. They considered the influence of the radiation on the ambient gas in the encapsulated transistor as well as the bulk changes within the devices.

In 1964, Hughes and Giroux (*33*) reported on the degradation of commercially available MOS transistors that had been subjected to ^{60}Co irradiation. They found that both the transconductance (g_m) and the channel conductance degraded after exposure to 10^6 rads. Even at low doses (60 rads) some degradation of the output characteristics was observed. No explanations for the mechanisms causing the degradation were given.

In 1965, Kooi (*34*) reported on the effects of ionizing radiation on the surface properties of oxidized silicon. He noted that absorbed radiation left a positive charge in the oxide, with the charge equilibrium being affected during continued irradiation because new trap centers were formed in the oxide or at the surface of the silicon. He found that the surface potential of the oxide was altered whenever the charged oxide was illuminated with UV radiation. (Kooi concluded that electrons flowed from the silicon substrate to the oxide.) He observed that the method of oxide preparation strongly influenced the irradiation effects. A distinction was made between the behavior of "wet" and "dry" oxides which contain different defect distributions in the oxide. He found that new defect (or trapping) centers were generated during irradiation. It was found that the "dry" oxides were not as susceptible to electron capture at the charged centers (recombination) as the "wet" oxides were whenever the oxide was illuminated with ultraviolet. Thermal annealing of the trapped charge was found to be effective using temperatures between 300° and 500°C.

In 1965, Szedon and Sandor (*35*) reported results on the irradiation of n-type capacitors (1600-Å thick) using 10- to 20-kV electrons. (They did not indicate if the MOS capacitor gate was shorted to the substrate during irradiation.) They concluded that the incident electrons induced a positive charge in the oxide near the silicon interface. They found that the charge could be removed by thermal annealing at temperatures up to 200°C for 15 min.

In 1965, Green *et al.* (*36*) described the observation of reversible changes in transistor characteristics caused by electron-beam irradiation in a scanning electron microscope (16 keV). They reported a marked decrease in the transistor beta and changes in the reverse-bias leakage current and breakdown voltages. They suggested that the incident radiation produced ionization in the oxide with the trapped charge causing a permanent change in the surface potential of the silicon. Partial recovery by thermal annealing at temperatures up to 250°C was reported.

In 1965, Snow *et al.* (*37*) reported on ion transport phenomena in SiO_2. They deposited alkali ions at the metal–oxide interface and then studied the transport of the ions through the oxide as a function of time, temperature, and applied voltage. For a positive gate bias, the number of ions accumulated at the oxide–semiconductor interface was found to be a function of the square root of time with an ultimate trend toward saturation. They observed that the ions resided in the vicinity of one interface or the other and were rapidly transported across the remainder of the oxide whenever diffusion occurred.

In 1965, Deal *et al.* (*38*) reported on the effectiveness of the MOS capacitance–voltage method for studying MOS systems. They found this method to be a powerful and extremely versatile one in the study of semiconductor surfaces, particularly for irradiated MOS structures. This method has become one of the most commonly used techniques for determining surface-charge densities, surface potentials, ion migration in the oxide, surface mobilities in inversion layers, and impurity redistribution in thermal oxidation.

In 1966, Grove and Snow (*39*) found that irradiated MOS structures show an excess charge buildup that follows a square root of applied gate bias dependence. For positive gate bias, the saturation charge was reported to be independent of oxide thickness. The saturated charge value was reported as $Q_{sat} = 3 \times 10^{18}$ cm^{-3}.

In 1966, Hofstein (*40*) identified several ionic species as being responsible for the charging and relaxation effects in MOS structures having large applied electric fields. He found that the relaxation of charges trapped in the oxide was influenced by thermal as well as electric field bias. He reported that the charging–discharging asymmetry was caused by ionic trapping at the metal–oxide interface and that fast recovery was tied to ionic hydrogen.

In 1966, Zaininger (41) reported on the irradiation of MOS structures using high-energy (125 keV and 1 MeV) electrons. He concluded that the incident electrons generated hole–electron pairs with subsequent capture of the holes by stationary traps. In his descriptive model, he stated that some recombination of electrons and holes will occur, but that trapped holes account for the charge buildup. The radiation-generated electrons move out of the oxide. For no bias on the gate, electron–hole recombination dominates and no significant excess positive charge is retained in the oxide. With applied gate biases, the electrons rapidly move out of the oxide. Charge buildup occurs near the oxide–silicon interface for positive gate biases. Charge buildup occurs near the metal–oxide interface for negative gate biases. He concluded that the traps must be located more than about 4 eV above the valence band of the oxide, based on illumination studies using UV radiation. Temperature annealing was found to render the oxide essentially charge-free.

In 1966, Zaininger (42) also noted that the induced charge in the oxide monotonically increased until an incident flux of approximately 10^{12} electrons/cm^2 had been reached when irradiating with 0.1- to 1-MeV electrons. Above that value, the curve became linear for the next two orders of magnitude. He also showed that the density of induced oxide charged states follows the form shown in Fig. 1.

Zaininger noted an asymmetry in the curve with polarity of gate bias. The effect was tremendously increased for positive gate bias. He explained

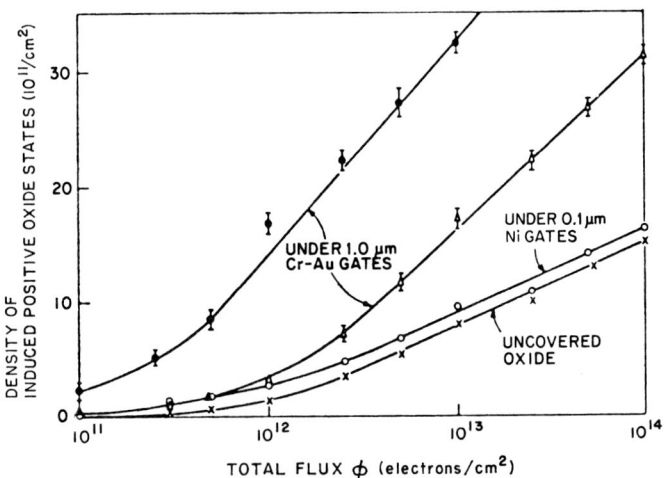

FIG. 1. Density of radiation-induced positive states in the oxide for various gate thicknesses and gate compositions versus incident dose: 0.1 MeV, ●; 1 MeV, △, ○, ×. [See Fig. 1 in Zaininger (41).]

these results as being due to induced positive charge at or near the oxide–silicon interface for positive gate bias and due to induced charge at or near the metal–oxide interface for negative gate bias. He proposed that this process was due to the generation of electron–hole pairs in the oxide with the electrons moving either out of the oxide or recombining with holes. The holes diffuse in the oxide and many of them are captured into stationary traps. Gate bias during irradiation was reported to influence the spatial distribution of the trapped holes. The induced oxide states were found to follow the form shown in Fig. 2. Zaininger reported that for positive gate bias the large induced charge at the oxide–silicon interface was due to electrostatic interaction at the interface. For negative bias, the charge was introduced at the metal–oxide interface and, owing to the distance from the silicon interface, most of the image charge appeared in the metal and hardly any effect was observed by MIS-type measurements.

In 1966, Barry and Page (43) reported on the effect of low ionizing dose levels on transistor characteristics. They found that the threshold voltage monotonically increased with radiation dose as shown in Fig. 3a. They concluded that the charge distributions within the oxide followed those shown in Fig. 3b for positive, negative, and zero gate bias during irradiation. They also showed the small-dose radiation sensitivity as a function of gate voltage after successive large doses for positive gate bias, as shown in Fig. 3c. It is significant to note that the curves show a minimum in the negative gate-bias region with the minimum moving downward and to the left as the

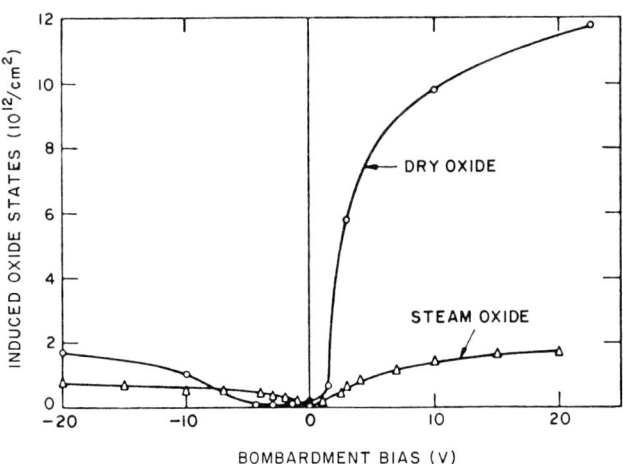

FIG. 2. The density of radiation-induced positive states versus gate bias during irradiation. [See Fig. 13 in Zaininger (42); copyright 1966, IEEE, New York.]

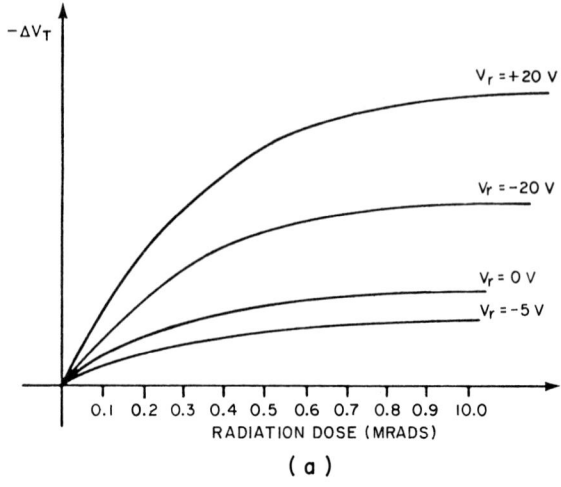

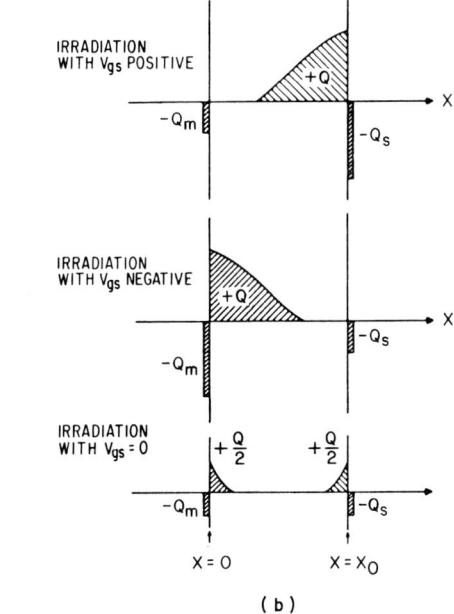

Fig. 3. Radiation-induced behavior for MOS field-effect transistors. (a) Shift of gate threshold voltage versus radiation dose for various gate biases (see Fig. 3 in Ref. *43*); (b) proposed charge distributions in the oxide following irradiation for three gate-bias conditions (see Fig. 5 in Ref. *43*); (c) threshold voltage sensitivity following irradiation for three dose values. [See Fig. 6 in Sommerfeld (*3*); copyright 1966, IEEE, New York.]

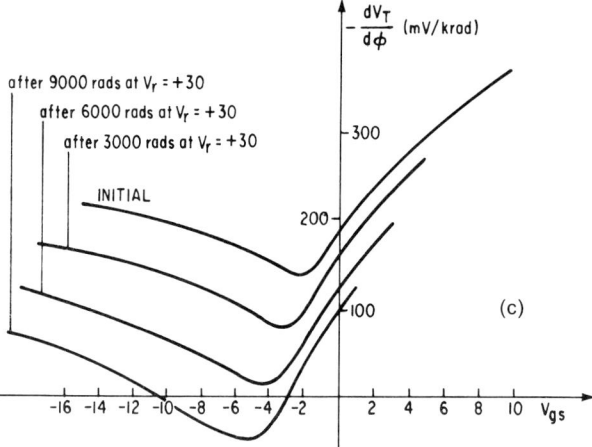

Fig. 3 (Continued)

radiation is increased. The four curves show the same general slope in both the left and right directions from the minima.

In 1967, Snow et al. (44) reported the effects of various ionizing radiations on MOS structures. They considered high- and low-energy electron (20 keV and 1 MeV), X-ray, and UV radiation effects on the charge buildup in SiO_2. Their model was based on trapping of holes generated in the oxide by the absorbed radiation. They found a high density of hole traps near the oxide–silicon interface, which decays rapidly with distance into the oxide. They studied thermal annealing effects and found oxide space-charge annealing occurred when using UV light.

Throughout the 1970s, so many excellent papers were published in the technical journals on radiation effects in MOS structures that it is not possible to reference them all. So, for the sake of brevity, we mention those publications that have had a direct bearing on our work or represent significant aspects of radiation effects in SiO_2 systems.

In 1971, Powell and Derbenwick (45) used UV illumination to show that positive charge could be induced at the oxide–silicon interface even when the UV was absorbed near the gate electrode. They also observed that under irradiation with positive (gate) bias, positive space charge was formed near both interfaces. This was found by etch-back experiments. Mobile holes were responsible for the effect and were trapped at the oxide–silicon interface.

Derbenwick and Gregory (46) made measurements on dry oxides and arrived at an oxide thickness-cubed dependence for flat-band shift. Later, Hughes et al. (47) used electron-beam, VUV, and corona charging to determine that the dependence was not cubic but was quadratic.

Buxo et al. (48) reported the results for irradiation of MOS transistors in what they termed a "four-parameter" model (surface barrier height, mobility, number of traps, and the equivalent carrier temperature). In their model, the radiation-induced positive charges in the oxide formed in two principal locations at each heterojunction of the MOS sandwich.

Boesch and McGarrity et al. have published a number of papers (49–51) on hole-transport phenomena, charge relaxation, charge yield, and dose effects in SiO_2 systems. They performed irradiation with UV, ^{60}Co and energetic electron beams (52). C. T. Sah (53) has contributed much insight into the origin of surface states and oxide charges. The IBM group (54–60) has published extensively on the location of charges in SiO_2 films, electron trapping, and the effects of processing. R. C. Hughes et al. (61, 62) have studied time-resolved hole transport of excess holes in amorphous SiO_2 as a function of electric field and temperature. H. L. Hughes et al. (63, 64) studied the effect of oxide impurities on radiation sensitivity and radiation-induced surface states. G. W. Hughes et al. (47, 65, 66) reported on the oxide thickness dependence of the flat-band shift, the contribution of donor and acceptor states to the total flat-band and threshold voltages, and how to stabilize and increase processing yields.

Gregory and Gwyn (67) reported on the radiation-induced degradation of semiconductor performance. Gwyn (68) developed a "hybrid" recombination model and described a numerical device analysis technique to calculate various parameters and externally measured characteristics. In his model he assumed a rectangular charge distribution at each interface with the electric field equal to zero in between. A 1-nsec pulse of ionizing radiation was absorbed in the oxide and time integration was carried out over the next 2 nsec.

We began our study of the characteristics of MOS devices that are subjected to ionizing radiation nearly a decade ago. We modeled the transient behavior of MOS structures (69–71) for applied voltage step functions and calculated the resulting distributions of charge in the oxide. The results of this model were modified and applied to cases where various quantities of ionizing radiation were absorbed in the oxide for a wide range of applied gate biases (72–74). The details of the model and the resulting conclusions are presented in this article.

II. Modeling Considerations

There are only a few observables that can easily and meaningfully be monitored to determine the effects of irradiation on MOS structures. Some of these are what we will term direct observables, meaning that they can be monitored directly either during or after termination of each irradiation increment. Others are what we will term indirect observables, since they cannot be observed without additional irradiation increments and/or processing of the device.

Some examples of direct observables are (1) the current flowing into the gate during an irradiation transient, (2) the position and shape of the C–V and G–V plots after the irradiation has been terminated, and (3) direct microscopic inspection of the device. Some examples of indirect observables are (1) the effect of postirradiation annealing, (2) the history dependence under successive increments of irradiation, and (3) the memory effect.

For the purpose of developing a model, measurements involving the direct observables are usually more productive, since direct observables provide a more intimate contact with the physical processes themselves. Hence, by concentrating on the direct observables, the investigator should be in a better position to deduce the basic physical processes that ultimately control the radiation-induced changes in device characteristics. The indirect observations are also valuable, but more for "tweaking" the finer details of the model and as a check on its overall validity than as a means for deducing the basic model structure required.

A typical C–V plot for one of our MOS capacitors before and after it had been irradiated is shown in Fig. 4. Curve 1 shows the C–V characteristics of this device before irradiation, and curve 2 shows the characteristics of the same device after irradiation, with $V_{gr} = +1.3$ V, where V_{gr} is the gate bias voltage applied during radiation. The most readily apparent change in the

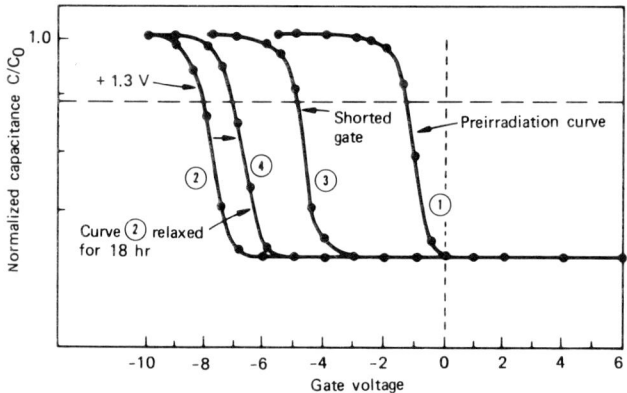

FIG. 4. C–V curves for p-type MOS devices. Curve 1, Preirradiation curve; curve 2, gate bias of $+1.3$ V and irradiation dose of approximately 10^8 rads; curve 3, shorted gate to substrate followed by dose of approximately 5×10^7 rads; curve 4, an 18-hr relaxation of curve 2 with shorted gate to substrate.

equilibrium properties of the device after electron irradiation is a simple shift in the C–V curves along the bias voltage axis. Since there was no observable distortion in the *shape* of the C–V curves after irradiation, it was concluded that the readings are not influenced by surface states. (Some investigators have reported distortion as well as shift in the C–V plots (*35*), *42*, *53*) but the subject of irradiation-induced surface states is outside the scope of this present discussion since the mechanisms by which radiation-induced surface states are formed are still largely unclear (*75*).)

The radiation-induced flat-band shift S may be characterized quantitatively by the equation

$$-S(V_{gr}) = V_{fb} - V_{fb0} \qquad (1)$$

where V_{fb} is the flat-band voltage after irradiation at the particular gate bias V_{gr} and V_{fb0} is the preirradiation value of flat-band voltage. After the radiation has been turned off, the radiation-induced charge distribution in the oxide remains essentially "frozen in." This induced charge density produces image charges in the semiconductor, which account for the measured flat-band shift. We have found that the induced charge distribution is relatively stable and that the induced flat-band shift decreases only at a slow rate over a period of several years at room temperature.

Considerable direct information can be gained by not only observing the distortion (when it exists) and shift in the C–V and the G–V plots immediately after irradiation has been terminated, but also by monitoring these plots for short- and long-term relaxation effects over periods of days, weeks, months, and even years. However, in attempting to interpret these data with a view toward formulating a model, one must always be careful to distinguish between a number of possible ranges of irradiation parameters. Failure to do this in the past has led to the use of certain models in areas where they are, at best, only marginally acceptable and the rejection of other models even in those areas where they actually do serve a useful purpose.

There are a number of irradiation parameters which can have an important influence on the response of MOS structures to irradiation. The cumulative dose could, perhaps, be considered one of the most important of these. We find it is extremely important to distinguish between three unique ranges of cumulative dose: the low-, medium-, and high-dose ranges. The nature of the internal electric fields and charge distributions in the oxide for each of these ranges can be quite different. Thus, the physical processes which dominate in each of these ranges can also be quite different. Many experiments have been reported in each of these ranges whereby various investigators have tried to relate the results of their own work in a given range to the results reported by other investigators working in a different range. However, failure to recognize that the mechanisms involved in one range may differ substantially from those of another means that many of the comparisons were not actually valid after all. As a result, data which should have brought about better understanding instead led to further confusion.

We wish to stress that there is no sharp division between the three ranges of cumulative dose, since each range merges gradually with the other ranges. However, there are some broad guidelines that can be used for definitions. For our present purposes we will define the three ranges roughly as follows.

A. Low-Dose Range

The low-dose range covers the range of values of cumulative dose over which no appreciable amount of flat-band shift is detected. Of course, under actual experimental circumstances, some small but nonnegligible amount of shift must usually be present in order to determine that some radiation-induced changes have in fact occurred. Therefore we can only apply this definition in the broadest sense. However, in practice, irradiation has actually been terminated at the point where a nonnegligible flat-band shift was first detected (76).

In the absence of an appreciable flat-band shift, we surmise that only an insignificant amount of charge could have been trapped inside the insulator. By Gauss' law, this would imply that the electric field is uniform and that there is no substantial energy-band curvature inside the insulator. Thus, we see that models that rely on the assumption of a uniform electric field must either exclude bulk traps in the oxide or else be restricted to the low-dose case.

B. High-Dose Range

The high-dose range is defined as that range of values of cumulative dose for which the flat-band shift under a constant gate bias is independent of additional increments of irradiation dose. This definition applies to those situations where the charge, field, and current distributions have reached a state of dynamic equilibrium. Throughout this range there is no appreciable history dependence in the flat-band shift. That is, after each change in gate bias, even a relatively small amount of additional incremental dose will cause the flat-band voltage to quickly approach the value which is characteristic for the new value of gate bias and which is independent of all previous irradiation history. Once again, the boundary between high- and medium-dose ranges is not rigidly delineated.

C. Medium-Dose Range

In the medium-dose range, the flat-band voltage will be significant, but has not reached a dynamic equilibrium value. Except in the complete absence of bulk states, the presence of a significant shift implies that there is a large concentration of charges in the insulator. Thus, according to

Gauss' law, the fields must be nonuniform. In the medium- as well as the high-dose range the electric field throughout a large portion of the insulator can differ by orders of magnitude from the field value one would calculate from the ratio of applied gate voltage and oxide thickness. If the total radiation-induced charge is large enough (as it is for some typical experimental situations), it is quite possible that the total internal field may actually oppose the applied field in some regions of the insulator. Thus, even for the medium-dose range, one cannot assume that holes are swept continuously across the insulator in one fixed direction, since holes may actually be moving in different directions at different locations.

D. Other Factors

According to the definitions given above, we see that irradiation of an MOS structure is somewhat similar to a transient in an electrical circuit. In the case of irradiation, the independent variable is cumulative dose rather than time. The physical variables we are concerned with are quantities such as flat-band shift rather than voltage or current. In this analogy, the low-dose range is analogous to the initial few moments of the transient response and the medium-dose range is comparable to the continuation of the transient, whereas the high-dose range corresponds to the steady-state part of the total response.

In an electrical transient, the initial response will often be strongly affected by various parameters which do not actually represent the overall circuit or topological configuration itself. For example, transients in an RLC circuit (especially if it contains nonlinear elements) will not generally be recurrent unless the initial charge on capacitors, the initial current through inductors, the initial phase angles of the signal sources, and the initial magnetization of inductor cores is carefully reset to the same values each time the transient is initiated. Even small variations in some of these initial conditions can cause drastic variations in the overall nature of the transient response. The steady-state response, on the other hand, usually depends only on the intrinsic topological properties of the circuit and sources themselves, and will hardly be affected by even gross variations in the initial conditions.

If the irradiation of an MOS structure is considered to be analogous to an electrical transient then, as in an *RLC* circuit, the steady-state response (or high-dose range) is a more direct indicator of the actual physical processes which govern the development of the charge and field distribution than is the initial transient (or low-dose range). The low-dose range is sensitive to what we consider less important details, such as defect density, bulk physical stress at the interface, as well as minor variations in laboratory processing times, temperatures, etc.

Models formulated exclusively from data pertaining to the low-dose range would, in all probability, be valid only in that range, since the other ranges probably have more complex charge and field distributions. On the other hand, models formulated from data taken at the high-dose levels must by nature be able to account for complex charge and field distributions. Hence, the uniform-field case found either in the absence of oxide bulk traps or at low cumulative dose values would already be included as a simple special case. It is easier to generalize a model based on dynamic equilibrium effects so as to include the lower dose regions than it would be, in principle, to generalize a model in the opposite sense. Thus, for the purpose of formulating a model we find that the high-dose range is an especially important area to study, and it is relatively easy to extend a high-dose model into the lower dose regions.

At this point, we must be careful to distinguish between two equally important but widely divergent goals. For production-oriented applications, radiation-induced changes in devices are considered strictly detrimental. For these applications it is desirable merely to vary the processing parameters systematically in a relatively large number of runs until radiation sensitivity has been minimized. The final product is then called "radiation hardened." (Hardened devices are defined as those whose radiation sensitivity has been minimized by processing and/or operation conditions.) For such applications, it is important to know what parameters affect radiation sensitivity but it is only marginally important to actually understand the physical processes governing radiation effects. For other applications, however, the goal is to exploit the radiation-induced changes in device characteristics rather than minimize them. There are some potential device applications in which radiation effects could be utilized in a practical manner (77). In this case, it is not adequate to merely vary the fabrication parameters until some desired effect is seen. Instead, it is critical that the actual physical mechanisms which give rise to the changes in the electronic behavior of the devices be thoroughly understood. (Unfortunately, most of the more recent research and development resources have typically been channeled into developing radiation-hardened devices with very few going to understanding basic physical mechanisms.) In order to design systems which put radiation effects to practical use, one must have not only good models covering each of the three ranges of cumulative dose, but also a single, comprehensive model whose validity extends over all ranges.

In formulating a model, the objective is to find a convenient way to design devices for a variety of conditions of operation. A model can be formulated by first studying the experimental data for the dose range to be considered and then, based on that knowledge, by making an educated guess as to the specific physical mechanisms that dominate the device behavior in that range. These mechanisms would then be analyzed mathematically to determine whether or not the proposed model actually agrees with the experimental data from which it was developed. The model must also agree with any new experimental tests which might be suggested by mathematical analysis of the model.

Some models might give accurate predictions over a wide range of experimental conditions while other models might give reasonable results over a narrow range of conditions only. There is nothing inherently "wrong" with a model which works only in a narrow dose range. As long as it gives accurate results in some well-defined range, it must be considered a valid model. Sometimes, because of simplicity, such a model might even be more useful than a more comprehensive model. Some models will fail to agree at all with experimental findings and such models, of course, must be discarded. Only rarely does one expect to be able to formulate a model that works well under all conditions. (Even Ohm's law is a model whose accuracy is generally restricted to a limited range of linearity where the mobility is independent of the electric field.) Ultimately, one hopes for a radiation model that gives a reasonably good prediction of flat-band voltage V_{fb} for any arbitrary combination of gate bias V_{gr} and dose rate, $R(t)$.

The key to successful modeling is to determine which are the most critical physical mechanisms out of all those that might conceivably be operating in the system. In the case of irradiated MOS devices, there are many mechanisms influencing the evolution of charge distributions. Some of these will be crucial for inclusion in the model, whereas others that have only a minor effect on the results can be omitted from the model without a significant loss in accuracy.

We have already noted how important it is to specify the cumulative dose range in which the model is to be applied. Another important factor is the physical nature of the traps in the insulator. Are they bulk traps or interface traps? Are they preexisting (inherent) traps or are they traps created by the radiation itself? There are many other important factors, as well. The three-dimensional diagram in Fig. 5 shows that many unique combinations are possible with these few considerations alone.

In Fig. 5 the horizontal direction is split up into three regions, corresponding to the low-, medium-, and high-dose ranges, respectively. The vertical direction is split into three layers. The lowest layer corresponds to models based on the presence of interface traps alone while the upper layer corresponds to the case of bulk traps alone. The middle layer therefore includes those models which assume the simultaneous involvement of both bulk and interface traps. The third dimension has been divided up into three slices as well. In this case, the front slice corresponds to models incorporating preexisting traps alone, the back slice corresponds to models where only traps created by the irradiation are included, and the middle slice corresponds to models which incorporate both preexisting and radiation-induced traps. All together, there are 27 different possibilities. Unless one has a highly sophisti-

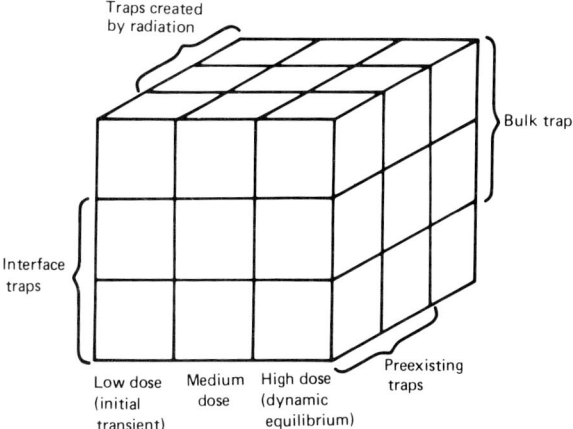

FIG. 5. Global view for modeling various traps and trapping mechanisms for low-, medium-, and high-dose ranges.

cated global model that simultaneously incorporates all of the subcubes in Fig. 5, one is forced to formulate unique models for various convenient separate combinations of the 27 different cases.

Only three dimensions have been shown in Fig. 5. When other important physical mechanisms are considered as well, the three-dimensional figure becomes a multidimensional block and the total number of possibilities becomes enormous. Obviously, we do not expect to readily find a simple model which is able to simultaneously treat a much larger number of possibilities. One must often be content with simple models that provide fast, reasonably accurate results with minimal effort. Simple models representative of this case will be discussed in Section IV.

When greater accuracy or validity over wider ranges of conditions is absolutely necessary, one must make use of numerical simulation by computer analysis. This additional generality is not achieved without cost, since a comprehensive computer model is more difficult and requires more time to develop. One such model will be reviewed in Section V.

For the purpose of modeling it is important to consider not only the spatial distribution of trapped centers but the energy distribution of traps as well. For example, electron and hole traps could either be located at certain discrete energy levels or spread out over a whole range of energies between the valence and conduction bands of the insulator.

In order to develop an understanding of the physical processes involved in radiation-induced changes in MOS structures, it is important to determine which dominant mechanisms govern charge generation and transport. In order to determine these prevailing mechanisms it is not sufficient to have a model that works only over a limited range of values of V_{gr}. Unless the model matches the experimental results in great detail over a *wide* range of values of V_{gr} for *both* polarities of voltage, it will not be possible to sort out the contributions from various mechanisms. Ideally, if the model predicts results closely matching a fairly complex pattern of measured data over the whole range of positive and negative values of V_{gr}, then it is likely that the model is a good one and that the proper contributions from all the various mechanisms have been finely tuned in the model. For completeness, one might want to include the effect of stress-induced (*78*) traps as well as mobile traps—those able to move about from one place to another in the insulator under various influences such as fields, stress, etc.

Although the dose rate is not a significant parameter in modeling radiation-induced changes in MOS devices, the energy of the incident radiation is important and does have an influence on the nature of the results. As was noted in the introduction, high-energy electrons and ^{60}Co gamma rays can produce lattice damage and displacements. These damage sites influence the density and nature of the oxide traps. Thus, proper allowance for time-dependent trap creation and possible self-annealing must be provided for in any comprehensive model.

The nature of the mobile carriers in the insulator and the semiconductor as well as the energy-band structure of the two heterojunctions can have a significant influence on the reliability of the model. Although the carrier mobility in virgin SiO_2 is fairly well known, the mobility in irradiated oxide depends on a number of elusive factors. In particular, experiments have been done to determine the mobility of holes and/or mu–tau product, but numbers differ in some cases by many orders of magnitude (*76*). The widely divergent experimental results may be explained in terms of accurate representations of the specific quality of the irradiated oxide. Thus, models for use in the high-dose range would require mobility values differing markedly from those in models intended for use in the low-dose ranges.

Another important consideration in modeling this physical system is the field dependence of the mobility (*79*). Certain data that support the geminate recombination theory could be alternatively explained by invoking the concept of scatter-limited velocity in SiO_2. At the present time this problem cannot be considered fully resolved.

A complete model needs to allow for trapping cross sections that might be velocity- and/or field-dependent. With the intense fields which are often encountered in irradiated oxides, the model might need to include the effect of impact ionization (*80*). Therefore in the case of electron-beam irradiation with energetic primary electrons special considerations are merited.

There are a number of other physical processes that might need to be included in a truly complete model. For example, when the field near one interface or another becomes large enough, tunneling into the insulator can occur. Tunneling arises with intense applied fields and/or when the charge density near the interface is sufficiently large (81). Concentrations of surface charge larger than approximately 4×10^{12} cm^{-2} would "encourage tunneling of electrons from the silicon, annihilating the positive charge" (81). Thus, tunneling could play a significant role in maintaining the dynamic equilibrium condition that characterizes the high-dose irradiation cases. The Schottky effect can also become important under these conditions.

For large radiation fluences, the oxide may heat up. It is important, therefore, to adjust the radiation flux to prevent thermal annealing from occurring during irradiation. Oxide heating is typically not of any significance in the absorption of low-dose-rate ^{60}Co and X-ray radiation. With high-flux electron beams or corona charging, more caution must be exercised. Also, in those experiments where no gate electrode is used, caution must be exercised to prevent erratic or unpredictable results that come from unknown floating equivalent gate potentials during irradiation.

In some irradiation experiments the gate contacts have been left unconnected during irradiation. However, when the gate floats in this manner the device will develop a substantial self-bias that can affect the results of irradiation significantly when electron beams are used. The actual value of this self-bias is unknown and, since it depends on many variables, the floating gate conditions should generally be avoided.

Another situation which must be considered, since it amounts to a substantial effect, is the zero-gate-bias flat-band shift (shorted gate to substrate). In some of the earlier articles on irradiation of MOS structures, it was stated that V_{fb} was essentially unshifted by radiation when the gate was shorted during irradiation (41). (It is important to recognize that even when $V_{gr} = 0$, the electric field within the oxide can be quite large.) However, as more exact measurements were reported, it was seen that the zero-volt flat-band shift amounted to a rather large value (often as much as 20 V or more) and could not be ignored after all. Figure 6 shows a curve where the flat-band voltage is 14 V at the end of an irradiation at $V_{gr} = 0$ V. The presence of a 14-V shift under zero-bias conditions must be considered a significant constraint imposed upon any proposed model. If the proposed model is unable to account for a large zero-bias flat-band shift, then that model must be considered questionable for all but the low-dose range.

Another situation that ought to be considered in the formulation of a useful model is the nature of the generation process for mobile carriers. In the past, it has often been assumed that mobile carriers were created by ionizing radiation only as hole–electron pairs—a process which certainly would be present during irradiation. However, since trapping of carriers by trapping centers is considered essential to most models, then the generation

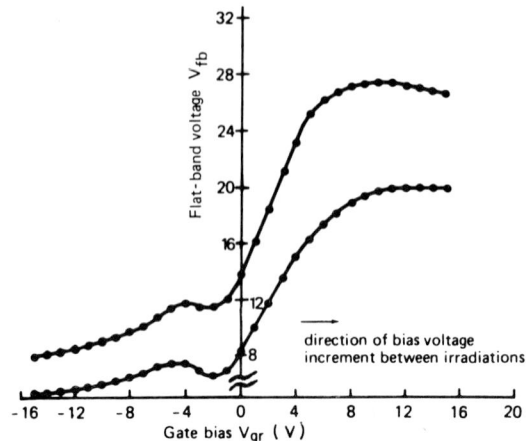

FIG. 6. Flat-band voltage versus applied gate bias during irradiation V_{gr}. Voltage increments between irradiations started at extreme left end for both curves. Bottom curve irradiated at 3×10^8 rads/point followed by irradiation for upper curve at 3×10^8 rads/point (500-Å-thick oxide); both curves from same device.

of mobile carriers by these same trapping centers must also be considered (71). For example, the generation of hole and ionized-acceptor pairs and/or the generation of electron and ionized-donor pairs is a mechanism which could contribute in a nonnegligible manner to the formation of the charge distribution that is ultimately responsible for the measured shift in flat-band voltage.

III. GENERAL RELATIONSHIPS BETWEEN V_{fb} AND $V_{gr}(D)$

Experimentally, it is found that the flat-band voltage after irradiation depends on the gate-bias voltage V_{gr} applied to the gate during irradiation as well as on the cumulative dose level. In irradiation research, the most common practice is to pick a value of V_{gr} and hold it constant during the entire irradiation. However, in a typical integrated circuit application (a computer in a satellite, for example) V_{gr} would certainly vary. It is not sufficient, therefore, to distinguish only between three ranges of cumulative dose. One must also specify the time dependence of V_{gr} and dose D, as well as the incident beam energy and fluence, in order to uniquely describe a given irradiation experiment.

However, as we have found, the response of ultraclean, dry-oxide MOS devices to irradiation is essentially independent of dose rate, at least for dose rates up to 10^7 rads (Si)/sec. In other words, the functional dependence of D and V_{gr} on time is not important. The values of all the direct observables seem to be uniquely determined by the cumulative dose D_C and by the function $V_{gr}(D)$ for all $D < D_C$. Barring unusual effects in the extremely high dose-rate cases, it is the functional relationship between V_{gr} and D alone, rather than the instantaneous values of these parameters, that determines the actual response of the device to its radiation environment. (However, in cases where the irradiation is administered in 30-nsec bursts, such as was done by Maier and Tallon (86), a "photovoltaic" bias effect occurs. In such cases, rate effects can be important.)

This principle is illustrated by the experimental data shown in Fig. 7. (All of the data in Fig. 7 were obtained from the same sample.) For this

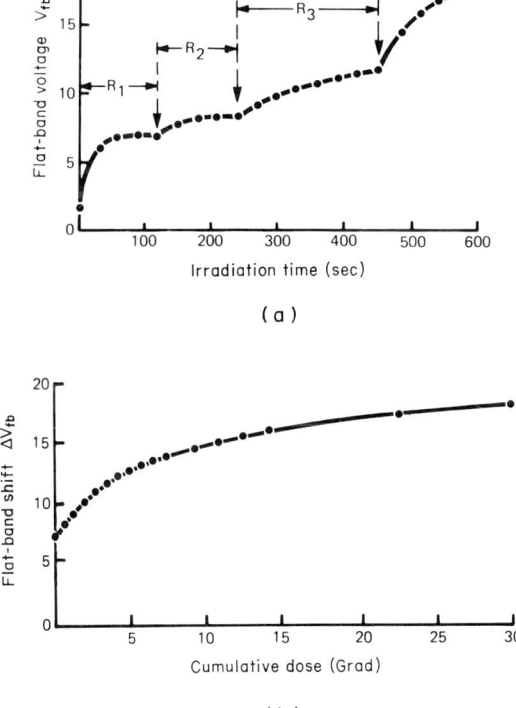

FIG. 7. Flat-band voltage versus cumulative dose for $+1$-V gate bias (500-Å-thick oxide). (a) Flat-band voltage versus irradiation time, $R_1 \approx 10$ rads/sec, $R_2 \approx 7 \times 10^5$ rads/sec, $R_3 \approx 10^7$ rads/sec, and $R_4 \approx 10^8$ rads/sec; (b) Flat-band voltage versus dose.

sample, the gate was held at a fixed value of $+1$ V during irradiation. At the end of each dose increment (indicated by dots in the figure), the gate bias was removed, the flat-band voltage was measured, and the gate voltage was returned to $+1$ V. The next irradiation increment was then initiated as quickly as possible. After five data points at a fixed dose rate R_1, as indicated in the figure, the dose rate was increased to R_2 in order to generate the next segment of the curve, containing four data points. This process was repeated for dose rates R_3 and R_4.

The curve in Fig. 7a shows cusps at each of the points where the dose rate was changed. At first glance, it might appear that this is a dose-rate effect. But in fact it is not. The cusps appear only because in Fig. 7a the flat-band voltage is plotted versus irradiation time. When the identical experimental data is replotted against cumulative dose rather than time, then as shown in Fig. 7b the cusps disappear. Thus, by choosing to display flat-band voltage versus time one might be inclined to believe there is a dose-rate effect when, as Fig. 7b clearly shows, the results really depend only on the cumulative dose.

In a general experiment, one might choose to vary V_{gr} in some completely arbitrary manner while D increases monotonically (but not necessarily at a constant rate) with time. For modeling purposes, it would be much too difficult to interpret the experimental results for an arbitrary functional variation $V_{gr}(D)$. It is more convenient, instead, to treat the arbitrary variation of V_{gr} as if it consisted of sequential irradiation in increments using discrete values of V_{gr}. If the increments are chosen small enough, then a sequential irradiation could, in principle, be made to correspond to the arbitrary $V_{gr}(D)$ as accurately as desired.

Figure 8 illustrates two basic types of experimental operations which are commonly done in irradiation of MOS devices. The curve labeled "type 1" represents data for irradiation over a large incremental dose at a constant value of V_{gr}. We shall call this a type 1 irradiation. In the curve labeled "type 2" the gate bias is varied over a given range in some continuous manner. As described earlier, it is often more convenient to vary the voltage in uniform steps, as indicated by the staircase, rather than to vary it continuously as in the sawtooth curve. The flatband voltage would, presumably, be measured after each dose increment. This process of cycling the gate voltage back and forth in uniform steps between sequential irradiation increments will be called a "type 2" irradiation. It is essential to distinguish between type 1 and type 2 irradiation experiment, since different aspects of device behavior are exhibited in the different types of experiment. Failure to recognize this important difference has led in the past to misconceptions concerning the usefulness of certain models.

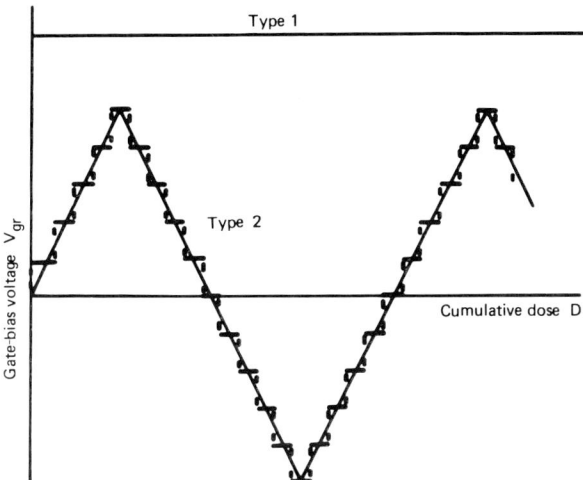

FIG. 8. Basic types of gate-bias conditions used during irradiation. Top curve labeled "type 1" represents constant gate bias. Lower sawtooth curve labeled "type 2" represents gate bias varied in discrete increments (positive upward, negative downward.)

A. Type 1 Irradiation

Figure 9 shows how the flat-band voltage is typically related to the cumulative dose for type 1 irradiation experiments at a variety of fixed values of V_{gr}. We can draw some conclusions from these curves concerning the general characteristics of type 1 irradiation. First, we observe that there is a significant difference in the basic features in the low- and the high-dose

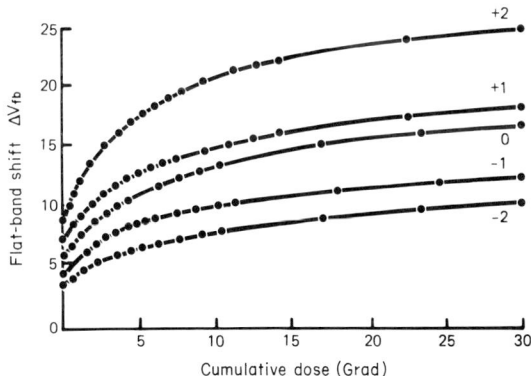

FIG. 9. Measured flat-band voltage for five values of fixed gate bias during irradiation (500-Å-thick oxide). Five devices used to obtain data.

ranges. In the low-dose range, the curves are close together and steep. In fact, for very low doses (not shown in Fig. 9), the curves are almost identical. [Our electron-beam system was designed for medium- and high-dose fluences. See Sze (*82*), p. 479, for a sample of low-dose irradiation data.] Conversely, at the high-dose end, the curves appear to be flatter and somewhat uniformly spaced, signifying the approach to dynamic equilibrium. The data of Mitchell (*83*) supports this general assessment. [See, for example, the curves for $V_{gr} = +2, +5$, and $+10$ V for dose values <0.1 Mrad in Fig. 39, Chapter 9 of Sze (*82*).]

In order to see more clearly the ramifications of these general features of the experimental data, it will be convenient to redraw the curves of Fig. 9 using a "piecewise linear" approximation. In this manner, we can emphasize the outstanding features of the low- and the high-dose ranges and deemphasize the details of the transition region between these two extremes. The type 1 irradiation curves shown in Fig. 9 can be redrawn as a highly idealized schematic curve, as shown in Fig. 10a. This is done to facilitate the analysis of the response of MOS devices to type 1 irradiation. A similar approach was taken by Boesch and McGarrity (*51*), as shown in their Fig. 7A. An analysis by means of these idealized curves will, of course, impart a significant alteration to certain aspects pertaining to the medium-dose region. But, at the same time this piecewise linear approach allows certain fundamental features of the response, which have previously been hard to distinguish, to emerge from obscurity and to be understood more easily.

In Fig. 10a, all the curves for an irradiation dose less than 7 units lie on the same straight line regardless of gate bias. As the dose is increased, the curves for the various values of gate bias split off one at a time and approach their respective high-dose asymptotes. The high-dose asymptotes in this piecewise linear approximation correspond to the nearly level region of the curves in Fig. 9. These horizontal regions are shown to be evenly spaced. [The curves shown in Fig. 39, Chapter 9 of Sze (*82*) support the choice for even spacing.] The curve for $V_{gr} = -3$ V has a smaller flat-band voltage than any other members of the family of curves. This minimum-voltage value agrees with experimental results (*74*).

B. *Type 2 Irradiation*

Figure 10b shows the same schematic device behavior as in Fig. 10a except that in this case V_{fb} is plotted with V_{gr} as the abscissa. Figure 10b derives directly from Fig. 10a. The horizontal parts of the curves in Fig. 10b correspond to the sloping parts of Fig. 10a, and *vice versa*. Although the families of curves in both figures are completely equivalent so far as information content is concerned, the two alternative forms of displaying the data enable one to gain valuable insight into irradiated MOS device behavior under both

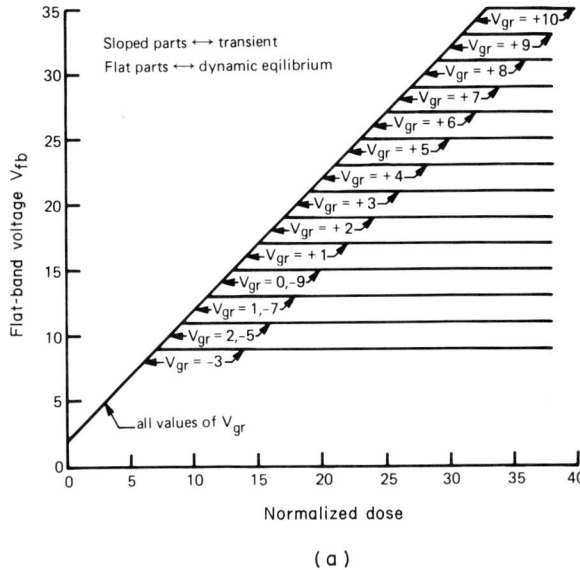

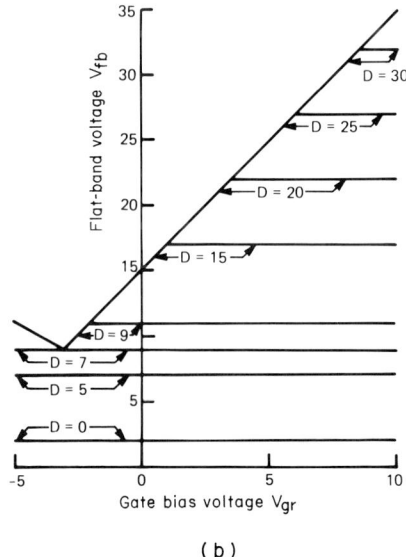

Fig. 10. Idealized schematic curves for analyzing certain fundamental features. (a) Flat-band voltage versus normalized dose for various fixed gate-bias values; (b) flat-band voltage versus gate bias for various cumulative doses.

irradiation conditions. This approach makes it possible to resolve what appear to be conflicting ideas in the literature.

From Fig. 10b we see that for low values of cumulative dose (e.g., $D \leq 15$ units) the flat-band voltage will essentially be independent of V_{gr}—except, possibly, for a small range of voltages near $V_{gr} = 0$. Conversely, for the high-dose case (e.g., $D = 30$ units) in the range $-3 \text{ V} < V_{gr} < 10 \text{ V}$, the flat-band voltage will be a strictly linear function of V_{gr}.

C. Applications

In their studies of oxide thickness dependence, Hughes et al. (47) and Derbenwick and Gregory (84) irradiated MOS devices using a constant value of applied field. Since they used different oxide thicknesses, it was also necessary for them to use different values of V_{gr} in order to maintain a constant initial electric field. The conclusions concerning thickness dependence require that the behavior be truly independent of V_{gr}.

Derbenwick and Gregory concluded that "There appears to be no difference between the constant field and constant voltage irradiations for either the flat-band or voltage shifts" (84). These conclusions are consistent with the $D = 15$ curve in our Fig. 10b. We conclude, therefore, that their irradiation would fall near the low-dose range or, in other words, in what we term the transient part of the response.

Freeman and Holmes-Siedle (87) described a model for predicting radiation effects in MOS devices based on a simple charge-sheet buildup in the oxide. The growth curves in their Fig. 4 are very much like our Fig. 10b except that they are plotted on logarithmic scales.

Curves like $D = 15$ units, together with simple models to explain the voltage independence, have appeared in the literature (47, 84). Curves like $D \geq 30$ units have also appeared in the literature (42, 77), but of course this type of curve requires a totally different model (74). Although the difference between these two modes of behavior (transient versus steady-state) is quite striking, it is only rarely that any attempt has been made to distinguish between low- and high-dose cases in the literature (85). One possible reason for the failure to recognize or comment on the differences is that by far the majority of published data fall in the low- to medium-dose (i.e., initial transient) category.

In the past, there has been a tendency to treat the physical models corresponding to the two separate cases (low- and high-dose) as if they were somehow in competition with one another to obtain acceptance. One of the models for the low-dose type of behavior (to be discussed in Section IV) has, in fact, been almost universally accepted as valid even though it fails to account for the high-dose type behavior, whereas models based on high-dose behavior have not been widely accepted even though they do agree with much of the experimental data. However, as Fig. 10a and b so graphically show, the two types of models are not mutually exclusive at all: They are complementary! The two divergent modes of behavior do not actually represent either conflicting data or conflicting models but, rather, they represent two extreme ranges of cumulative dose values. It should not be claimed that a specific model based on a high-dose behavior is inappropriate because it differs from an accepted model which is based only on the low-dose behavior. Instead, it is necessary to recognize that both types of models might be valid; only in different ranges of D. Both types of model are useful, each in its own way. To really be complete, of course, it would be necessary to have a model that simultaneously describes the complete spectrum of behavior—from the low-, through the medium-, and including the high-dose ranges—and accounts for the behavior when $V_{gr} < 0$. One such model (74) that has been proposed will be discussed in Section V.

As we have seen, the dynamic irradiation response at fixed V_{gr} (i.e., type 1) yields a certain class of data typified by Fig. 10a. As useful as this kind of information truly may be, certain important features of the radiation-induced changes in the electrical behavior of MOS devices probably would not have been detected at all from an exclusive reliance on type 1 experiments. However, these features are easily detected in type 2 experiments. Once again we encounter a situation where two different experimental methods have often been treated as if they were in competition, whereas in actual fact they were complementary! We will now consider the nature and some of the ramifications of type 2 irradiation experiments. Here, too, it is crucial that we be aware that there will be dramatic differences between the low- and high-dose cases.

Figure 11 shows typical experimental results for type 2 irradiation in the three different ranges of cumulative dose. The arrows on the curves indicate the polarity of the voltage increments in V_{gr} between successive irradiation increments. Once again each data point represents the flat-band voltage taken immediately after the termination of an increment in irradiation. After the resulting new value of V_{fb} had been determined, the gate voltage was set to the next value called for in the sequence and then the electron beam was reapplied to commence the next dose increment.

First, consider the characteristics unique to the medium-dose range. Run 1 in Fig. 11a shows the results of an irradiation experiment in which the initial irradiation was administered at $V_{gr} = 0$. This figure shows that V_{fb} was increased after each irradiation increment starting with the first data point at $V_{gr} = 0$, continuing through each of the data points up to the maximum value, $V_{gr} = +4$ V. All of the data points in run 1 were obtained at a dose of 1.5×10^8 rads/point. After the data point at $V_{gr} = +4$ V had been obtained, the gate voltage was set at $V_{gr} = +5$ V and run 2 (with negative V_{gr} increments) was begun. In order to facilitate the taking of data,

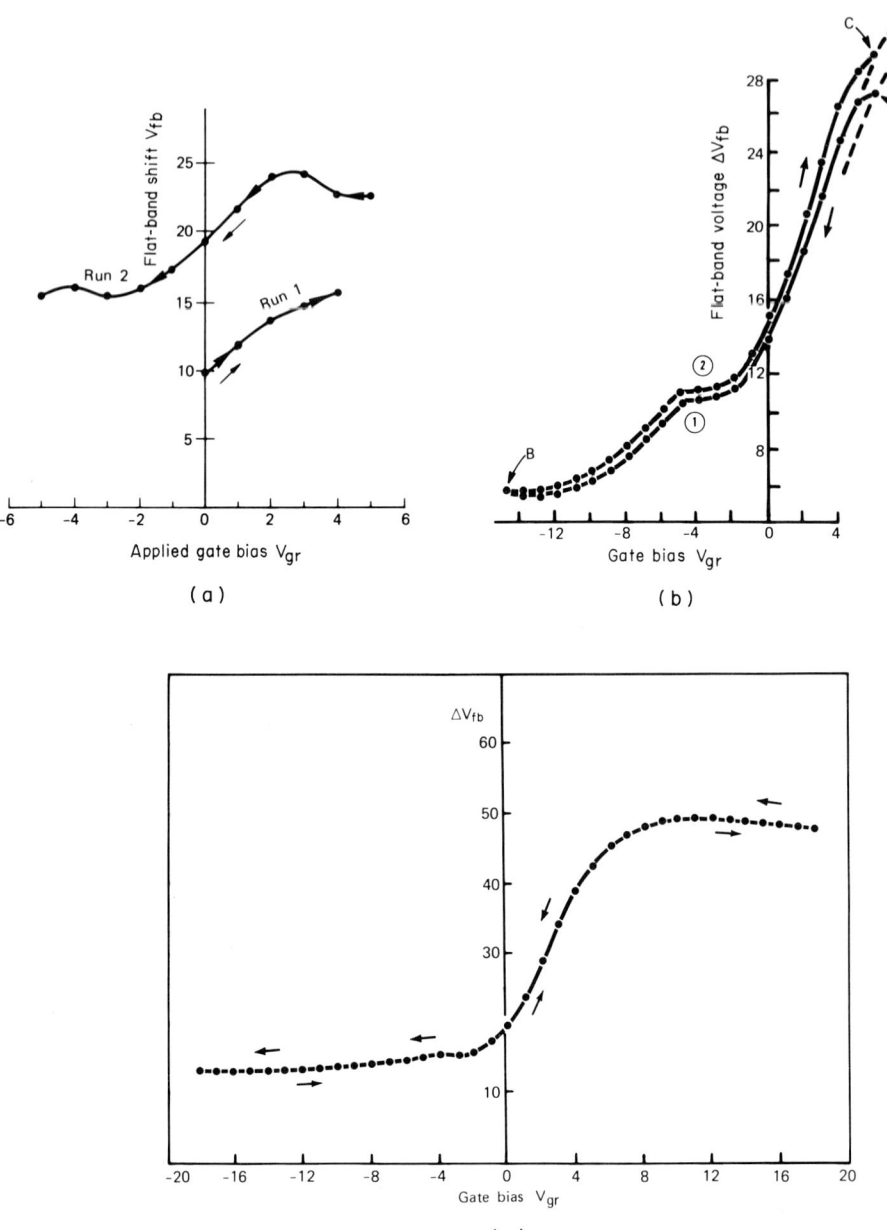

Fig. 11. Measured flat-band voltage versus gate bias. (a) Run 1, 1.5×10^8 rads/point; run 2, 1.5×10^9 rads/point (500-Å-thick oxide); (b) 10^9 rads/point; (c) 2.5×10^9 rads/point (700-Å-thick oxide).

run 2 was performed at a dose of 1.5×10^9 rads/point. (It had already been determined that the results were independent of dose rate.) As might be expected, the flat-band voltage at $V_{gr} = +5$ V was greater than those previously obtained at $V_{gr} = +4$ V. And, as one also might expect, subsequent data points at $V_{gr} = +4$ V and $+3$ V also showed that the flat-band voltage increased with increasing cumulative dose. For subsequent doses, (i.e., those for $V_{gr} < +3$ V in run 2) we observe a new effect that did not appear in the type 1 irradiation. That is, the flat-band voltage *decreased* rather than increased, even though the cumulative dose had increased! (As will be shown later, this does not conflict with the type 1 data but, rather, enhances it.) Type 2 data, therefore, shows that V_{fb} sometimes decreases even as the cumulative dose increases. This fact could not be directly determined by type 1 data alone. However, it is an essential fact since it shows dramatically that valid models for the irradiation-induced change in MOS characteristics *must* allow for V_{fb} to decrease as well as to increase, even though D continues to increase monotonically with time. Some widely accepted models, unfortunately, do not conform to this requirement! Since the decrease occurred even while V_{gr} remained positive, it must have been the decrease in *magnitude* of applied field alone, rather than the polarity of the field, that caused the flat-band voltage to decrease. As the negative increments in gate bias were continued, the measured flat-band voltage continued to decrease for a while, until at $V_{gr} = -3$ V a point was reached where the flat-band voltage reached a minimum value.

In Fig. 11b, we show a curve for data belonging slightly above the boundary between the medium- and high-dose ranges. (In type 2 irradiation, it is not always easy to specify a unique dose range since the device has received a different cumulative dose by the time it reaches the endpoint than it had at the initial point of the curve.) Notice that at these higher dose levels the two branches of the curves are rather close together, and the relationship to the linear behavior in the high-dose portion of the curves in Fig. 10b is becoming more apparent.

In this case, the total difference in flat-band voltage at the initial point A and final point C is only about 2 V even though the device has accumulated 5×10^{10} rads more by the final point than it had at the initial point. Thus we see that as the device acquires more and more cumulative dose the various branches of the type 2 curves begin to fall closer and closer to one another and to approach a general asymptotic structure similar to the curve for $D = 30$ in Fig. 10b. In other words, for these higher dose levels the flat-band voltage after irradiation has almost the same value whether the previous gate voltage was greater than or less than the present value of V_{gr}. Furthermore, as we will see later, approximately the same value is found for flat-band voltage for both type 1 and type 2 irradiations performed at the same value of V_{gr} and D. Notice, again, that in curve 1 the flatband voltage is *decreasing* during each additional irradiation increment.

Note that, after the two branches of the curve have been traced out from A to B to C, a total increase in cumulative dose amounting to 5×10^{10} rads has brought about only a 2-V difference between the initial (A) and final (C) values of V_{fb}. This small difference is found even though V_{fb} went all the way from 27 V at point A down to about 5 V at point B and back up to 29 V at point C during the sequence where V_{gr} was cycled from $+6$ V to -15 V and back again to $+6$ V. Thus, even after the flat-band voltage moved down and up again by more than 20 V (or approximately 400%), the final value of V_{fb} differed from the initial value by only about 7%.

Another significant feature to observe is that the minimum which appeared in the curve of Fig. 11a is not seen at all in Fig. 11b. This minimum is often quite pronounced at the lower dose levels. However, as shown in Fig. 12, it grows progressively less significant as the cumulative dose continues to increase, until at the dose levels in Fig. 11b and c it has become more like a narrow plateau than a minimum. For those dose levels where the minimum does exist, the minimum flat-band voltage increases with increasing radiation dose and also moves to more negative values of gate bias. As will be shown in Section IV, there is reason to believe that the minimum is related to the transition between accumulation and inversion conditions in the silicon.

The variation in value and position of this minimum as D increases may be seen by comparison of the curves in Fig. 12. A similar nonzero minimum effect is observed in gate threshold voltage shift curves for irradiated MOS transistors. See, for example, Fig. 6 of Barry and Page (*43*) and Fig. 2 of Maier and Tallon (*86*). The existence of what we have termed an NZNZ minimum has already been reported (*72*).

Curves 1 and 2 in Fig. 11b represent the case where V_{gr} was stepped back and forth between the two extreme values, $+6$ and -15 V, in 1-V increments. Consider next a different, but related case where V_{gr} is switched directly from $+6$ to -15 V and back again to $+6$ V without any intermediate values. Let us assume that when the cycle begins, the device is in the conditions corresponding exactly to point A in Fig. 11b. Let us assume, further, that the irradiation times are adjusted so that the total cumulative dose at the end of the two subsequent irradiation increments at -15 and $+6$ V are exactly the same as the values of cumulative dose indicated for points B and C, respectively, in Fig. 11b. The flat-band voltage is measured after each of the irradiations at -15 and $+6$ V, and when these results are plotted directly on Fig. 11b, it is found that these two new data points will fall directly on top of the original data points B and C, respectively. This demonstrates quite dramatically the nearly total lack of history dependence in the results.

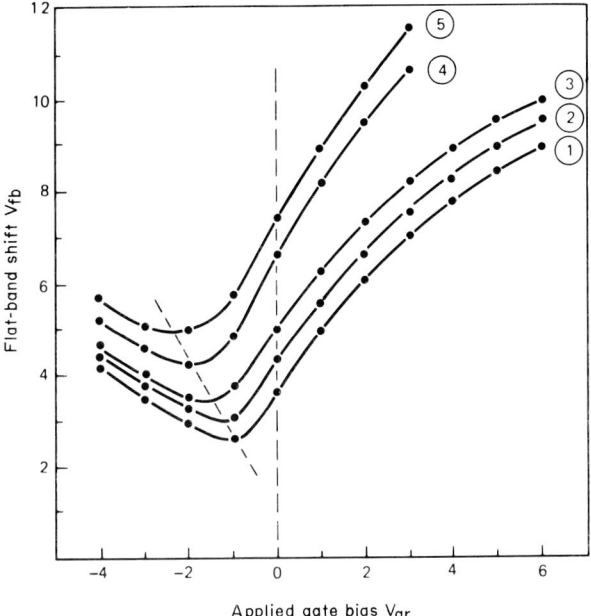

FIG. 12. Measured flat-band voltage versus gate bias at low and medium cumulative dose values. Note relocation of minimum for each curve as cumulative dose was varied:

Curve	Dose rate (rads/sec)	Dose per point (rads)	Cumulative dose (rads)	Time per point (sec)
1	10^4	2.4×10^6	2.6×10^7	240
2	2×10^4	2.4×10^6	2.6×10^7	120
3	1.2×10^5	3.6×10^6	4.0×10^7	30
4	1.7×10^6	5.3×10^7	4.7×10^8	30
5	1.7×10^6	5.3×10^7	9.6×10^8	30

Furthermore, when one performs type 1 irradiations at +6 and −15 V, each with the same respective doses as indicated for points A, B, and C, then, once again the corresponding data points fall directly on top of the previous points for type 2 irradiations. (The locus of curves obtained by type 1 techniques is indicated by dashed lines in Fig. 11b.)

The data in Fig. 11c show the very high dose level case. As indicated in the figure, for these high-dose cases, V_{gr} can be stepped back and forth over and over again without any observable change in structure or position of the curve. Thus, at these high dose levels a condition of dynamic equilibrium has been achieved in which, due to a balance between opposing physical processes, the charge distribution in the oxide is unaffected by further increases in cumulative dose. Note that, even though the shape of the curve has stabilized, the actual value of V_{fb} varies periodically (between the high value at the right end of the curve and the low value at the left end) as V_{gr} is cycled repeatedly back and forth between the two extreme values. Thus, the actual distribution of charges in the insulator is variable, depending on the value of V_{gr}, but is essentially independent of D (in this dose range). We conclude from these results that once a minimum value of cumulative dose has been achieved, the positive charges already induced in the oxide are rapidly redistributed during each irradiation to achieve some kind of balance between competing physical processes. A proposed model capitalizing on this redistribution will be described in Section IV.

Although the charges themselves can be almost completely removed from the oxide by thermal annealing, we find that only a small additional dose of irradiation is needed to totally restore the device to its dynamic equilibrium state corresponding to the given total dose. In other words, even though annealing has removed the trapped charges, the device "remembers" that it has previously been irradiated and what the previous total dose was.

It has been reported that the so-called "memory" of the damage can be recovered rapidly by restoring the original applied bias during irradiation (*88*). Such a conclusion is a gross oversimplification of a rather complex combination of processes that accompany irradiation of an MOS structure. It is known that the absorbed radiation creates lattice-site changes in the bulk oxide as well as at the interfaces (*79, 89*). However, as we have reported (*90*), irradiation of devices where the absorbed dose exceeds 10^8 rads revealed that the flat-band shift was uniquely determined by the *final* irradiation conditions *only*. The structures did not lose their memory of cumulative dose but did lose memory of all *previous* gate-bias history. The complete understanding of the mechanisms involved in describing this phenomenon is tied to a rather complex charge redistribution process mentioned above. A proposed model that capitalizes on this so-called memory effect will be given in Section IV. The model to be presented in Section V fully accounts for observed memory effects.

In type 2 experiments, V_{gr} can be varied by the investigator in a completely arbitrary manner as D continues to increase monotonically with time. The overall structure of the resulting plots of V_{fb} versus $V_{gr}(D)$ will be uniquely determined by the specific functional relationship between V_{gr} and D that was chosen. Consider, for example, the specific function $V_{gr}(D)$ shown in Fig. 13a. It is instructive to determine the flat-band voltage at the end of each irradiation increment for this example. In order to simplify the analysis, we will assume that history-dependence is totally nonexistent, so that the flat-band shift will be entirely determined by the piecewise-linear family of approximate curves shown in Fig. 10a and b and by the chosen form of $V_{gr}(D)$.

By reading the value of V_{fb} from Fig. 13b that corresponds to the specified combination of V_{gr} and D at the end of each horizontal line segment in Fig. 13a, one can easily construct the hypothetical (but realistic) curve shown in Fig. 13c. Since Fig. 13b is a piecewise-linear approximation to the actual curves, we should expect that Fig. 13c likewise will be a piecewise-linear approximation to the curves which would be obtained in an actual experiment using the specific function $V_{gr}(D)$ defined in Fig. 13a.

By tracing out the curve in Fig. 13b on a point-by-point basis, one can easily verify that the line segments labeled a, b, f, and g in Fig. 13c all correspond to points on the sloping line in Fig. 13b. In other words, these particular line segments are traced out while the combination of V_{gr} and D falls outside the dynamic equilibrium range. Conversely, the line segments labeled d, e, and h in Fig. 13c correspond to data points in Fig. 13b where the device is in the dynamic equilibrium conditions. The segment labeled c is traced out during the transition between the tilted and the horizontal segments in Fig. 13b.

As the first six data points are traced out, the dose is increasing but the applied gate voltage is also increasing. Hence, the device is being pulled farther away from dynamic equilibrium conditions with each increase in V_{gr}. For the next five data points (segment b) the dose still continues to increase, but now, as shown in Fig. 13a, the gate voltage is decreasing with each step; hence, as can be seen by the dashed line in Fig. 13b, the device is drawing closer to dynamic equilibrium. Finally, at $V_{gr} = 0$ V the device falls exactly at the transition point between the nonequilibrium and the dynamic equilibrium modes. For subsequent irradiation increments the cumulative dose is great enough that for relatively small values of V_{gr} the device is able to sustain dynamic equilibrium. However, after V_{gr} again begins to increase with each step (see Fig. 13a), the "safety margin" for dynamic equilibrium decreases, until at $V_{gr} = +4$ V the device once again is pulled into the sloping segment of the device curves in Fig. 13b. This causes a kink to appear in the curve between segments e and f of Fig. 13c even though, as shown in Fig. 13a, the sequence of values of V_{gr} and D have both continued to increase in a uniform manner. The nondynamic equilibrium behavior exhibited in segment f would continue indefinitely if it were not for the sudden switch from positive to negative increments in gate-bias voltage occurring at $V_{gr} = +10$ V and $D = 27$ units as shown in Fig. 13a. In segment g, the device is once more being brought closer to a dynamic equilibrium mode, since V_{gr} is decreasing between irradiation increments. Finally, for segment h dynamic equilibrium is actually achieved. Note that

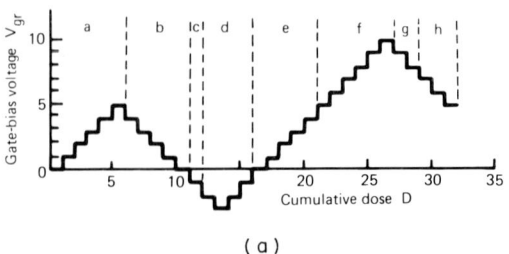

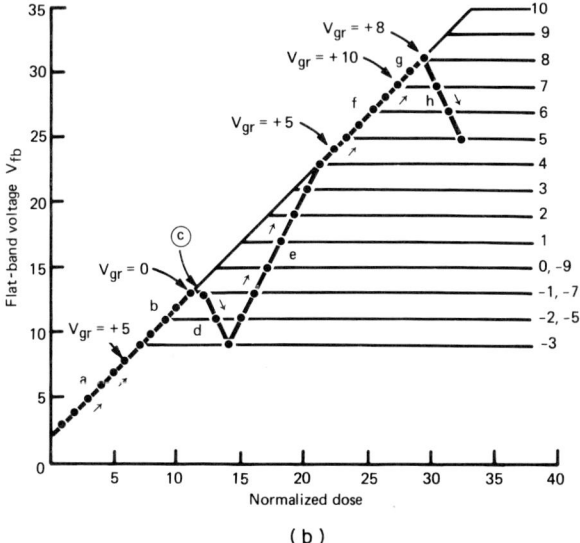

FIG. 13. Idealized schematic curves showing functional relationships between flat-band voltage, gate bias, and cumulative dose. (a) Stepwise variation of gate bias versus cumulative dose; (b) flat-band voltage versus normalized dose for given gate-bias values; (c) flat-band voltage versus gate bias for particular dose increments.

if V_{gr} continued indefinitely to decrease by 1 V at each irradiation step, then the extension of line h would fall exactly on top of line segment e. In general, for this approximation the slope for the nonequilibrium segments is exactly equal to one-half of the slope for the segments corresponding to dynamic equilibrium behavior. Obviously if the corners were to be rounded off in the piecewise linear approximation of Figs. 10a and b in order to make the curves more realistic, then the corresponding corners in the results shown in Fig. 13c would likewise be rounded off a bit.

Results similar to the curves shown in Fig. 13c are actually observed

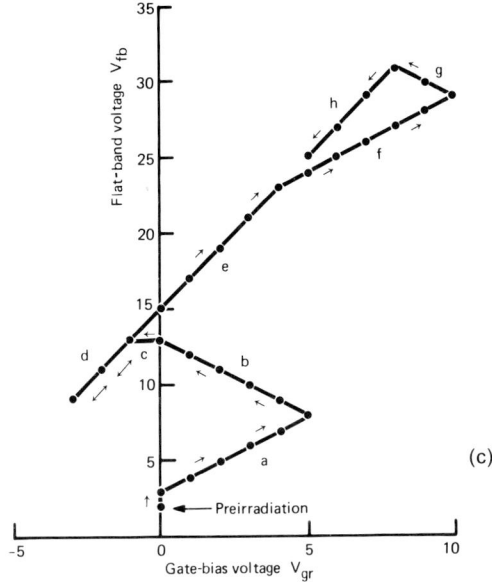

Fig. 13. (Continued)

experimentally. In Fig. 14 we show data obtained from a 500-Å device. This irradiation sequence started at the left-hand end of curve 1 ($V_{gr} = -1$ V) and proceeded toward the right, using positive increments in V_{gr} until the end point at $V_{gr} = +16$ V had been reached. Note that there is a rather sharp break in the slope of curve 1 at the point $V_{gr} = +1$ V. This break in the slope is comparable to the cusp at the juncture between line segments e and f in the upper curve of Fig. 13c. After reaching the end point in curve 1 of Fig. 14, V_{gr} was returned to -1 V and further irradiation increments were applied to trace out curve 2. Notice how curve 2 extends smoothly from the linear segment of curve 1 at the cusp. Both the value and the slope of curve 2 match the linear segment of curve 1 precisely at this point. This smooth connection between the two portions of the curves is analogous to the smooth extension of line segment h (high dose) into line segment e (low dose) in Fig. 13c. In the past, the occurrence of this sharp cusp in the experimental data has been a mystery, but now we see that it is just a natural consequence of the general nature of MOS device response as emphasized for easy analysis by the piecewise-linear approximation in Fig. 10.

It should be pointed out that in a type 2 experiment some finite amount of radiation must first be incident upon a device before there can be any radiation-induced change in flat-band voltage. If, for example, the next dose increment after a change in the gate-bias voltage is vanishingly small, then

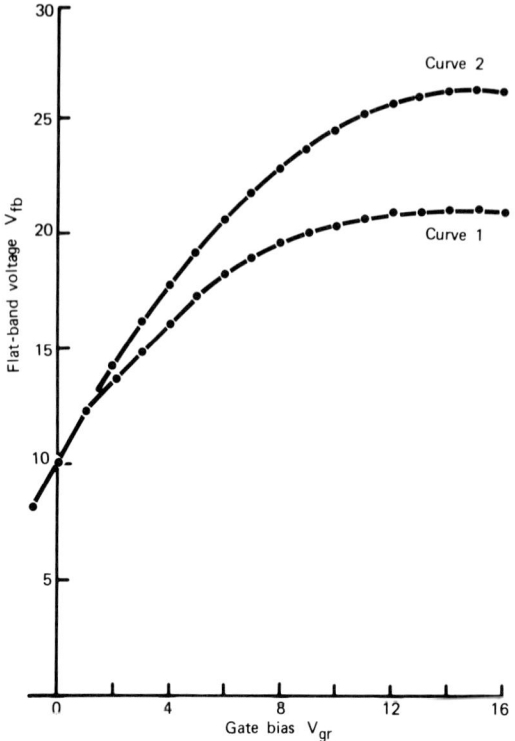

FIG. 14. Measured flat-band voltage versus applied gate bias for 500-Å-thick oxide. Curve 1, dose 4×10^7 rads/point; curve 2, dose 4×10^8 rads/point.

the charges will remain frozen in their previous locations and V_{fb} will not be able to change at all. This means that our previous statements concerning the absence of any history dependence cannot be considered to be an *absolute* pronouncement. Therefore, it is important to determine exactly the conditions under which history dependence can or cannot be neglected. In order to define these limits, the following experiments involving combinations of type 1 and type 2 irradiation were performed (*90*).

Numerous MOS devices were irradiated to cumulative doses of at least 5×10^{10} rads. Families of curves were generated for these devices showing flat-band shift versus cumulative dose while the gates were kept at fixed gate-bias values. For the devices we used it was determined how much flat-band shift occurred as a function of cumulative dose. For each device, a fixed gate-bias range from -10 to $+20$ V was used to obtain the family of monotonically increasing flat-band curves.

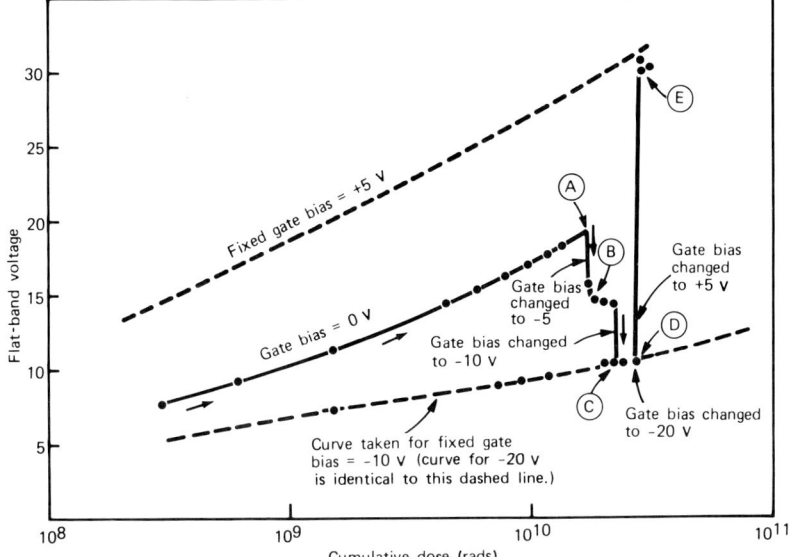

FIG. 15. Measured flat-band shift versus dose for three gate-bias values. Bias was held at 1 V until reaching point A; then the bias was changed as indicated; $W_{ox} = 500$ Å.

Figure 15 shows some of the results of this aspect of our study. The two dashed curves represent measured flat-band shifts for two different applied gate biases. The upper curve represents data taken for a gate bias of $+5$ V. The lower curve represents data taken for a gate bias of -10 V. (All experimental runs for gate biases between -10 and -20 V produced identical results, as would be expected from the flat nature of the curve for this voltage range in Fig. 11c.) The figure shows data over three decades of cumulative dose starting at 10^8 rads.

An MOS capacitor was biased at $V_{gr} = 0$ V and irradiated with prescribed radiation increments to produce the monotonically increasing curve at the center of the figure. When the device had absorbed 1.7×10^{10} rads (point A) the gate bias was changed to -5 V. Then the device was given an additional radiation increment of 3×10^8 rads. At the completion of that increment, the device achieved the same flat-band value it would have had if it had been maintained with a gate bias of -5 V (on the curve located approximately halfway between $V_{gr} = 0$ and $V_{gr} = -10$ V but not shown on this figure). This is shown as point B.

Several more increments of 3×10^8 rads were absorbed by the device to determine if any further flat-band shift would occur. As can be seen, no significant change occurred and the device had, essentially, reached a dynamic equilibrium value.

At point C the device was biased at -10 V and given another 3×10^8-rad dose of ionizing radiation. The flat-band value now dropped to the $V_{gr} = -10$ V curve *as if* it had received a total cumulative dose of 2.5×10^{10} rads at a fixed value of $V_{gr} = -10$ V. A few more incremental doses of 3×10^8 rads each caused the device to follow the $V_{gr} = -10$ V curve to the right.

At point D the device was biased at $+5$ V and given a dose of 3×10^8 rads. As expected, the flat-band value now moved upward and became fixed at the dashed curve at E for $V_{gr} = +5$ V. Further irradiation with $V_{gr} = +5$ V caused the device to monotonically increase in flat-band shift toward the right on this dashed curve.

In summary, the approach to dynamic equilibrium conditions comes about by applying the new gate bias and then irradiating the device with an increment that is on the order of 1% of the total dose that the device has ever received. However, if the cumulative dose is less than approximately 10^8 rads, then the redistribution mechanism (to be discussed in Section IV) is incomplete and the 1% rule does not hold.

After the radiation study on a number of devices had been completed, they were placed in light-tight containers with the gates externally shorted to the substrates. At the end of given repeated time intervals, the flat-band voltages were measured. Subsequent plotting of the resulting data showed that the flat-band voltage monotonically decreased in time (*91*). In all the cases we studied, there was a two-component relaxation effect that occurred over a time period of minutes to months. We followed an observed "fast" component for the first segment of the data as well as the "slow" component which seemed to last for months. We attributed the flat-band voltage decrease to room temperature annealing effects.

This two-component relaxation behavior was also found in the simulation. One explanation for this complex behavior involves curvature in the band structure of the oxide following irradiation. More details of the distribution of trapped holes and electrons in the oxide, and the resulting potential minimum and corresponding barrier effects will be described in Section V. Both fast and slow components that matched the experimental data qualitatively were observed in the simulation, but we did not explore this behavior extensively.

A two-component relaxation effect was reported by Lindmayer (*92*). He attributed the fast component to electron flow across the Si–oxide interface. He explained the slow relaxation component in terms of a rising barrier process and the number of internally trapped electrons. We agree with his conclusions.

IV. Simple Descriptive Models

To achieve maximum accuracy in modeling radiation effects in MOS structures it is necessary to formulate the problem in terms of a complete set of coupled nonlinear differential equations. These differential equations must include terms representing each and every one of the physical mechanisms that are thought to be active in the MOS structure during irradiation. Furthermore, each of the terms must be expressed in the specific form that most accurately describes the particular mechanism it represents.

Although such a high degree of accuracy may occasionally be useful, the difficulties involved in setting up such a complex set of nonlinear equations, not to mention solving them numerically, may be prohibitive. For this reason it is usually more practical to make a trade-off between accuracy and simplicity. For example, if a slightly less accurate answer is acceptable, then some of the physical mechanisms which contribute only in a minor way to the overall response may be omitted from the differential equations. In this case the answers would still be acceptable but a numerical solution to the simpler set of equations would presumably be much less involved. Such a model, in which some of the less important mechanisms have been omitted but which is still accurate enough to require numerical solution by computer, will be discussed in some detail in Section V.

For many purposes (e.g., device design) it is not practical to use a model so complex that it requires extensive numerical solution. In cases where quick, closed-form answers to design problems are needed, it is adequate to take an entirely different approach—to let some highly simplified conceptual model be substituted for the differential-equation model. In this section we will briefly review a number of highly simplified models which have been proposed in the past as reasonably accurate representations of irradiated MOS structures.

Some of the earliest proposed models were based on the concept of dynamic equilibrium. These early models were useful to some degree in understanding what happens to MOS devices under irradiation, but they failed to agree with certain important aspects of the observed behavior. Thus emphasis gradually seemed to shift away from the early dynamic equilibrium concepts toward models involving transient phenomena. These also provided insight into some aspects of the observed behavior but contained serious deficiencies. For example, the dose range used for experimental verification of these proposed models fell neither in the low- nor the very-high-dose ranges. Strictly speaking, therefore, neither dynamic equilibrium nor transient models were appropriate for describing the observed results. Models formulated for the medium-dose range would have been much more appropriate. Recently, dynamic equilibrium models have enjoyed a resurgence in popularity. Our work was done in the high-dose range, and a dynamic equilibrium model was proposed (*90, 93, 94*) that provided a degree of success in reproducing the general shape of the V_{fb} versus $V_{gr}(D)$ curves over a substantial range of both positive and negative values of gate bias. Generally speaking, all of the simple models fall into one or the other of two main categories. One category is that of dynamic equilibrium models. The other is that of models based on the capture of mobile carriers as they sweep across the insulator under the influence of the applied electric field. For convenience, dynamic equilibrium models will be referred to as *DE models*, whereas models involving carriers that are captured during transit will be called *CDT models*. First, we will review the CDT models.

A. Captured during Transit (CDT) Models

It is assumed that the incident radiation induces mobile hole and electron pairs. The carriers in the oxide move rapidly toward either the Si–SiO$_2$ interface or the M–SiO$_2$ interface, depending on their charge and on the polarity of the applied voltage V_{gr}. In the CDT models it is assumed that holes encounter trapping centers as they move at a relatively small speed in the direction of the applied field, whereas electrons move in the opposite direction with relatively high mobility through an oxide which contains essentially no electron trapping centers. Due to the high mobility and the lack of electron traps, the electrons rapidly leave the oxide. However, some of the holes stay behind where they attach to hole traps in the oxide.

In these models it is also assumed that as long as the concentration of charged traps at position x and time t, $N_T^+(x)$, is very much less than the total concentration of trapping centers $N_T(x)$, then $N_T^+(x)$ will be proportional to the total number of holes that have been swept past that position prior to time t. For a sufficient irradiation dose, of course, the trapping centers could become completely filled with holes. In such a case the charge distribution in the oxide would become fixed but not necessarily uniform.

It has been common to assume that $N_T(x) = 0$ except for a very narrow region (100 or 200 Å wide) near the Si–SiO$_2$ interface, where a very dense concentration of interface traps would exist. Although there seems to be a considerable amount of experimental evidence in support of this assumption, there also appears to be substantial evidence supporting the case for trapping centers distributed throughout the oxide material. As we will show later, models based on a distribution of trapping centers throughout the oxide layer are also consistent with the data, whereas CDT models fail to agree with data obtained in type 2 experiments.

Nevertheless, in order to present a brief review of the CDT model it will be necessary to assume that interface traps only are present and active in the oxide. In this case, if $V_{gr} > 0$, holes would be swept toward the semiconductor and some of them would be trapped at the interface. When the irradiation is terminated, these trapped holes would be "frozen" into their position and would induce image charges in the semiconductor. The image charges would manifest themselves by shifting the flat-band voltage toward more negative values.

On the other hand, if $V_{gr} < 0$, then holes would all be swept *away* from the trapping centers and the traps at the Si–SiO$_2$ interface would remain neutral. Thus irradiation at positive gate bias would induce a negative increment in V_{fb}, whereas irradiation with negative V_{gr} would not produce an appreciable shift in V_{fb}. For irradiation at small but nonzero values of V_{gr} there would be only a very small applied field in the oxide, and hence the holes and electrons would tend to recombine in the same general area where they were generated (geminate recombination). In this latter case, there would be only a few holes that reach the interface traps and the induced flat-band shift would be minimal.

From the above review of the CDT models, one predicts that the flat-band voltage would be vanishingly small for irradiation at $V_{gr} < 0$ and would have some value independent of V_{gr} for irradiation at $V_{gr} \gg 0$. (It is important to note that, according to the CDT model, the flat-band voltage for $V_{gr} \gg 0$ must be a monotonically *increasing* function of cumulative dose.) For small positive values of V_{gr} there would be only a small shift in V_{fb}. Thus, the CDT models predict that for type 1 irradiation the relationship between V_{fb} and V_{gr} should be essentially as shown in Fig. 16. For sufficiently large values of cumulative dose, the flat-band voltage for $V_{gr} \gg 0$ might tend to approach some fixed value corresponding to the case where the available traps are saturated by holes.

Curves having the same general structure as those shown in Fig. 16 are sometimes obtained experimentally for $V_{gr} \geq 0$ from type 1 irradiation. (For

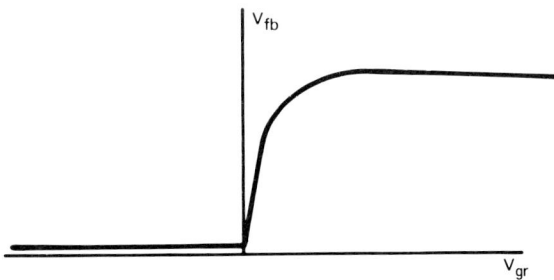

FIG. 16. Idealized flat-band voltage versus gate bias as predicted by the simple CDT model. Curve has relatively small value in negative gate-bias region and shows monotonically increasing values in the positive region.

$V_{gr} \leq 0$, however, the data differ substantially from those shown in Fig. 16.) In fact, the CDT models have had considerable success in predicting the oxide-thickness dependence of the radiation-induced flat-band shift. According to the theory, holes trapped in a thin layer near the Si–SiO$_2$ interface will impart a contribution to the flat-band shift proportional to the oxide thickness W_{ox} through the last factor in the equation

$$-S = (qN_T^+/\epsilon_{ox})W_{ox} \qquad (2)$$

where N_T^+ is the surface density of trapped holes. The term N_T^+ could also be dependent on oxide thickness if the radiation is able to penetrate into the bulk of the oxide. If, for example, the radiation penetrates through the oxide all the way to the Si interface, then the total number of holes that have been generated and have swept past the layer of traps at the interface will also be proportional to the first power of the oxide thickness. In this case the net flat-band shift will depend quadratically on W_{ox}. On the other hand, if the radiation consists of photons in the vacuum ultraviolet, then the radiation will only penetrate a small distance into the outer surface of the oxide and the number of holes swept past the interface traps will be totally independent of oxide thickness. In this case, the net flat-band shift will have only a linear dependence on W_{ox}.

The oxide-thickness dependence predicted by the CDT models has been observed experimentally by Hughes, Powell, and Woods (HPW) (47). For 1-MeV electron irradiation, HPW observed a quadratic dependence and, for the vacuum ultraviolet, they also observed the linear dependence expected in agreement with the predictions of the CDT models. Thus, it might appear that the CDT models are suitable for use in modeling radiation-induced changes in MOS structures. However, although they were successful in predicting the correct dependence on W_{ox}, the CDT models fail to agree with certain other aspects of the experimental data.

First of all if, as in the context of reference 47, we assume that N_T^+ continues to increase as long as holes continue to be swept past the traps at the interface (i.e., assuming that the system is not in dynamic equilibrium), then V_{fb} should also continue to increase along with the cumulative dose D. However, from both their vacuum ultraviolet and corona data HPW calculated $N_T^+ = 2.1 \times 10^{12}$ cm^{-2}. But this density is approximately the same value that Woods and Williams quote (81) for the maximum possible concentration because of the onset of tunneling at about this concentration of surface charges. If this were the case, then N_T^+ would have necessarily reached a *stable* value independent of the cumulative dose and in such a case the CDT models could not, strictly speaking, be operating at these dose levels. Actually, under such tunnel-limited operation the system would be operating in a state of dynamic equilibrium rather than in the CDT mode.

In this regard, it is interesting to note that in order to maintain a constant initial value of electric field in the oxide, it was necessary for HPW to apply a voltage which was itself proportional to W_{ox}. When the data in Fig. 2 of reference 47 are replotted versus V_{gr} rather than W_{ox}, it is found that

$$-S = -0.7 + 1.0 V_{gr} \tag{3}$$

Thus, except for a constant offset voltage (possibly due to work function differences) the flat-band voltage has adjusted itself during irradiation so as to be *exactly equal* to the applied gate voltage. This appears to be more like a dynamic equilibrium than a CDT situation. It is rather surprising to find $dS/dV_{gr} = -1.0$ in Eq. (3) since in the context of the CDT models N_T^+ should vary with cumulative dose and hence should not give such a slope except through a highly improbable coincidence. It is not clear from the data of reference 47 whether this is truly a coincidence or whether in fact the data represent some situation in which dynamic equilibrium is achieved when the flat-band shift precisely equals the applied gate voltage!

Second, as was pointed out earlier, V_{fb} must be a monotonically increasing function of D for any irradiation according to the CDT models. However, as was shown in Fig. 11 in Section III, the flat-band voltage sometimes *decreases* when the applied field is reduced by a small amount even though the *direction* of the applied field has not changed. This presents a problem for the CDT models, since it only allows for V_{fb} to increase as the interface continues to trap a fraction of all the holes that sweep past it. There is no mechanism provided in the CDT models for N_T^+ to decrease merely because the field intensity has been slightly reduced.

A third problem with the CDT models is that they fail to account for either the relatively high value of flat-band shift at $V_{gr} = 0$ or the observed minimum in the curves for V_{fb} versus V_{gr}. Thus, as we have seen, the CDT models have been able to accurately describe the observed shape of the curve for V_{fb} versus V_{gr} in the low-dose ranges for $V_{gr} \geq 0$ and have accurately described the oxide-thickness dependence. However they fail to describe certain critical aspects of the experimental data, as described above. By using type 1 and type 2 irradiation techniques in a cooperative manner, we have been able to test the CDT models more fully and to determine some of their strengths and weaknesses.

B. Dynamic Equilibrium (DE) Models

Some of the earliest models proposed were based on dynamic equilibrium concepts. Many aspects of the results obtained from these early models were in good agreement with the corresponding experimental data. A good review of a few of these early models can be found in the paper by Snow et al. (44). In these early DE models, it was often assumed that there was a uniform concentration of hole trapping centers in the oxide but that only those traps located in a single layer near the silicon interface would be occupied. The width or height of the layer would grow until most of the applied voltage V_{gr} appeared across the layer itself. Thus the electric field elsewhere in the oxide would be correspondingly reduced to an almost negligible value and hole–electron pairs subsequently generated in these zero-field regions would recombine before they could be captured by any trap sites. One such model (39) was able to correctly predict the polarity of gate bias required to induce a large charge in the semiconductor and the eventual saturation of this charge with increasing radiation dose. However, this model also predicted a square-root dependence of V_{fb} on V_{gr}, whereas later data revealed that there was actually a linear dependence. (It was later shown that the earlier data implying a square-root dependence had been obtained under conditions such that the cumulative dose fell below the values required for true dynamic equilibrium.)

MacDonald and Everhart (95) later proposed a model in which the traps were distributed in energy as well as in position. In their model, the total charge increased quadratically with the applied voltage, which in turn gave rise to a linear relationship between V_{fb} and V_{gr} in agreement with the experimental observations. However, the model appears to limit the expected range of linearity to about 3 or 4 V whereas the observed range of linearity sometimes extends to nearly 10 V (88). Various other models having a variety of assumed distributions of trapping centers were also proposed. Some of these (96) had the desired wide range of linearity but were difficult to justify on the basis of reasonable physical mechanisms.

We proposed a simple two-layer approximate model intended specifically to describe the *linear* range of the flat-band shift of irradiated MOS devices (93) (see, for example, Fig. 6). The model was based on dynamic equilibrium concepts and thus was intended to be valid only in or near the high-dose region. The model did predict a linear relationship between V_{fb} and V_{gr} over the necessary range of V_{gr}. Furthermore, the model was able to provide estimates of the width of the inner layer of trapped charge, and this value was found to be consistent with the measured width (54) of less than 100 Å as determined by vacuum ultraviolet, etch-back experiments, and 20-keV x rays.

It has been stated (77) that the range of V_{gr} for which this two-layer model would allow full linearity is much smaller than that actually observed experimentally. However, (1) no justification was given for this opinion and (2) we disagree with the opinion.

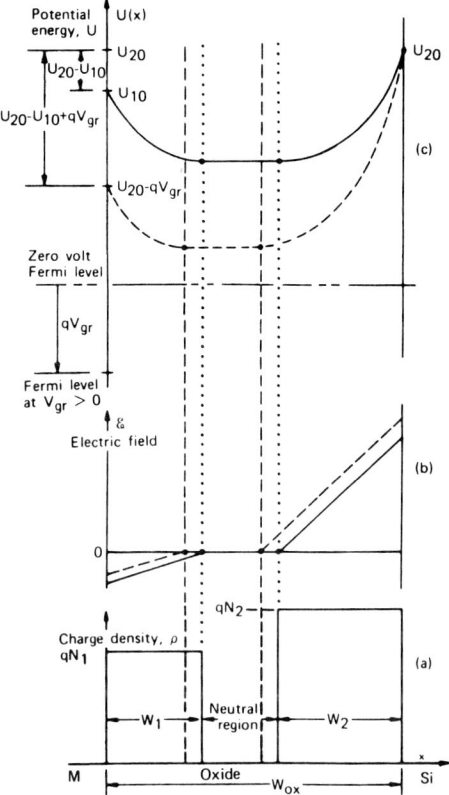

FIG. 17. Various two-layer distributions in the oxide region for two different gate-bias voltages. (a) Charge density distribution; (b) electric field distribution; (c) potential distribution.

According to the two-layer model (see Fig. 17) there is a uniform concentration N_1 of trap sites near the metal–insulator interface, but only those traps located in a layer of width W_1 adjacent to the metal interface are ionized. The model also calls for a uniform concentration of trap sites N_2 near the semiconductor–insulator interface, but only those traps within a layer of width W_2 adjacent to the semiconductor interface are ionized. As we showed in reference 93, since the vacuum potential must be continuous throughout the entire device, the layer widths must be

$$W_1(V_{gr}) = W_{10} - \alpha V_{gr} \qquad (4)$$

$$W_2(V_{gr}) = W_{20} + \beta V_{gr} \qquad (5)$$

where W_{10} and W_{20} are the equilibrium layer widths for $V_{gr} = 0$, and where

$$\alpha = \epsilon_{ox}[qN_1^{1/2}(N_1^{1/2}W_{10} \pm N_2^{1/2}W_{20})]^{-1} \qquad (6)$$

$$\beta = \pm\epsilon_{ox}[qN_2^{1/2}(N_1^{1/2}W_{10} \pm N_2^{1/2}W_{20})]^{-1} \qquad (7)$$

If we ignore work-function differences, then the flat-band voltage due to a positive charge distribution $\rho(x)$ in the oxide may be obtained from the expression:

$$V_{fb} = -(1/\epsilon_{ox}) \int_0^{W_{ox}} x\rho(x)\,dx \qquad (8)$$

For the two-layer charge distributions shown in Fig. 17a, Eqs. (4)–(8) yield the linear relationship

$$-V_{fb} = A + BV_{gr} \qquad (9)$$

where

$$A = q[N_1 W_{10}^2 + N_2 W_{20}(2W_{ox} - W_{20})]/2\epsilon_{ox} \qquad (10)$$

$$B = [N_2^{1/2} W_{ox}/(\pm N_1^{1/2} W_{10} + N_2^{1/2} W_{20})]^{-1} \qquad (11)$$

where q is the electronic charge, and ϵ_{ox} is the dielectric constant of the insulator material.

From Eq. (9) we see that the two-layer model predicts a strictly linear relationship between flat-band voltage and gate-bias voltage. According to this model, it is the *combined* influence of the two layers that is responsible for the linearity of the segment of the curve in Fig. 6 in the range $-1 \text{ V} < V_{gr} < +5 \text{ V}$. We also see from Eqs. (9) and (10) that the two-layer model predicts a nonzero value of flat-band shift even when the gate is shorted to the substrate during irradiation. Although this shorted-gate flat-band shift is seldom small enough to actually be considered negligible, it has often been ignored and sometimes even denied in the literature. The two-layer model shows that this effect is a natural consequence of the energy-band structure of the MOS system. An extension of the two-layer model also accounts for the minimum in the flat-band shift curves, since according to Eq. (5), layer 2 vanishes when $V_{gr} \leq V_{min}$, where $V_{min} = -W_{20}/B$. Since W_1 continues to depend on V_{gr} even when $W_{20} = 0$, the result is that the flat-band shift is an increasing function of $|V_{gr} - V_{min}|$, in qualitative agreement with the data shown in Fig. 12.

Another attribute of the two-layer model is that it provides a prediction concerning the oxide-thickness dependence of V_{fb} after irradiation. In Eq. 11, we see that if $(N_1/N_2)^{1/2} W_{10} \ll W_{20}$ and if W_{ox} is very much larger than W_{20}

then the slope B will be approximately equal to W_{ox}/W_{20}. For 500-Å-thick devices, taking $W_{20} = 100$ Å (54), this gives $B = 5$ V/V, which differs by only about a factor of 2 from the measured value of 2.2 V/V from Fig. 6. Furthermore, if V_{gr} is chosen proportional to W_{ox} (as it was in reference 47), with B approximately proportional to W_{ox}, then Eq. (9) shows that $-(V_{fb} + A)$ will depend *quadratically* on the oxide thickness. Thus, even though the two-layer model was intended only for order-of-magnitude calculations, it still agrees very well with the experimentally observed dependence on W_{ox}, the slope B, the minimum in flat-band shift curves, and the 0-V flat-band voltage A. As was discussed above, this quadratic dependence of V_{fb} on W_{ox} has also been explained using a CDT model in which trap centers in the bulk were excluded. Thus, we have two entirely different models, both of which have the proper dependence on W_{ox}. Unlike the CDT model, however, the two-layer model also allows for both increasing and *decreasing* values of V_{fb} during type 2 irradiation.

There is another very promising model that also involves a two-layer charge distribution and also predicts an approximately linear region in the V_{fb} versus V_{gr} curves. This model can be derived as an extension and modification of the work on mobile ions in oxides by Marciniak et al. (97) and by Tangena et al. (98). In addition to covering the linear region, the model also predicts results outside the linear region that resemble the curves in Fig. 11b.

The limiting cases of our work included an equilibrium charge distribution basically resembling the charge distribution found in Marciniak and Przewlocki (97) and Tangena et al. (98). We calculated the radiation-induced flat-band shift in MOS structures for such distributions. Marciniak and Tangena considered the static distribution of mobile ions in the dielectric layer for a constant total number of charges in dynamic equilibrium distributions under the combined influence of drift and diffusion mechanisms. They did not, however, calculate the flat-band shift. Their results gave a closed-form solution to a relatively simple case. In a previous publication we presented a non-closed-form (i.e., computer-simulation) solution for a more extensive case for a similar MOS configuration. Since some aspects of our results for radiation-induced charge generation and transport in the dielectric resemble the main results of Marciniak and Tangena (97, 98), it occurred to us that there might be a fundamental reason for the striking similarities in the charge distributions.

Although we used a similar model, it must be emphasized that the charges in one case (no irradiation) are inherently mobile, whereas in the other case (irradiation) the charges are immobile until the oxide is irradiated. This is a slight departure from our previous conclusions, but we wish to present an alternative view, both for the sake of completeness and because it has considerable merit.

Marciniak and Tangena assumed that the charge motion is governed by drift and diffusion and obtained the following results (expressed in our notation),

$$N_T^+(x) = \frac{\epsilon_{ox}}{q\beta^2 W_{ox}} \left[\frac{\phi^2}{\cos^2\{\phi(x/W_{ox}) + \arctan[(\beta^2 W_{ox} E_0)/\phi]\}} \right] \quad (12)$$

$$E(x) = \frac{\phi}{\beta^2 W_{ox}} \tan\left[\phi\left(\frac{x}{W_{ox}}\right) + \arctan\left(\frac{\beta^2 W_{ox} E_{ox}}{\phi}\right) \right] \quad (13)$$

Where ϵ_{ox} is the dielectric constant of the oxide, q is the electronic charge, $\beta^2 = \mu/2D$, μ is the charge mobility, D is the diffusion constant, $\phi = \alpha\beta W_{ox}$, E_0, and E_0' are, respectively, the electric field and its gradient at the metal–oxide interface, and $\alpha^2 = E_0' - \beta^2 E_0^2$. It is assumed that μ and D satisfy the Einstein relation $D/\mu = kT/q$ at the elevated electron temperature T.

In order to maintain a constant total number N of charges per unit area, we must satisfy the constraint

$$N = \int_0^{W_{ox}} N_T^+(x)\, dx \quad (14)$$

which, with Gauss' law, leads to

$$E(W_{ox}) - E_0 = qN/\epsilon_{ox} \quad (15)$$

When Eqs. (13) and (15) are combined they yield

$$E_0 = \left(-\frac{qN}{2\epsilon_{ox}}\right)\left\{1 + \left[1 + \frac{\left(\frac{2}{W_{ox}}\right)\left(\frac{qN}{2\epsilon_{ox}}\right)\left(\frac{\phi}{\tan\phi} - \frac{\phi^2}{\beta^2 W_{ox}^2}\right)}{\beta^2(qN/2\epsilon)^2}\right]^{1/2}\right\} \quad (16)$$

Equation (16) is a constraint imposed on E_0 and ϕ due to the constancy of N. By integrating the electric field in Eq. (13) from $x = 0$ (metal interface) to $x = W_{ox}$ (silicon interface), and setting this result equal to $(V_{WF} + V_{gr})$, where V_{WF} is the work-function difference between the metal and the semiconductor, we obtain

$$\int_0^{W_{ox}} E(x)\, dx = V_{WF} + V_{gr} \quad (17)$$

After substituting Eq. (13) into Eq. (17) and solving for E_0, we obtain

$$E_0 = \frac{1}{\beta^2 W_{ox}} \left[\frac{\cos\phi - \exp[-\beta^2(V_{WF} + V_{gr})]}{\sin\phi/\phi} \right] \quad (18)$$

Equation (18) represents an additional constraint imposed on E_0 and ϕ due to the definition $E = -\nabla V(x)$. Equations (16) and (18) are simultaneous equations in the two parameters E_0 and ϕ that have been evaluated. The flat-band shift is given by Eq. (8). When the integration is carried out, using the previous equations, we obtain the following:

$$-V_{fb} = -(V_{WF} + V_{gr}) + (\phi/\beta^2)\tan\{\phi + \arctan[\beta^2 W_{ox} E_0/\phi]\} \quad (19)$$

Knowing the values of E_0 and ϕ from the simultaneous solutions of Eqs. (16) and (18), it is possible by means of Eq. (19) to plot flat-band voltage versus applied voltage. [Actually, the form of Eq. (19) involving trigonometric functions is only valid when $\alpha^2 \geq 0$. When $\alpha^2 < 0$, the trigonometric functions of Eqs. (12)–(19) become hyperbolic functions. These calculations are similar to those given in Eqs. (12)–(19) and are not elaborated upon here.] The result of the calculations for both positive and negative values of α^2 is shown in Fig. 18. Basically, the trigonometric form shown in Eq. (19) produces the nearly linear segments of the curves in Fig. 18, whereas the hyperbolic form produces the remaining parts of the curves.

Figures 18a,b, and c, respectively, show curves for three different values of N. In addition, for Figs. 18a and b we show the curves for three different values of electron temperature. The curves labeled 1 correspond to $\beta^2 = 0.25$, the curves labeled 2 to $\beta^2 = 0.5$, and the curves labeled 3 to $\beta^2 = 0.75$. One notes that the linear segments of the curves are much steeper at the low temperatures than at the high temperatures. The intermediate-temperature curves (labeled 2 in the figure) have all been replotted on the same axis in Fig. 18d for easy comparison. One notes the resemblence between the curves in Fig. 18d and the curves in Fig. 6.

The flat-band condition is indicated with dashed lines in Fig. 18. For points on the curves which lie above the flat-band line, the device is in inversion during irradiation. For points lying below these lines the device would be in accumulation during irradiation. Thus there is a range of 2 or 3 V in the vicinity of the dashed lines in which there is a transition between inversion and accumulation. We feel that this may be the source of the minima and/or the plateaus in the curves of Figs. 6, and 11a,b. In Fig. 18 all the curves are drawn as if the device always stayed in the inversion mode. The necessary modification of the model to incorporate the transition from inversion to accumulation has not yet been attempted. We expect, however, that the net effect of the transition would be to shift the portions of the curves lying below the dashed lines to the left by approximately 1 or 2 V. This would result in a plateau-like structure similar to that seen in Fig. 11b.

Rockstad (77) proposed a model for the flat-band shift in irradiated MOS devices based on dynamic equilibrium analysis. In his model there was only a single layer of charged traps in the oxide, located near the semiconductor–insulator interface. For the linear region of $V_{fb}(V_{gr})$, the width of this layer

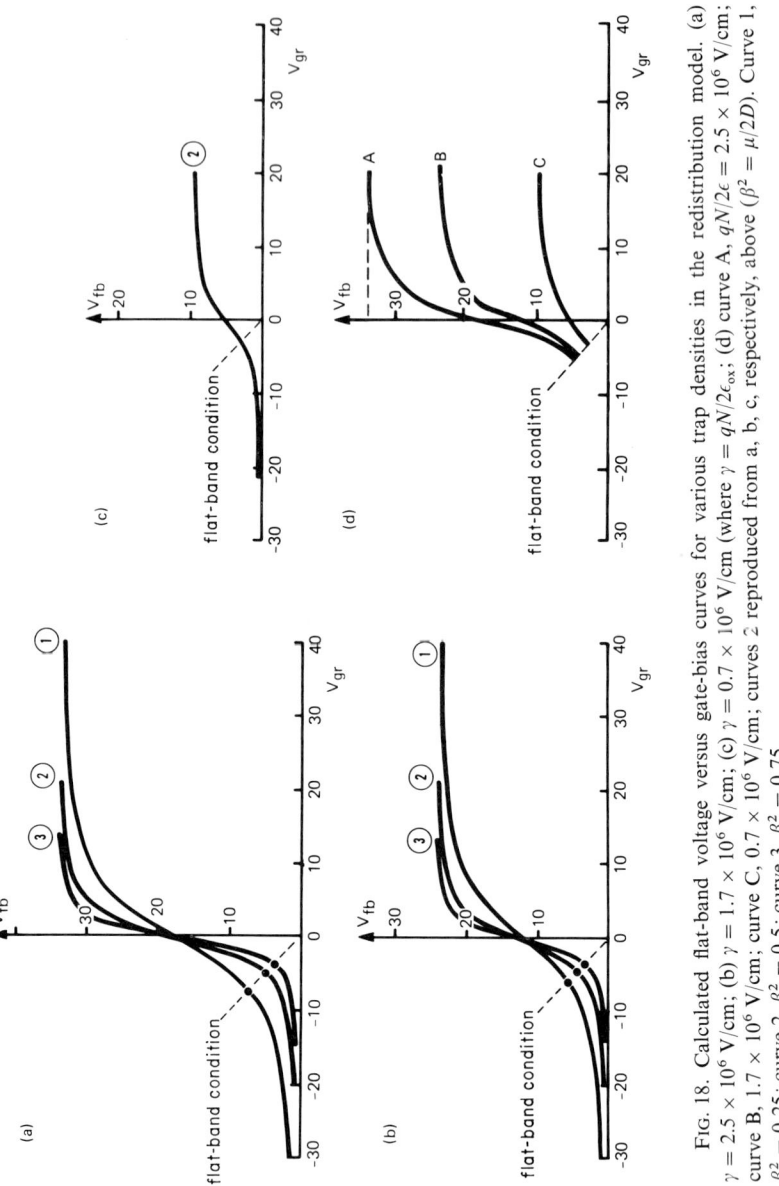

FIG. 18. Calculated flat-band voltage versus gate-bias curves for various trap densities in the redistribution model. (a) $\gamma = 2.5 \times 10^6$ V/cm; (b) $\gamma = 1.7 \times 10^6$ V/cm; (c) $\gamma = 0.7 \times 10^6$ V/cm (where $\gamma = qN/2\epsilon_{ox}$; (d) curve A, $qN/2\epsilon = 2.5 \times 10^6$ V/cm; curve B, 1.7×10^6 V/cm; curve C, 0.7×10^6 V/cm; curves 2 reproduced from a, b, c, respectively, above ($\beta^2 = \mu/2D$). Curve 1, $\beta^2 = 0.25$; curve 2, $\beta^2 = 0.5$; curve 3, $\beta^2 = 0.75$.

was assumed to be constant, whereas the concentration consequently varied linearly with gate bias. Although Rockstad's one-layer model is consistent with a linear relationship between V_{fb} and V_{gr}, it postulates a charge density proportional to V_{gr} and, consequently, implies that the flat-band shift must be negligible for irradiation with shorted plates. This is in conflict with a considerable amount of published experimental data (*41, 42, 63, 68, 77, 84, 86, 99*). It is not clear why the oxide charges should behave in the manner that Rockstad assumed. He gave no indication of any physical mechanism which would tend to make the charges behave in this rather unusual way.

We have attempted in this section to briefly review some of the simple descriptive models that have been used for quick analysis of radiation-induced effects in MOS structures. Simple models are often useful in obtaining first-order predictions in the initial phases of design, but for greater accuracy a more comprehensive model including second-order effects is needed. One such model is reviewed in the next section.

V. COMPLETE COMPUTER SIMULATION

The MOS system that was simulated is shown in Fig. 19. In the simulation (*74*), the oxide absorbed increments of penetrating "radiation" while the gate was held at a fixed bias potential, V_{gr}. After each increment of radiation dose, the flat-band voltage was "measured" and the gate bias was then "set" to the next value in preparation for the subsequent dose increment.

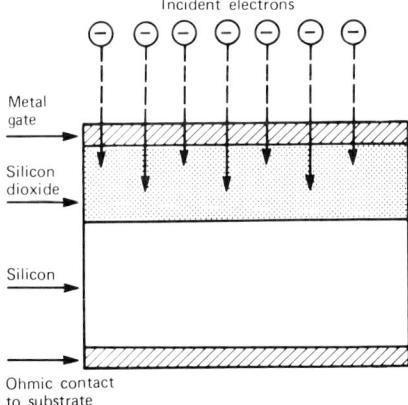

FIG. 19. Cross section of the MOS devices investigated. Incident electrons interact in the oxide region only.

To simulate device behavior, a mathematical model for the charge dynamics in both the oxide and semiconductor regions was constructed. The following assumptions were made:

(1) The incident electrons generate electron–hole pairs uniformly throughout the oxide at a rate G.

(2) Irradiation induces latent (inherent) trap sites of uniform concentration N_T which may be either neutral or positively charged. Once trap sites have been created they are not readily removed, even by annealing at moderate temperatures, although the actual occupancy statistics may vary with position and/or other irradiation parameters in the oxide.

(3) The dominant trapping mechanism of the electrons and holes occurs via trap centers rather than via direct band-to-band recombination.

(4) Electron–hole pair generation, due to the incident electrons, occurs only in the oxide region. (Although the computer simulation permits pair generation in the silicon substrate, this phenomenon was not included in order to focus on the fairly common situation in which primary electrons of less than 20-keV energy interact with the oxide. We previously found that the inclusion of pair generation in the semiconductor region did not appreciably change the general nature of the results.)

A. Describing Equations

With the aforementioned assumptions, the mathematical model was formulated using a consistent set of describing equations for the total structure.

In the oxide region we have the following expressions. Invoking the continuity of electrons and holes gives

$$\partial n/\partial t = D_n \nabla^2 n + \mu_{n,\text{ox}} \nabla \cdot (n\mathbf{E}) - R_n + G \qquad (20)$$

$$\partial p/\partial t = D_p \nabla^2 n - \mu_{p,\text{ox}} \nabla \cdot (p\mathbf{E}) - R_p + G \qquad (21)$$

Poisson's equation can be written as

$$\nabla^2 U = -\rho/\epsilon_{\text{ox}} \qquad (22)$$

where U is the electrostatic potential and

$$\rho = q(p - n + N_T^+) \qquad (23)$$

The rate equation for traps becomes

$$\partial N_T^+/\partial t = R_p - R_n \qquad (24)$$

where

$$R_p = \sigma_p v_{th} N_T^0 p \qquad (25)$$

$$R_n = \sigma_n v_{th} N_T^+ n \qquad (26)$$

with

$$N_T = N_T^+ + N_T^0 \qquad (27)$$

In the semiconductor region we have the following expressions. Invoking the continuity of electrons and holes gives

$$\partial n/\partial t = D_n \nabla^2 n + \mu_n \nabla \cdot (n\mathbf{E}) - [(n - n_0)/\tau_n] \qquad (28)$$

$$\partial p/\partial t = D_p \nabla^2 p - \mu_p \nabla \cdot (p\mathbf{E}) - [(p - p_0)/\tau_p] \qquad (29)$$

Poisson's equation may be written as

$$\nabla^2 U = -\rho/\epsilon_{sc} \qquad (30)$$

where

$$\rho = q(p - n + N_D - N_A) \qquad (31)$$

B. Boundary Conditions

Equations (20)–(31) represent a mathematical description of charge dynamics in the oxide and semiconductor regions of the device. These coupled nonlinear differential equations were solved subject to certain boundary conditions imposed by the nature of the energy-band structure at the two heterojunction interfaces.

The energy-band diagram corresponding to the simulated cases under equilibrium conditions is shown in Fig. 20. The simulation showed that the positive charge distribution present in the oxide is nonuniform. This charge profile forces the energy bands in the oxide to be concave upwards as shown. As this diagram also shows, under typical conditions both types of mobile carriers would have to overcome a large potential barrier in order to pass into the oxide region from either the semiconductor or the metal. (Under extreme conditions, the potential barrier would be so thin that tunneling into and out of the oxide (*81, 100*) might possibly occur.) The role of the potential barriers which oppose carrier flow into the oxide, was modeled as follows. At the silicon–oxide interface, the condition becomes

$$J(n)_{Si \to ox} = J(p)_{Si \to ox} = 0 \qquad (32)$$

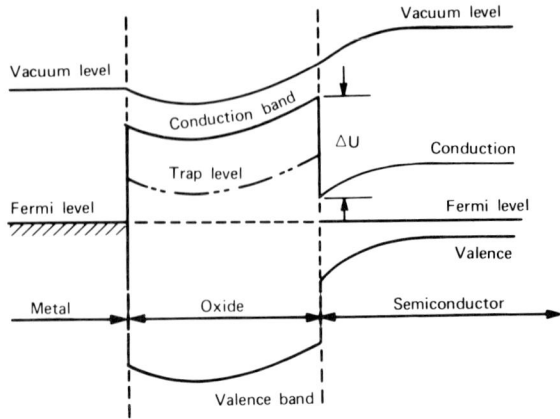

Fig. 20. Energy-band diagram for the entire MOS structure. The band structure in the oxide acquired the shape drawn due to charged bulk traps for $V_{gr} = 0$ during irradiation.

At the metal–oxide interface the condition becomes

$$J(n)_{M \to ox} = J(p)_{M \to ox} = 0 \tag{33}$$

These equations may impose an overly stringent condition on possible photoemission and hot-electron processes (101–104). However, in the interest of simplicity, and since this simulation agreed so well with experiment, we chose to use Eqs. (32) and (33). Improved accuracy would of course be obtained by a suitable simple modification in these boundary conditions.

On the other hand, carriers that are already inside the oxide and which arrive at either of the two interfaces can freely flow over the potential barriers from the conduction band of the oxide into the lower energy levels available in either the metal or semiconductor regions. Thus, the potential barriers tend to assist the passage of mobile carriers out of the oxide and into the metal or semiconductor regions, respectively. The role of the barriers in assisting the escape of mobile carriers from the oxide region was modeled as follows. At the silicon–oxide interface the condition is

$$J(n)_{ox \to Si} = -q\mu_{n,ox}E_{eff}n_{ox} \tag{34}$$

$$J(p)_{ox \to Si} = q\mu_{p,ox}E_{eff}p_{ox} \tag{35}$$

At the metal–oxide interface the condition is

$$J(n)\big|_{ox \to M} = q\mu_{n,ox} E_{\text{eff}} n_{ox} \quad (36)$$

$$J(p)\big|_{ox \to M} = -q\mu_{p,ox} E_{\text{eff}} p_{ox} \quad (37)$$

where $E_{\text{eff}} = \Delta U/q\, \Delta x$ is an effective field that represents the above-mentioned tendency for the potential barrier to assist in the escape of carriers from the oxide. It is obtained by dividing the height of the potential barrier $\Delta U/q$ by the characteristic distance Δx over which the barrier potential varies.

Due to the large fields created by the potential barriers at both interfaces, the electron and hole fluxes out of the oxide region consist almost entirely of drift currents, with the diffusion currents being very small in comparison. This fact defines a diffusion boundary condition at the oxide–silicon interface: the diffusion current crossing this boundary must be zero. This assumption allows for a stable numerical solution of the continuity equations for electrons and holes in the oxide region using boundary conditions based on sound physical principles. Ample justification for the use of this boundary condition is supplied by the results.

To satisfy Maxwell's equations we imposed the additional boundary condition that

$$\nabla \cdot \mathbf{D} = \rho_s \quad (38)$$

where ρ_s is the charge density at the silicon–oxide interface.

We did not include the work-function difference (*105*) between the silicon and the metal. (This amounts to nothing more than a constant horizontal displacement of the flat-band curves.) The semiconductor substrate contact was taken as the reference point and was assigned an electrostatic potential of 0 V.

C. Comments

The structure of this model is more general than the one proposed by Frankovsky *et al.* (*106*) for the following reasons:

(1) The *total* device was modeled using appropriate boundary conditions at the interfaces.

(2) The capture rates are nonlinear functions involving the products of n, p, N_T^+ and N_T^0 that must be solved self-consistently for the electron and hole concentrations. The capture time in the oxide was not assumed to be constant in this model (*74*). (In this regard, it is important to note that

HSR time constants (*19–21*) are generally valid only under steady-state conditions (*71*). In our approach, no time constants need be assumed because the correct time-dependence occurs automatically due to the nature of the equations employed.)

(3) The difference in capture time between the electrons and holes is contained in the self-consistent solutions. We did not assume that the holes are immediately captured after being generated. (Keeping in mind the degree of accuracy sought here, we did not restrict the simulation to electron and ionized trap pair creation.)

(4) We used the calculated values of $n(x)$, $p(x)$, $N_T^+(x)$, and N_T^0 to determine the resulting flat-band voltage shift via Eq. (8).

(5) The nonequilibrium nature of the physical system was automatically taken into account by using the rate equation approach. It was therefore unnecessary to invoke Fermi–Dirac statistics or hot-electron concepts.

Numerical techniques were developed (*69*) to solve the complete set of nonlinear differential equations. In a previous publication (*71*) similar techniques were applied to the transient analysis of an MOS capacitor.

D. Results of the Computer Simulation

Equations (20–(31) were solved self-consistently by computer simulation subject to the specified boundary conditions. The results were expressed in terms of the flat-band and potential brought about by electron-beam irradiation. Essentially, two different cases were modeled. One was the fixed gate-voltage case (type 1). The other was a case where the gate voltage was varied between each discrete dose increment (type 2). A locus of points giving the flat-band voltage versus gate voltage V_{gr} and D was obtained.

Table I lists the parameter values and/or ranges used. These values provided the best agreement with experimental data.

Case 1. Fixed Gate Voltage (Type 1)

Figure 21 shows the computer-simulated time evolution of the flat-band and shift for three different gate biases during irradiation. These curves reflect the growth of positively charged traps in the oxide during irradiation as each gate potential value was held constant. These computer-simulated curves show a structure similar to measured flat-band data in Fig. 22 using 500-Å-thick MOS structures. The dose rate used to obtain the experimental data in Fig. 22 was 2×10^7 rads (Si)/sec. No leveling off of the curves was observed up to a cumulative dose of 3×10^{10} rads.

The type 1 curves in Fig. 21, which were obtained by simulation, dramatically show that the flat-band voltage for $V_{gr} = -4$ V lies above the

TABLE I

LIST OF PARAMETERS USED IN THE SIMULATION TO OBTAIN THE "BEST FIT" FOR THE EXPERIMENTAL CURVES

Parameter	Symbol	Value
Oxide:		
Oxide thickness	W_{ox}	560 Å
Oxide trap concentration	M_T	10^{17}–10^{19} cm^{-3}
Generation rate	G_n, G_p	10^{17}–10^{18} cm^{-3} sec^{-1}
Electron mobility in oxide	$\mu_{n,ox}$	2×10^{-6} cm^2 V^{-1} sec^{-1}
Hole mobility in oxide	$\mu_{p,ox}$	0.8×10^{-6} cm^{2}V^{-1} sec^{-1}
Capture cross section of traps	$\sigma_{n,ox}; \sigma_{p,ox}$	10^{-18} cm^2
Interface barrier potential	$\Delta U/q$	4 V
Relative oxide permittivity	ϵ_{ox}	3.9
Substrate:		
Bulk doping (p type)	N_A	10^{15} cm^{-3}
Electron mobility (substrate)	μ_n	1000 cm^2/V sec
Hole mobility (substrate)	μ_p	400 cm^2/V sec
Electron diffusion coefficient	D_n	25.9 cm^2/sec
Hole diffusion coefficient	D_p	10.7 cm^2/sec
Temperature	T	300°K
Trap doping	N_T	0.5–1.0×10^{15} cm^{-3}
Relative silicon permittivity	ϵ_{si}	11.7
Substrate thickness	L	1.5–2.0 µm
Capture cross section of traps	$\sigma_{n,p}$	1.0×10^{-15} cm^2

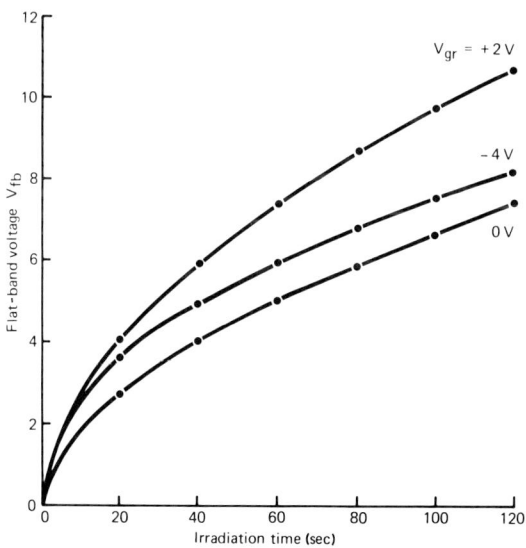

FIG. 21. Simulated time evolution of flat-band voltage with irradiation time for three fixed gate-bias values: $W_{ox} = 560$ Å; $G = 10^{17}$ cm^{-3} sec^{-1}.

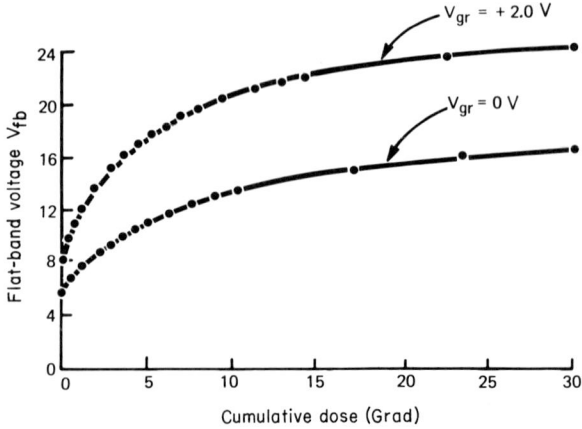

FIG. 22. Measured flat-band voltage versus cumulative dose for two fixed gate-bias values.

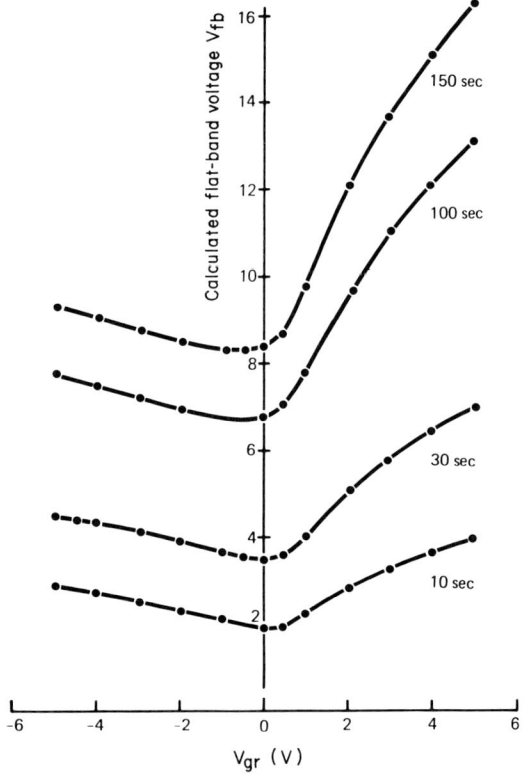

FIG. 23. Calculated flat-band shift versus gate bias for four simulated irradiation time intervals (500-Å-thick oxide). The curves have not been corrected for work function differences. Irradiation: type 1.

curves for $V_{gr} = 0$. If the type 1 data shown in Fig. 21 are replotted in a type 2 format, as is done in Fig. 23, then it is readily seen that the model consistently agrees with the data shown in Fig. 12 for the positive and negative regions of V_{gr}. (The minima in Fig. 23 are located near $V_{gr} = 0$ because the work-function difference was not included in the computer simulation.) Curves that exhibit this type of behavior have been published by numerous investigators, but no satisfactory model was given to explain this effect. We presented (94) a one-layer model that qualitatively explained the origin of this effect. It is a simple descriptive model and is not quantitatively rigorous; nevertheless, it matched the experimental data rather well. The results in Figs. 21 and 23 show that the computer-simulated model fully accounts for these characteristics quantitatively.

Figure 23 contains a family of curves for the simulated flat-band voltage V_{fb} with irradiation time as a parameter. Each point on a given curve represents the same value of total dose received by previously unirradiated devices.

The experimental curves shown in Fig. 24 have much the same general shape as the simulated curves. It is particularly apparent that the linear region on the measured curves is similar to the linear region on the computer-simulated curves. The linear behavior can be explained using a simplified two-layer model, which of course is inherently embodied in the more complete computer simulation (74). The curves show that a minimum flat-band shift occurs somewhere in the vicinity of $V_{gr} = -1$ V for curve 1. The minimum moves to the left with additional radiation. This is consistent with the simulation curves in Fig. 23.

The rather substantial flat-band shift for zero gate bias is the result of the formation of two positively charged layers—one near the silicon surface and one near the metal electrode. However, the layer situated nearest the silicon has the greatest influence on the flat-band shift. The formation of these two layers is related to the presence of image charges in the adjacent regions and is a natural consequence of the band structure of the metal, oxide, and silicon materials. Inspection of the trap energy levels in Fig. 20 shows the tendency for the formation of two charge layers.

Case 2. Sequential Gate Bias (Type 2)

The flat-band shift for simulated irradiation using repeated sequential gate-bias values is shown in Fig. 25a. To obtain the curve, the initial bias was set at zero. Then 30 sec of simulated irradiation was applied to the device. The gate bias was then increased by $+2$ V and another simulated dose of irradiation was applied. This procedure was repeated until the curve in the positive gate-bias region was traced out.

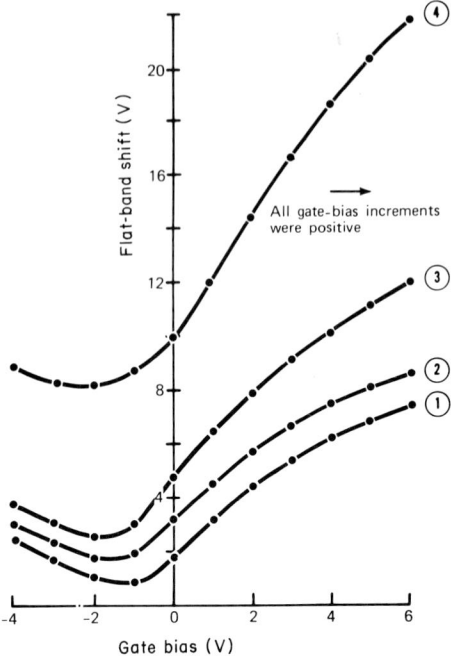

FIG. 24. Measured flat-band voltage versus applied gate bias for four different cumulative dose values (500-Å thick oxide). Irradiation, type 2:

Curve	Dose/point (rads)
1	2.4×10^6
2	4×10^6
3	5×10^7
4	3×10^8

This routine was repeated in the negative gate-bias region starting at $V_{gr} = 0$. Negative 2-V bias increments were now used with no previous irradiation included. (The direction of gate-bias increments is indicated in Fig. 25 by arrows.) The bias voltage was taken over a range sufficient to show departure from linearity toward a plateau for large positive bias (73) and sufficient to show the peak for large negative bias that was reported in reference 72. These features of the curves are due to the transition from inversion to depletion modes that occur in this gate-bias range.

The experimental data that correspond to type 2 irradiation are shown in Fig. 25b. The curves in the positive and negative gate-bias regions were done in the same sequential irradiation directions as the simulated data in

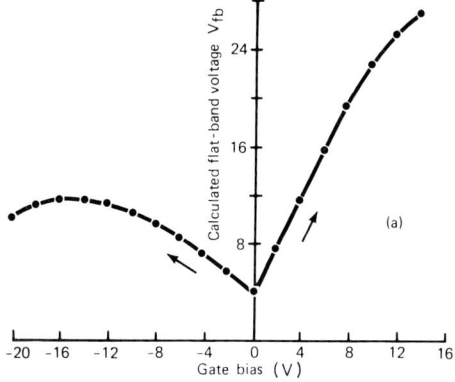

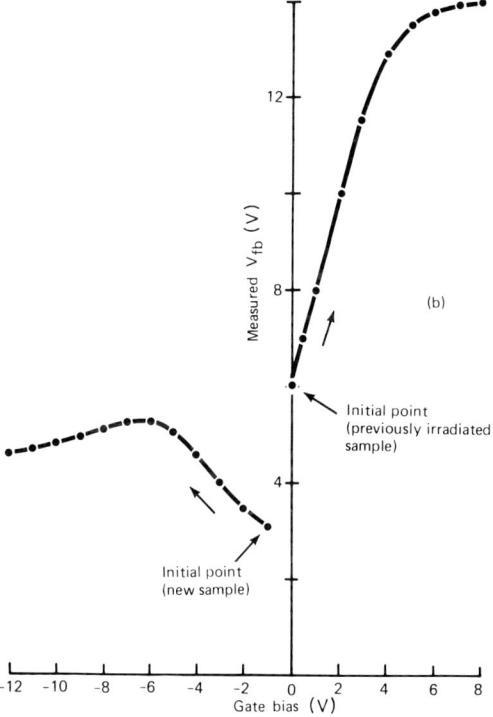

Fig. 25. Flat-band voltage versus applied gate bias for sequential (type 2 irradiation. (a) Calculated flat-band voltage with 30 sec/point. Calculations for positive and negative gate-bias regions commenced at 0 V. (b) Measured flat-band voltage with dose fixed at 5×10^7 rads/point. Curve in positive-bias region had 6-V offset due to prior irradiation. No initial offset existed in the negative-bias region.

Fig. 25a. However, the negative gate-bias region of Fig. 25b is approximately 3 V lower than its counterpart in Fig. 25a. This 3-V offset is present because the device used for $V_{gr} < 0$ had not received any previous irradiation. The initial flat-band value at the onset of irradiation for the device used in the positive gate-bias region was 6 V. The shapes of the curves agree in both bias regions. For comparison, note the shape of the curves in the positive gate-bias region of Fig. 1, Churchill et al. (93). The slight difference is easily accounted for by the fact that the MOS devices used to obtain curves of this type sometimes had received small amounts of prior irradiation. Both halves of the figure, however, start out at zero gate bias.

We also observed that there was a direct correlation between the curves in Fig. 25 and the corresponding curve in Fig. 23. That is, for a given total accumulated dose the final flat-band shift was independent of the history of the device and (barring excessive changes in V_{gr}) was determined only by the gate bias applied during the final irradiation increment. For example, the flat-band shift for $V_{gr} = 4$ V in Fig. 25a is slightly less than 12 V for a total radiation time of 90 sec. The corresponding point on the 100-sec curve in Fig. 23 is 12 V for $V_{gr} = 4$ V.

Nondynamic equilibrium effects were also obtained from the simulations, as shown in Fig. 26. Here, curves of flat-band voltage are shown for two

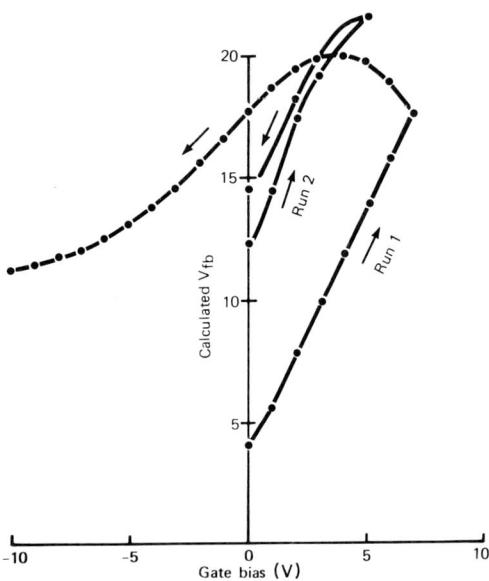

FIG. 26. Calculated flat-band voltage versus gate bias. In run 1, the dose was not sufficient to bring about dynamic equilibrium. In run 2 (an order-of-magnitude greater dose than run 1) dynamic equilibrium was approached.

different dose rates (i.e., generation rates). In run 1 the initial bias was set at zero. Subsequent bias values were increased in +1-V steps after each 30-sec dose increment. When the +7-V bias point had been calculated, the gate bias was decreased in 1-V steps after each successive 30-sec dose.

The curves show that for run 1 ($G = 10^{17}$ cm^{-3} sec^{-1}), the curve for negative gate-bias steps did not completely retrace the curve obtained for positive steps. At this relatively low rate a 30-sec dose was not sufficient to bring the system into dynamic equilibrium (i.e., a steady state). However, for run 2 ($G = 10^{18}$ cm^{-3} sec^{-1}) the curve for negative steps more nearly retraced the curve for positive steps. The 30-sec dose increments at this higher rate were nearly sufficient to bring about dynamic equilibrium at each value of applied gate bias. The transient changes in flat-band voltage, therefore, were essentially over in 30 sec at an irradiation-induced generation rate of 10^{18} hole–electron pairs/cm^3 sec, but not at the lower dose rate. (Actually, it is not dose rate, but rather the dose per irradiation or cumulative dose per point, that determines, whether the system is able to reach dynamic equilibrium during a given dose increment.)

Once the cumulative dose was large enough to produce dynamic equilibrium, additional dose increments at a fixed gate bias induced no further flat-band shift. However, with a change in bias voltage, there was typically a short transient fluctuation while the system rapidly approached dynamic equilibrium at the new stable point.

The experimental curves corresponding to the simulation runs of Fig. 26 are shown in Figs. 27 and 28. As was done in the simulation, integral positive gate-bias steps were used to obtain the bottom segments of the curves and integral negative gate-bias steps were used to obtain the top segments (i.e., positive 1-V steps in going from point A to point B at dose D_1 in Fig. 27, and negative 1-V steps in going from C to D at dose D_2.)

Although the data plotted in Figs. 27 and 28 were not originally taken for the purpose of comparison with Fig. 26, subsequent plotting of these points showed general close agreement in the shapes of the curves. This agreement is particularly evident for the top region of Fig. 27 and the top region of the run 1 curve in Fig. 26 for a generation rate $G = 10^{17}$ cm^{-3} sec^{-1}. A well-defined maximum in this top region occurs when proceeding to the left from point C, since not enough holes were generated by the irradiation to fill the available traps for equilibrium. In other words, the average concentration of occupied traps continues to increase until the competing processes for trapping and emission of holes approach dynamic equilibrium. (This kind of equilibrium should not be confused with the case in which all the traps are ionized.)

The experimental curve in Fig. 28 should now be compared with run 2 in Fig. 26. (The initial flat-band value at $V_{gr} = 0$ was large since the device

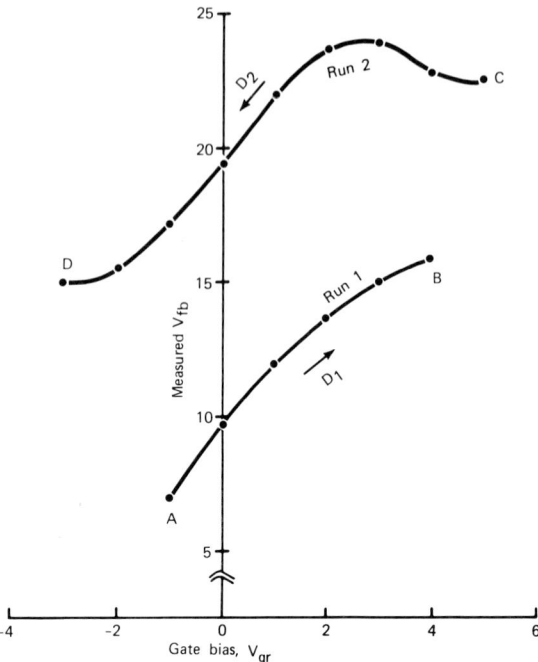

FIG. 27. Measured flat-band shift versus applied gate bias showing that dynamic equilibrium was not reached for each dose value received (run 1, 1.5×10^8 rads/point, run 2, 1.5×10^9 rads/point).

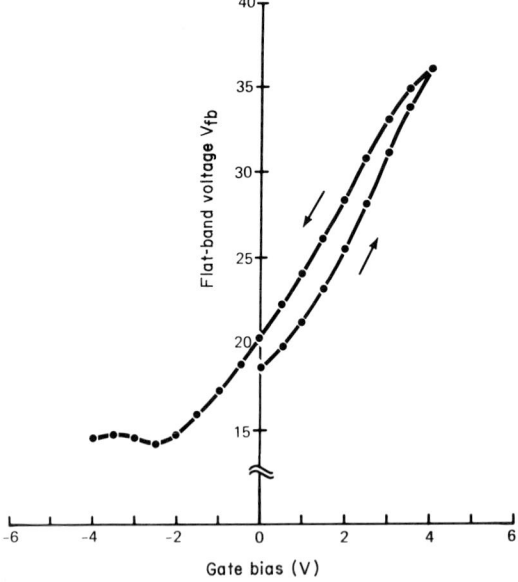

FIG. 28. Measured flat-band shift versus gate bias. Dose was 1.5×10^9 rads/point. Cumulative dose was large enough to approach dynamic equilibrium.

used had previously been irradiated.) Again, however, the structures of the two curves compare reasonably well. The irradiation conditions used to obtain the data for Fig. 28 correspond to a situation where there are enough available holes in the oxide for redistribution depending on the electric field magnitude and direction. The "effective trapping delay" that causes the openness of the curve in run 1 of Figs. 26 and 27 no longer exists when the more intense irradiation conditions for Fig. 28 are used. In other words, the hysteresis-like conditions are removed when the cumulative dose in the oxide is increased.

E. Discussion of the Simulation

Figure 29a–29d shows the electrostatic potential, electric field, trap density, and hole–electron density for a 30-sec dose with the gate bias $V_{gr} = 0$. As shown in Fig. 29, the electric field remains very small in the middle region of the oxide and is large in magnitude at the interfaces. Further, the field is negative near the metal contact but is positive near the silicon interface. These findings are consistent with our published two-layer model (*93*) and have a bearing on the annealing process for aluminum-gated SiO$_2$ devices (*58, 63, 107, 108*).

The charge density for trapped holes of Fig. 29 also exhibits a "two-layer" distribution, being quite large at both interfaces but small in the middle

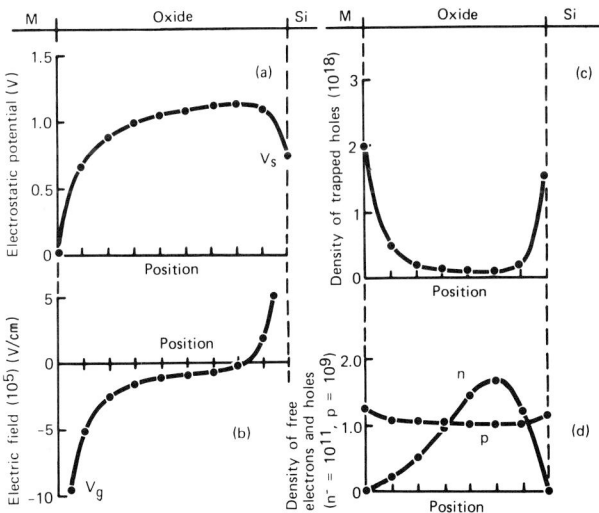

FIG. 29. Various calculated distributions in the oxide for $V_{gr} = 0$ and 30-sec dose. (a) Potential, (b) electric field, (c) density of trapped holes, (d) density of free electrons and holes.

region. This is a consequence of the boundary conditions imposed at the interfaces and the free-electron and hole distributions shown in Fig. 29d. It was already noted that the magnitude and thickness of these charged layers functionally depend on gate bias, trap concentration, and other irradiation parameters.

The densities of both electrons and holes are small compared to the density of charged traps. This was the case for all parameter values used. The electron and hole concentrations shown in Fig. 29d represent free carriers in the conduction and valence bands of the oxide, respectively, whereas the density of charged traps in Fig. 29c reflects the density of the trapped holes only. Although the densities of electrons and holes are very small compared to the charged-trap density, the final density of charged traps is extremely sensitive to these quantities. In this particular case, the electric field confines the electrons to the middle region of the oxide, whereas the free holes are almost uniformly distributed throughout the region. This is most likely the origin of the relaxation effects which are described in Section IV.

There is close agreement between the appearance of the trapped charge and electric field distributions found here and the equilibrium ion distribution and accompanying electric field reported by Marciniak (*97*) and Tangena (*98*). Their approach consisted in allowing a constant number of mobile ions to drift and diffuse in the oxide without generation or recombination. In our model we permitted the activation of latent trap sites and the movement of electrons and holes to change the occupancy at a given gate bias V_{gr}. As described in Section IV, when the flat-band shift for the charge distributions found by Marciniak and Tangena is calculated, the results (see Fig. 18) are similar to our curve shown in Fig. 11c.

The curves shown in Fig. 30 suggest that the distribution of trapped charges is determined primarily by the gate bias. For large positive bias (curve 1), a one-layer distribution exists near the silicon–oxide interface. For small potentials (curve 2), a two-layer distribution emerges. Finally, for negative gate biases (curve 3), a one-layer distribution is again dominant, except that for this polarity the layer is found near the metal–oxide interface rather than the silicon–oxide interface.

At much larger gate-bias values, the trapped charges reside in well-defined single-layer configurations. This can be readily seen in Fig. 31, where the distributions of charged traps for gate biases of $+10$ and -16 V are shown. It is important to note the accompanying distributions of free electrons, free holes as well as the electrostatic potential and electric field. The case of zero gate-bias is included for comparison. The distributions very much resemble those obtained from the redistribution model discussed in Section IV [compare with Fig. 2 in Ref. (*109*)]. We have previously reported the essential features of the transition between one-layer and two-layer behavior (*74, 93, 94*).

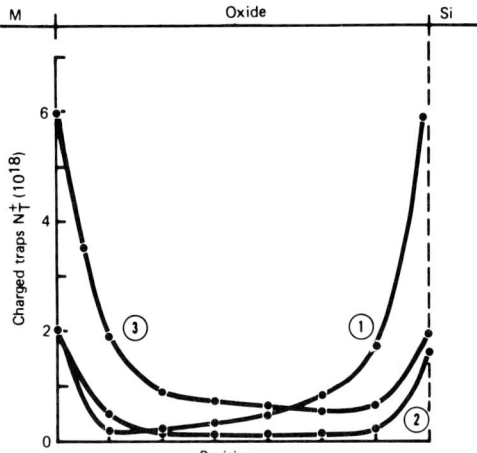

FIG. 30. Calculated distributions of charged traps in the oxide for +6-V gate bias (curve 1), zero gate bias (curve 2), and −6-V gate bias (curve 3).

Many simulation runs were carried out using a large range of values for the work functions, the electron–hole generation rate, the electron and hole mobilities, and the capture cross sections of traps in the oxide. In particular, the generation rate G was varied from 10^{17} to 10^{20} electron–hole pairs/cm^3 sec in the oxide. The total number of traps created as a result of the primary radiation was varied from 10^{17} to 10^{20} cm^{-3}. [When all the traps are ionized, the flat-band shift saturates. The experimental analog of this was reported by us (85).] The capture cross sections for holes and electrons were varied separately, but they were both kept within the range 10^{-13} to 10^{-18} cm^2, while the hole mobility was varied from 4×10^{-7} to 4×10^{-5} cm^2 V^{-1} sec^{-1}. The metal–oxide and oxide–semiconductor work functions were separately varied from 0 to 4 V.

The simulation results were matched to the experimental data by selecting a set of parameters that produced a "best-fit" results; they are listed in Table I. Some of these parameters may differ from the generally accepted nominal values. These parameters are a strong function of the preparation of the oxide (62) and are highly dependent on the electric field in the oxide (79). In fact, Hughes (79) argued that the yield of the electron–hole pairs to ionizing radiation, the field carrier injection from the electrodes into the oxide, ion motion, field-induced electron traps, and avalanche multiplication are highly dependent on the electric field. However, it is not likely that electron–hole pairs are created by intrinsic impact ionization (80) in our experimental work since the electric field is less than 6×10^7 V cm^{-1}.

In a study of electron beam-induced conductivity (EBIC), Taylor (110) used a best-fit analysis to estimate electron mobilities in SiO$_2$ for various

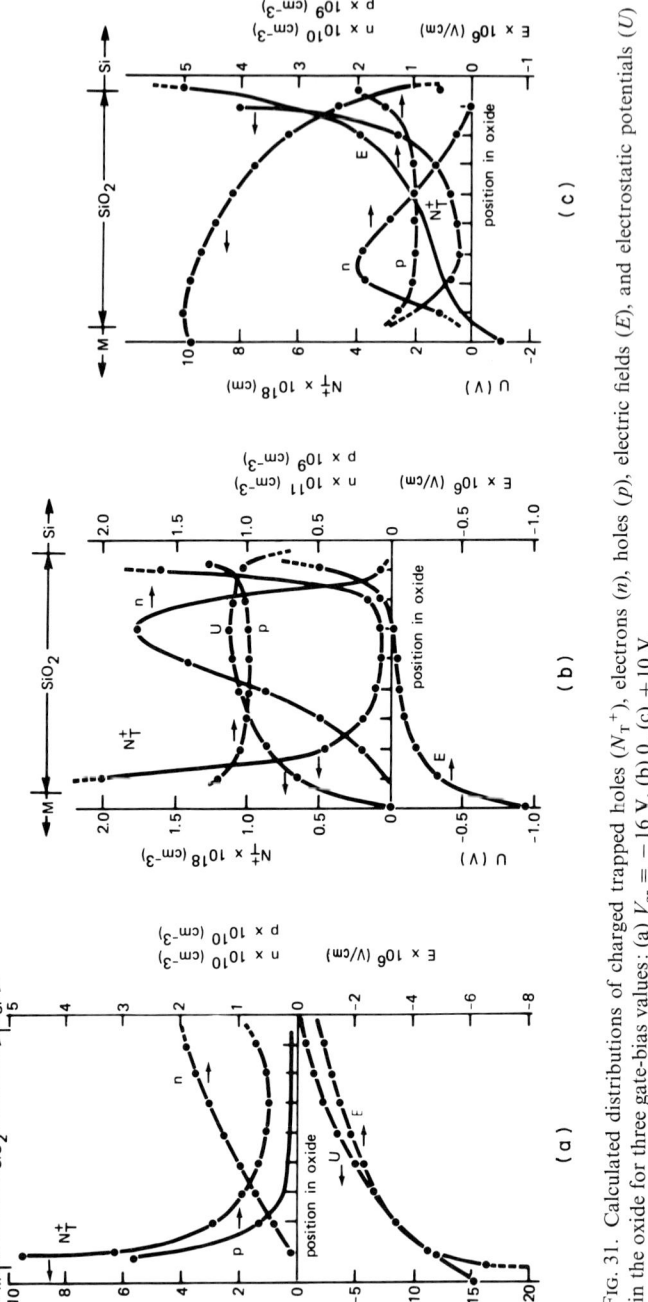

FIG. 31. Calculated distributions of charged trapped holes (N_T^+), electrons (n), holes (p), electric fields (E), and electrostatic potentials (U) within the oxide for three gate-bias values: (a) $V_{gr} = -16$ V, (b) 0, (c) +10 V.

electric fields. He obtained mobility values in the 0.1–0.6-cm^2 V^{-1} sec^{-1} range. These values are considerably smaller than values found by Hughes (*111*) who made time-of-flight measurements of the carrier drift mobility and lifetime. Mobility is highly dependent on defect concentration (intrinsic or radiation induced) as well as on electric field intensity in the oxide, as mentioned above.

Reported values for the mu–tau product in SiO_2 vary considerably. It was noted several years ago that the mu–tau product varied over more than five orders of magnitude (*76*). Variations this large were attributed variously to the material properties of the oxide, the experiments, and/or the analysis.

Hughes (*61*) discussed low intrinsic mobilities in general and noted that they are not unprecedented. In fact, as an example he cited that electrons in single-crystal sulfur have a mobility of 6×10^{-4} cm^2 V^{-1} sec^{-1} at room temperature and that no dispersive transport was present. He emphasized that this value is considerably lower than the generally accepted value for mobility in SiO_2 even though other parameters are similar. Certainly more work is needed in this area.

In view of the variations, and even discrepancies, in the experimental values for some of the physical parameters of the oxide, we fixed the simulation parameters as close to the nominal values as we could and still have the simulated curves match the shape of the experimental curves and give similar flat-band shift values. (Determining the sensitivity of the simulation output with variation of the parameter values at the input of the model can be tedious.)

Certainly the computer model may be extended by including other effects, as mentioned earlier, and the parameters can be further optimized to match the data. However, the goal was to analyze and synthesize the charge buildup characteristics of an MOS system. This was done by modeling the nonlinear equations describing the physics of the device while the structure was subjected to ionizing radiation. It is a simple model based on proven, conventional transport and trapping mechanisms.

There are four relatively direct ways to extend the model treated here: (1) inclusion of acceptor- as well as donor-type traps in the oxide, (2) allowing saturation of the mobility in the high-field regions, (3) inclusion of photoemission at the interfaces, and (4) inclusion of tunneling at the barriers. These changes would have the net effect of enhancing electron–hole recombination, which would require the use of a greater electron mobility in the simulation if the trapped holes are to accumulate more slowly. [In fact, trapping of this type in the model readily accounts for the negative values of flat-band shift, as shown in data similar to that of Rockstad (*77*).]

For perspective it is useful to compare the published results of various investigators. Figure 32 shows data from a number of authors (*42, 46, 63, 77, 85, 93, 112–114*) plotted together, where the ordinate is not flat-band

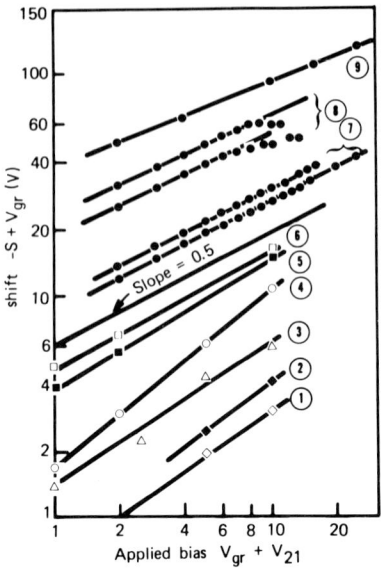

FIG. 32. Experimental results from various authors showing adjusted flat-band voltage versus adjusted gate bias. Data are from the following references: curve 1, Zaininger and Holmes-Seidle (42, 109); curve 2, Hughes, Baxter and Phillips (63); curve 3, Mitchell (110); curve 4, Derbenwick and Gregory (46); curve 5, Aubuchon (1 Mrad) (111); curve 6, Aubuchon (10 Mrad) (111); curve 7, Churchill et al. (500 Å) (85); curve 8, Churchill et al. (700 Å) (93); curve 9, Rockstad (77). Note line with slope 0.5.

shift but rather flat-band shift plus V_{gr}. This form for displaying the data, which results in linear plots and nearly equal slopes, is employed as a consequence of our analysis of the one-layer model (94). We observed that proper adjustment of both the abcissa and ordinate values yielded slopes nearly equal to 0.5 for the various published results.

The scope of the computer-simulated model is sufficiently significant and complete that inclusion of additional, questionable higher-order effects tends to obscure the main thrust of this work and would detract from the principle mechanisms that are operative during irradiation of SiO_2. Our objective was to present a simple model based on fundamental physical mechanisms that could assist in the analysis and design of MOS systems and provide insight into the behavior of the electrical characteristics following radiation exposure

VI. Discussion

Each simple descriptive model treated in Section IV has its own range of usefulness. Some of the models are valid in situations involving dynamic equilibrium, whereas others are valid in low-dose situations (initial transient). Models such as the two-layer one can, for example, fully account for the linear region of the flat-band voltage versus gate-bias curve shown in Fig. 6, but others, such as the redistribution model, can accurately describe the curve between both extreme ends of the positive and negative range of V_{gr}.

The more encompassing dynamic model described in Section V fully simulates the low, medium, and high cumulative dose range, which extend from initial transient to full dynamic equilibrium, and also correctly reproduces type 1 and type 2 irradiation data. For the case of type 2 irradiation, it reproduces the shape and position of the minimum in the curve and the large value of flat-band shift V_{fb} for $V_{gr} = 0$, as well as the linear region, the asymptotic behavior for the voltage independence at large positive gate bias, and the maximum in the curve for negative gate bias, as shown in Fig. 6. Some of the important factors that are reproduced by the computer simulation model have previously either (1) been overlooked by various investigators, (2) not been explained, or (3) been lightly brushed aside as inconsequential. (For example, the position of the minimum in the negative gate-bias region and its nonzero–nonzero behavior was quickly dismissed as related to high concentrations of interface states in the MOS samples being used (*112*).)

The computer-simulated model in Section V is more globally useful than the descriptive models in Section IV. For reasons of simplicity and ease of computation the model did not incorporate (1) electron tunneling at the interfaces, (2) high-field stress-generated surface and/or bulk defects, (3) photoemission effects, (4) Schottky emission, (5) secondary electron effects, (6) modification of the oxide parameters by localized high fields, or (7) impact ionization. However, it is relatively simple to extend the model to encompass these and other effects by adding the appropriate terms to the equations in Section V prior to commencement of the computer simulation.

The strength of the simulation model lies in its ability to reproduce and accurately represent the rather complex nature of the curves for type 1 data, as shown in Fig. 9, as well as for type 2 data, shown in Fig. 6. It reproduces the outstanding features of the curves shown in Fig. 25b even though there is a slight departure at large positive gate bias and the curve does not drop off as rapidly at *large* negative gate-bias values as evidenced in Fig. 25a.

Although the computer-simulated model given in Section V was based on bulk trapping centers in the oxide, a simple modification of the describing equations would readily permit the incorporation of a sheet layer of traps (*87, 116*) at the silicon interface or the dynamic bond modification (*117*) during irradiation. The final simulation results show excellent agreement with experimental data, even without allowance for traps at the interface. This suggests either that the bulk effects tend to dominate in the medium- to high-dose range or that an equivalence of surface states is found by means of the final trapped charge distributions in the oxide. With the inclusion of a uniform distribution of bulk traps in the oxide, the resulting charge distributions following radiation exposure take the form of a rather thin charge layer near the silicon interface when $V_{gr} = 0$, as shown in Figs. 30 and 31. Thus, the bulk-trap model is consistent with experiment when contrasted with the commonly assumed sheet-layer picture of the CDT models. Our simulation model fully explains type 2 irradiation experiments, in which the flat-band voltage decreases by a small amount when the gate-bias value is slightly decreased, followed by an additional radiation increment, whereas the sheet-trap models do not! No model that relies heavily on surface states to explain the observed behavior of MOS structures that are subjected to relatively large amounts of radiation can justify the large flat-band voltage for negative gate bias as well as for $V_{gr} = 0$. In fact, considerable confusion and contradictory conclusions concerning the degree of flat-band shift that results from $V_{gr} = 0$ was evidenced by the reported results dating from 1966–1978 (*42*, *44*, *77*, *113*).

Another argument supporting the inclusion of interface traps, as well as bulk traps, in the model comes from the reported work of Marquardt and Sigel (*118*). They employed electron spin resonance and etch-back techniques to identify radiation-induced defect centers and to determine their spatial distribution in thermally grown oxide films in an SOS configuration. They observed that the distribution of E' centers in oxide films irradiated to a dose of 10^8 rads (Si) consisted of a uniform bulk distribution plus a concentration buildup near the $Si-SiO_2$ interface. When a positive 10-V gate bias was applied during gamma irradiation, they found an order of magnitude increase in the E' center concentration. They reported that approximately 62% of the total E' centers that were formed and detected were distributed uniformly throughout the oxide.

In view of these various arguments, a completely accurate model would be based on the combination of the two types of traps rather than concluding, or assuming, that the charge is located in a sheet at the interface or only in the form of bulk traps. However, our use of bulk traps alone (which become positively charged as a consequence of some of the charges occupying a region very near the interface under irradiation) is more realistic than any attempt to use interface-like traps alone to explain the features of the experimental curves! In this latter case there is no allowance for a redistribution of the charged sheet into bulk states.

We believe it is a credit to the computer-simulated model that it gives results agreeing so well with such a wide variety (low-, medium-, and high-dose ranges) of experimental data. We know that the inclusion of photoemission and tunneling would bring a closer match at the higher positive and negative gate-bias regions and that a slight modification of the describing equations would permit the use of simulation parameters with values somewhat closer to those generally used. However, the model can be readily used to determine effective values of some of the parameters by successive iteration in the curve-fitting operation.

The main emphasis in our experimental effort was directed at the medium-to high-dose range. Although our irradiation equipment was primarily designed to study the high-dose range and would not permit measurement for dose increments below 10^6 rads, our data agreed with many published results in the overlap zone at the lower end of the medium-dose range. Most of the published results come from the low-dose range or (occasionally) from the overlap zone into the medium-dose range. However, we found that none of the existing models was adequate to explain the observed results in either a quite high dose range (up to 10^{12} rads) or a quite wide gate-bias range ($-30-+30$ V). In consequence we proceeded to develop a model—eventually the one given in Section V—that matched the data. We found no other model that was accurate over these wide ranges of dose and/or gate bias.

A versatile feature of the model we have presented is that the describing equations can be easily modified to include multilayered structures (MNOS devices, for instance). In its present form the model can easily be used to simulate the behavior of the III–V combinations which are of such great interest at present. For this latter case, it is simply a matter of inserting the corresponding values for the device parameters and including the appropriate boundary conditions.

Nomenclature

D Displacement vector, $C\ cm^{-2}$
D_n Diffusion constant for electrons, $cm^2\ sec^{-1}$
D_p Diffusion constant for holes, $cm^2\ sec^{-1}$
D_N Diffusion constant for redistribution model, $cm^2\ sec^{-1}$
D, D_C Incremental or cumulative dose, rads
E Electric field, $V\ cm^{-1}$
E_{eff} Effective electric field at the potential barrier, $V\ cm^{-1}$
E_0 Electric field at the metal–oxide interface, $V\ cm^{-1}$
E_0' grad E at the metal–oxide interface, $V\ cm^{-2}$

G Electron–hole pair generation rate, $cm^{-3}\ sec^{-1}$
$J(n)_{M\to ox}$ Current density for electrons moving from the metal to the oxide, $A\ cm^{-2}$
$J(n)_{Si\to ox}$ Current density for electrons moving from Si to oxide, $A\ cm^{-2}$
$J(p)_{Si\to ox}$ Current density for holes moving from the silicon to the oxide, $A\ cm^{-2}$
$J(p)_{M\to ox}$ Current density for holes moving from the metal to the oxide, $A\ cm^{-2}$
$J(n)_{ox\to Si}$ Current density for electrons moving from the oxide to silicon, $A\ cm^{-2}$

$J(n)_{ox \to M}$ Current density for electrons moving from the oxide to the metal, A cm^{-2}

$J(p)_{ox \to Si}$ Current density for holes moving from the oxide to the silicon, A cm^{-2}

$J(p)_{ox \to M}$ Current density for holes moving from the oxide to the metal, A cm^{-2}

N Total number of holes in the oxide, cm^{-2}

N_1, N_2 Concentration of trapped holes in layers 1 and 2, cm^{-3}

N_A, N_D Acceptor and donor concentration, cm^{-1}

n Free-electron concentration, cm^{-3}

n_0 Equilibrium concentration of electrons, cm^{-3}

n_{ox}, p_{ox} Free-electron–free-hole concentration in the oxide, cm^{-3}

N_T Uniform metalurgical trap density, cm^{-3}

N_T^0 Density of neutral traps, cm^{-3}

N_T^+ Trapped hole density, cm^{-3}

p Free-hole concentration, cm^{-3}

p_0 Equilibrium concentration of holes, cm^{-3}

q Electronic charge, C

R Dose rate, rad sec^{-1}

R_n Recombination rate for electrons, cm^{-3} sec^{-1}

R_p Trapping rate for holes, cm^{-3} sec^{-1}

S Flat-band shift, V

t Time, sec

V_{min} Value of V_{gr} for $W_2 = 0$, V

V_{fb} Flat-band voltage, V

V_{fb0} Preirradiation flat-band voltage, V

V_{WE} Semiconductor–metal work function difference, V

V_{gr} Fixed gate bias during irradiation, V

v_{th} Thermal velocity of carriers, cm sec^{-1}

W_{ox} Oxide thickness, cm

W_1, W_2 Widths of layers 1 and 2, cm

W_{10}, W_{20} Values of W_1 and W_2 at $V_{gr} = 0$, cm

x Position variable, cm

ΔU Potential barrier height, V

ϵ_{sc} Dielectric constant of the semiconductor, F cm^{-1}

ϵ, ϵ_{ox} Dielectric constant of the oxide, F cm^{-1}

μ_p, μ_n Mobility for holes and electrons in silicon, respectively, cm^2 V^{-1} sec^{-1}

$\mu_{n,ox}, \mu_{p,ox}$ Mobility of electrons and holes in the oxide, cm^2 V^{-1} sec^{-1}

ρ Charge density, C cm^{-3}

ρ_s Charge density at Si–SiO$_2$ interface, C cm^{-1}

σ_p, σ_n Trapping cross section for holes and electrons, cm^2

τ_p, τ_n Lifetime for holes and electrons, sec

ϕ_E Electrostatic potential, V

References

1. T. A. Wertime, *Science* **182**, 875 (1973); C. S. Smith, *Phys. Today* **18** (No. 12), p. 18 (1965).
2. F. Bloch, *Z. Phys.* **52**, 555 (1928).
3. A. Sommerfeld, *Z. Phys.* **47**, 1 (1928).
4. A. H. Wilson, *Proc. R. Soc. London, Ser. A* **133**, 458 (1931).
5. L. Prandtl, *Z. Angew. Math. Mech.* **8**, 55 (1928).
6. U. Dehlinger, *Ann. Phys.* **2**, 749 (1929).
7. J. Frenkel, *Phys. Rev.* **37**, 17 (1931).
8. F. Seitz, *Discuss. Faraday Soc.* **5**, 271 (1949).
9. J. W. Corbett, "Electron Radiation Damage in Semiconductors and Metals," *Solid State Phys. Suppl.* 7, p. 2. Academic Press, New York, 1966.

10. F. Dyson, "Disturbing the Universe," p. 108. Harper and Row, New York, 1979.
11. W. Schottky, *Z. Phys.* **113**, 367 (1939); **118**, 539 (1942).
12. N. F. Mott, *Proc. R. Soc. London, Ser. A* **171**, 27 (1939).
13. A. B. Phillips, "Transistor Engineering and Introduction to Integrated Semiconductor Circuits," p. 4. McGraw-Hill, New York, 1962.
14. J. Bardeen, *J. Appl. Phys.* **11**, 88 (1940).
15. R. de L. Kronig and W. G. Penney, *Proc. R. Soc. London, Ser A* **130**, 499 (1930).
16. Symposium held 30 April–2 May 1979, organized by Sir Nevill Mott, F.R.S. *Proc. Roy Soc. London, Ser. A* **371**, 1–177 (1980).
17. J. Bardeen and W. H. Brattain, *Phys. Rev.* **74**, 230 (1948).
18. W. Shockley, *Bell Syst. Tech. J.* **28**, 435 (1949).
19. R. N. Hall and W. C. Dunlap, *Phys. Rev.* **80**, 467 (1950).
20. R. N. Hall, *Phys. Rev.* **83**, 228 (1951); **87**, 387 (1952).
21. W. Shockley and W. T. Read, Jr., *Phys. Rev.* **87**, 835 (1952).
22. R. E. Davis, W. E. Johnson, K. Lark-Horovitz, and S. Siegel, *Phys. Rev.* **74**, 1255 (1948).
23. R. Ohl, *Bell Syst. Tech. J.* **31**, 104 (1952).
24. W. Shockley, *Proc. IRE*, **40**, 1365 (1952).
25. S. R. Hofstein and F. B. Heiman, *Proc. IEEE* **51**, 1190 (1963).
26. W. Shockley and G. L. Pearson, *Phys. Rev.* **74**, 232 (1948).
27. J. R. Szedon and R. M. Handy, *J. Vac. Sci. Technol.* **6**, 1 (1969).
28. J. C. Slater, *J. Appl. Phys.* **22**, 237 (1951).
29. R. B. Murray and A. Meyer, *Phys. Rev.* **122**, 815 (1961).
30. C. M. Nelson and J. H. Crawford, *J. Phys. Chem. Solids* **13**, 296 (1960).
31. W. D. Compton and G. W. Arnold, Jr., *Discuss. Faraday Soc.* **31**, 130 (1961).
32. D. S. Peck, R. R. Blair, W. L. Brown, and F. M. Smits, *Bell Syst. Tech. J.* **42**, 95 (1963).
33. H. L. Hughes and R. R. Giroux, *Electronics* **37**, 58 (1964).
34. E. Kooi, *Philips Res. Rep.* **20**, 596 (1965); **20**, 306 (1965).
35. J. R. Szedon and J. E. Sandor, *Appl. Phys. Lett.* **6**, 181 (1965).
36. D. Green, J. E. Sandor, T. W. O'Keefe, and R. K. Matta, *Appl. Phys. Lett.* **6**, 3 (1965).
37. E. H. Snow, A. S. Grove, B. E. Deal, and C. T. Sah, *J. Appl. Phys.* **36**, 1664 (1965); see also *J. Appl. Phys.* **35**, 2458 (1964).
38. B. E. Deal, A. S. Grove, E. H. Snow, and C. T. Sah, *Trans. Met. Soc. AIME* **233**, 524 (1965).
39. A. S. Grove and E. H. Snow, *Proc. IEEE* **54**, 894 (1966).
40. S. R. Hofstein, *IEEE Trans. Electron. Dev.* **ed-13**, 222 (1966).
41. K. H. Zaininger, *Appl. Phys. Lett.* **8**, 140 (1966).
42. K. H. Zaininger, *IEEE Trans. Nucl. Sci.* **ns-13**, 237 (1966).
43. A. L. Barry and D. F. Page, *IEEE Trans. Nucl. Sci.* **ns-13**, 255 (1966).
44. E. H. Snow, A. S. Grove, and D. J. Fitzgerald, *Proc. IEEE*, **55**, 1168 (1967).
45. R. J. Powell and G. F. Derbenwick, *IEEE Trans. Nucl. Sci.* **ns-18**, 18, 99 (1971).
46. G. F. Derbenwick and B. L. Gregory, *IEEE Trans. Nucl. Sci.* **ns-22**, 2151 (1975).
47. G. W. Hughes, R. J. Powell, and M. H. Woods, *Appl. Phys. Lett.* **29**, 377 (1976).
48. J. Buxo, D. Esteve, G. Enea, and A. Martinez, *Solid State Electron.* **15**, 1029 (1972); see also B. Andre, J. Buxo, D. Esteve, and H. Martinot, *Solid State Electron.* **12**, 123 (1969).
49. H. E. Boesch, Jr., F. B. McLean, J. M. McGarrity, and G. A. Ausman, Jr., *IEEE Trans. Nucl. Sci.* **ns-22**, 2163 (1975).
50. F. B. McLean, H. E. Boesch, Jr., and J. M. McGarrity, *IEEE Trans. Nucl. Sci.* **ns-23**, 1506 (1976).
51. H. E. Boesch, Jr., and J. M. McGarrity, *IEEE Trans. Nucl. Sci.* **ns-23**, 1520 (1976).
52. P. S. Winokur, J. M. McGarrity, and H. E. Boesch, Jr., *IEEE Trans. Nucl. Sci.* **ns-23**, 1580 (1976).

53. C. T. Sah, *IEEE Trans. Nucl. Sci.* **ns-23**, 1563 (1976).
54. D. J. DiMaria, Z. A. Weinberg, and J. M. Aitken, *J. Appl. Phys.* **48**, 898 (1977).
55. D. J. DiMaria, D. R. Young, W. R. Hunter, and C. M. Serrano, *IBM J. Res. Dev.* **22**, 289 (1978).
56. J. M. Aitken and D. R. Young, *J. Appl. Phys.* **47**, 1196 (1976).
57. D. R. Young, E. A. Irene, D. J. DiMaria, R. F. De Keersmaecker, and H. Z. Massoud, *J. Appl. Phys.* **50**, 6366 (1979).
58. D. R. Young, D. J. DiMaria, and W. R. Hunter, *J. Electron. Mater.* **6**, 569 (1977).
59. J. M. Aitken, *J. Non-Cryst. Solids* **40**, 31 (1980).
60. R. A. Gdula, *J. Electrochem. Soc.* **123**, 42 (1976); see also *IEEE Trans. Electron. Dev.* **ed-26**, 644 (1979).
61. R. C. Hughes, *Phys. Rev. B* **15**, 2012 (1977).
62. R. C. Hughes, E. P. Eernisse, and H. J. Stein, *IEEE Trans. Nucl. Sci.* **ns-22**, 2227 (1975).
63. H. L. Hughes, R. D. Baxter, and B. Phillips, *IEEE Trans. Nucl. Sci.* **ns-19**, 256 (1972).
64. L. L. Sivo, H. L. Hughes, and E. E. King, *IEEE Trans. Nucl. Sci.* **ns-19**, 313 (1972).
65. G. W. Hughes, *J. Appl. Phys.* **48**, 5357 (1977).
66. G. W. Hughes and G. L. Brucker, *Solid State Technol.* **22**, 70 (1979).
67. B. L. Gregory and C. W. Gwyn, *Proc. IEEE*, **62**, 1264 (1974).
68. C. W. Gwyn, *Oxides Oxide Films* **4**, 99 (1976).
69. T. W. Collins, Ph.D. Dissertation, Univ. of California, Davis (1973).
70. T. W. Collins and J. N. Churchill, *IEEE Trans. Electron. Dev.* **ed-22**, 90 (1975).
71. T. W. Collins, J. N. Churchill, F. E. Holmstrom, and A. Moschwitzer, *Adv. Electron. Electron Phys.* **47**, 267 (1978).
72. J. N. Churchill, T. W. Collins, and F. E. Holmstrom, *IEEE Trans. Electron. Dev.*, **ed-21**, 768 (1974).
73. F. E. Holmstrom, T. W. Collins, and J. N. Churchill, *Appl. Phys. Lett.* **24**, 464 (1974).
74. J. N. Churchill, F. E. Holmstrom, and T. W. Collins, *J. Appl. Phys.* **50**, 3994 (1979).
75. V. Y. Kiblik, V. G. Litovchenko, and R. O. Litvinov, *Sov. Microelectron.* **8**, 398 (1979).
76. O. L. Curtis, Jr., R. Srour, and K. Y. Chiu, *J. Appl. Phys.* **45**, 4506 (1974).
77. H. K. Rockstad, *J. Vac. Sci. Technol.* **15**, 1039 (1978).
78. A.-K. M. Zakzouk, *J. Electrochem. Soc.* **126**, 1771 (1979).
79. R. C. Hughes, *Solid-State Electron.* **21**, 251 (1978).
80. D. K. Ferry, *Solid State Commun.* **18**, 1051 (1976).
81. M. H. Woods and R. Williams, *J. Appl. Phys.* **47**, 1082 (1976).
82. S. M. Sze, "Physics of Semiconductor Devices," p. 364. Wiley-Interscience, New York, 1969.
83. J. P. Mitchell, *IEEE Trans. Electron. Sec.* **ed-14**, 764 (1967).
84. G. F. Derbenwick and B. L. Gregory, *IEEE Trans. Nucl. Sci.* **ns-22**, 2151 (1975).
85. F. E. Holmstrom, J. N. Churchill, and T. W. Collins, *Solid-State Electron.* **21**, 915 (1978).
86. R. J. Maier and R. W. Tallon, *IEEE Trans. Nucl. Sci.* **ns-22**, 2214 (1975).
87. R. Freeman and A. Holmes-Siedle, *IEEE Trans. Nucl. Sci.* **ns-25**, 1216 (1978).
88. D. K. Nichols, *IEEE Trans. Nucl. Sci.* **ns-27**, 1016 (1980).
89. A. G. Revesz, *IEEE Trans. Nucl. Sci.* **ns-24**, 2102 (1977).
90. T. W. Collins, F. E. Holmstrom, and J. N. Churchill, *IEEE Trans. Nucl. Sci.* **ns-26**, 5176 (1979).
91. J. N. Churchill, T. W. Collins, and F. E. Holmstrom, IBM Tech. Rep. TR 02.603, Dec. 7, 1973. General Products Div., San Jose, California.
92. J. Lindmayer, *IEEE Trans. Nucl. Sci.* **ns-18**, No. 6, 91 (1971).
93. J. N. Churchill, F. E. Holmstrom, and T. W. Collins, *Solid-State Electron.* **19**, 291 (1976).

94. J. N. Churchill, F. E. Holmstrom, and T. W. Collins, *IEEE Annual Conf. Nucl. Space Radiat. Effects, Santa Cruz, Calif., July 17–20, 1979.* Summaries of Papers. Paper P-15, **26A**, p. 169.
95. N. C. MacDonald and T. E. Everhart, Elect. Res. Lab., Univ. of Calif., Berkeley, Rep. No. ERL-66-16, Sept. 15, 1966.
96. A. J. Speth and F. F. Fang, *Appl. Phys. Lett.* **7**, 145 (1965).
97. W. Marciniak and H. M. Przewlocki, *Phys. Status Solidi* **24**, 359 (1974).
98. A. G. Tangena, J. Middelhoek, and N. F. de Pooij, *J. Appl. Phys.* **49**, 2876 (1978).
99. K. C. Aubuchon, *IEEE Trans. Nucl. Sci.* **ns-18**, 117 (1971).
100. R. H. Fowler and L. Nordheim, *Proc. R. Soc. London* **119**, 173 (1928).
101. A. M. Goodman and J. J. O'Neill, Jr., *J. Appl. Phys.* **37**, 3580 (1966).
102. A. M. Goodman, *Phys. Rev.* **152**, 780 (1966).
103. K. Hess, *Solid-State Electron.* **21**, 123 (1978).
104. T. H. Ning, *Solid-State Electron.* **21**, 273 (1978).
105. B. E. Deal, E. H. Snow, and C. A. Mead, *J. Phys. Chem. Solids* **27**, 1873 (1966).
106. F. Frankovsky, H. Protschka, and F. Zappert, *IEEE Trans. Nucl. Sci.*, **ns-15**, 140 (1968).
107. D. H. Phillips, *IEEE Trans. Nucl. Sci.* **ns-22**, 2190 (1975).
108. J. M. Aitken, *J. Electron. Mater.* **9**, 639 (1980).
109. V. P. Romanov and Yu. A. Chaplygin, *Phys. Stat. Sol.* (a) **53**, 493 (1979).
110. D. M. Taylor, *IEE Proc.* **128** (No. 3), 174 (1981).
111. R. C. Hughes, *Phys. Rev. Lett.* **30**, 1333 (1973).
112. A. G. Holmes-Siedle and K. H. Zaininger, *Solid State Technol.* 40 (1969).
113. J. P. Mitchell, *IEEE Trans. Electron. Dev.* **ed-14**, 764 (1967).
114. K. G. Aubuchon, *IEEE Trans. Nucl. Sci.* **ns-18**, 117 (1971).
115. A. G. Holmes-Siedle and I. Groombridge, *Thin Solid Films* **27**, 165 (1975).
116. A. R. Stivers and C. T. Sah, *J. Appl. Phys.* **51**, 6292 (1980).
117. K. L. Nagi and C. T. White, *J. Appl. Phys.* **52**, 320 (1981).
118. C. L. Marquardt and G. H. Sigel, Jr., *IEEE Trans. Nucl. Sci.* **ns-22**, 2234 (1975).

Point Defects in GaP, GaAs, and InP

U. KAUFMANN AND J. SCHNEIDER

Fraunhofer-Institut für Angewandte Festkörperphysik
Freiburg, West Germany

I. Introduction	81
II. Experimental Techniques for Point Defect Assessment	84
III. Donors and Acceptors	86
IV. Intrinsic Defects	96
A. The Isolated Gallium Vacancy in GaP	97
B. Vacancies in GaAs	99
C. Antisite Defects	103
V. 3d Transition Metals	109
A. Copper	112
B. Nickel	112
C. Cobalt	114
D. Iron	114
E. Manganese	116
F. Chromium	118
G. Vanadium	130
H. Ni–Donor Pairs in GaP and GaAs	131
I. 4d and 5d Transition Metals	134
J. 4f Rare Earth Metals	134
References	135

I. Introduction

Along with the elemental semiconductors, silicon and germanium, the III–V compounds are among the most intensively studied and best understood semiconductors (Willardson and Beer, 1965–1968; Madelung, 1964). Some of them have certain physical properties (higher mobility, larger band gap) superior to those of silicon. Since economic and efficient growth of the most important representatives—GaP, GaAs, InP, and their alloys—is also feasible, they have become indispensible materials for many device applications ranging from optoelectronic components (Bergh and Dean, 1976; Pilkuhn, 1981; Kressel and Butler, 1977; Casey and Panish, 1978) to microwave devices (Liechti, 1976) and high-speed logic circuits (Liechti, 1977; Zucca *et al.*, 1980). Apart from the basic physics point of view, the great interest which these materials are attracting is motivated by this technological importance.

Almost all applications of semiconductors are based on impurity effects rather than on the properties of the ideal crystal alone. Deliberately added foreign elements in small concentrations (1–100 ppm), so-called dopants, can alter the physical properties of the material tremendously in a desired fashion; e.g., they can decrease (or increase) the resistivity by many orders of magnitude or increase the luminescence efficiency. Unintentional trace impurities and other lattice imperfections in extremely tiny concentrations (less than 1 ppm) can also affect the material quality—often uncontrollably and in an undesired manner. One very important field of semiconductor research is therefore the identification of such point defects and their characterization with respect to the effects on material quality and device performance.

This article presents an overview of the experimental investigations of point defects in GaP, GaAs, and InP. No attempt will be made to cover the field exhaustively. In particular this is true of our discussion of shallow centers, in which only representative examples will be given. Special attention will be paid to deep-center identification.

With the exception of the nitrides all III–V compounds crystallize in the cubic zincblende structure, wherein each cation is tetrahedrally coordinated by four anions and vice versa. In a pure and otherwise perfect semiconductor the valence band is separated from the conduction band by a "clean" band gap. This means that single-electron states which could be occupied by electrons do not exist in this forbidden zone. Any disturbance—impurities, missing or misplaced atoms—of the regular array of lattice atoms will create electronic states which can lie within the "forbidden" zone. Their energetic position depends decisively on the type of imperfection. The occupation of these states by electrons is governed by the position of the Fermi level. Impurities, intrinsic lattice imperfections, and their complexes are called point defects. In dealing with semiconductors it is useful to classify the latter as shallow and deep defects (or centers). In the following we consider those defects shallow which have energy levels not significantly deeper than those of the conventional shallow acceptors and donors for which effective mass theory provides an adequate description. All other defects, having energy levels between those of the shallow donors and acceptors, are referred to as deep. In contrast to the shallow centers, the deep centers can have ionization energies comparable with half the band-gap energy, $E_g/2$.

A point defect in a semiconductor may be considered well understood if the following characteristics are known:

- electrical properties (thermal ionization energies, capture cross sections), accessible charge states;
- optical properties (optical ionization energies, internal transitions, optical cross sections, luminescence behavior);

chemical identity and symmetry;
vibrational properties and interaction with lattice phonons (electron–phonon coupling).

A full understanding should also include a theoretical model that is able to predict the relevant measurable parameters of the defect.

A successful theoretical description for shallow defects, effective mass theory, was developed 25 years ago (Kohn, 1957). In treating deep centers theorists are faced with severe problems. Special papers and detailed reviews describing the complexity of the subject and the various mathematical approaches to its solution are available (Bassani, 1974; Roitsin, 1974; Pantelides, 1978; Jaros, 1980; Hjalmarson *et al.*, 1980).

Many of the important *shallow* donors, acceptors, and isoelectronic centers in III–V compounds are well understood in the above sense. There are several reasons for this success. Shallow defects primarily control the most basic property of a semiconductor, its electrical conductivity, and thereefore they deserve special attention. They lend themselves equally well to the classical, macroscopic characterization techniques such as resistivity and Hall-effect measurements and to microscopic techniques like optical spectroscopy.

Except in a few—although very important—cases, e.g., the semiinsulating modifications of GaAs and InP, *deep* centers are of minor importance for control of electrical conductivity. However, they play a key role in the recombination properties of semiconductors (Mott, 1978; Queisser, 1978; Abakumov *et al.*, 1978). Since carrier capture and emission rates vary exponentially with level depths, minority carrier lifetimes can be governed by the density of deep defects even if their concentration is only a small fraction of the shallow dopant density. In certain cases, e.g., for fast switching devices, very short excess carrier lifetimes are a desirable feature. However, for most other applications, where device efficiency requires long minority carrier lifetimes, e.g., for light-emitting diodes, laser diodes, or solar cells, deep centers have the deleterious effect of reducing lifetime.

Deep centers are much less well understood than shallow defects (Queisser, 1971; Milnes, 1973; Mircea and Bois, 1979). This manifests itself already in terminology. Deep centers are often referred to as deep traps, deep levels, or deep states. These terms cannot be avoided if microscopic information about the defect involved is lacking. It must be kept in mind, however, that this terminology emphasizes only one aspect of a defect center, namely, its energy level within the band gap. It also underlines the ignorance about the center in question. However, despite the considerable experimental difficulties associated with the study of deep centers in III–V compounds, definite progress in defect identification has been made during the last decade. These developments are reviewed in Sections IV and V.

II. EXPERIMENTAL TECHNIQUES FOR POINT DEFECT ASSESSMENT

The experimental methods suitable for defect investigations can be divided into destructive, impurity-element-specific and nondestructive, defect-specific techniques. Among the former, mass-spectroscopy, atomic absorption, and nuclear methods like neutron activation analysis are the most important. They provide information about the *total* impurity content of a specific element but do not distinguish between different defect centers involving this element. Nevertheless, they can be very helpful for defect assessment if they are combined with defect-specific techniques.

The latter may be grouped into four categories: electrical, optical, magnetic (ESR, ODMR, NMR, ENDOR), and nuclear (Mössbauer, positron and μ^+ annihilation). In the following sections some general aspects of the corresponding techniques are outlined, but experimental details are not discussed.

The standard semiconductor assessment techniques, *resistivity* and *Hall-effect* measurements, were and are indispensible tools to establish the electrical properties of shallow donors and acceptors. Their application to deep centers is considerably more difficult: Deep defects in general have low solubility and their selective and deliberate introduction is difficult. Additional complications arise if the sample has a high resistivity.

Other steady-state electrical techniques, such as thermally stimulated current, thermally stimulated capacitance, photoconductivity, and photocapacitance (Bube, 1960; Sah *et al.*, 1970; Milnes, 1973; White *et al.*, 1976; Grimmeiss, 1977) proved to be more useful for deep-level studies. However, apart from being time consuming, most steady-state capacitance measurements suffer from the drawback that effects due to different deep states are difficult to separate. Straightforward data analysis and interpretation is possible only if one deep center is known to dominate in concentration.

Significant progress was achieved with the advent of *deep-level transient spectroscopy* (DLTS) (Lang, 1974; Miller *et al.*, 1977). This method combines the high sensitivity of capacitance measurements with the possibility of displaying different trapping centers as separate peaks on a temperature scale. The DLTS technique yields information about all important electrical parameters of a deep trapping center, i.e., its energy level, its capture cross sections, and its concentration. It is therefore suitable for routine quality assessment of device-grade material and thus has found widespread application. However, its usefulness for defect identification is limited, as is the case for other electrical techniques. Several modifications of the DLTS technique have been reported. They all use the basic DLTS principle, i.e., the concept of the rate window, but they differ in the trap-filling mode (electrical or optical) or in the type of transient monitored (capacitance or current). Optical trap filling, for instance, has to be used if the material under study is semiinsulating (Martin, 1980).

Optical spectroscopy is one of the most powerful techniques for point defect identification, especially if coupled with external perturbations such as uniaxial stress or magnetic fields. For the III–V compounds the useful wavelength regime extends from the visible to the far IR.

To a very large extent the present, detailed microscopic understanding of shallow impurities in GaP, GaAs, and InP has been derived from careful analysis of the near band-gap luminescence spectra of specially selected and doped samples (Dean, 1972; Queisser, 1976; White, 1979a; Dean and Herbert, 1979). For two reasons luminescence has been less successful in deep-center research. First, because of moderate to strong electron–phonon coupling, recombination at deep centers can to a large extent be nonradiative. Even if a deep luminescence is detected, the usual lack of sharp zero-phonon lines prevents unambigous interpretation. Crystal field bands of 3d impurities are an exception to this rule. The second point is of an experimental nature. In the near IR, down to about 4 μm where deep-center luminescence bands may occur, detector performance is inferior to that in the near-band-gap region of GaP, GaAs, and InP.

Defect assessment by optical spectroscopy in the medium-to-far IR region ($\lambda \geq 10$ μm) is usually done in absorption. Two important types of spectra occur in this energy range: The first includes local mode spectra consisting of sharp absorption lines due to local vibrations of impurity atoms and their near neighbors. The second corresponds to electronic hydrogen-like spectra that arise from internal transitions within shallow donors or acceptors.

Among the magnetic techniques, *electron spin resonance* (ESR) has so far been most successfully applied. For defect identification the method has turned out to be less powerful in III–V semiconductors than in silicon or II_B–VI compounds (Watkins, 1975). Nevertheless, it is still the only method that provides direct microscopic information about intrinsic defects. The most obvious reason why significant progress has been made only recently is the reduced ESR sensitivity. Since all group III and group V elements consist only of isotopes with nonzero nuclear spin, unresolved hyperfine interaction leads to ESR linewidths that are one to two orders of magnitude larger than in Si or II_B–VI compounds. This causes a sensitivity reduction of two to four orders of magnitude and it can also result in a loss of information. Apart from this inherent problem, severe experimental difficulties can arise, especially for *n*-type GaAs and InP. Because of the small ionization energies of shallow donors, efficient carrier freeze-out may not be achieved even at temperatures below 80 K. This results in undesired microwave losses and leads to an additional decrease in sensitivity. But despite these problems progress has been made. It has been repeatedly stated that ESR "does not work" for III–V materials. In view of the ESR results now available this statement is no longer tenable.

Whenever a defect exhibits a strong ESR spectrum, further information about its structure may be obtained by *electron nuclear double resonance* (ENDOR) spectroscopy. Recent results reported for 3d transition metal impurities in GaP and GaAs are encouraging.

There is also the possibility for *optical detection of magnetic resonance* (ODMR). With this technique, the "flipping" of the unpaired spin of a defect center is monitored by a change in its luminescent characteristics. In this way a direct correlation of a luminescent band with a specific paramagnetic defect may be possible. *Nuclear magnetic resonance* (NMR) studies of defects in III–V semiconductors are seriously limited by low solubility of most defects in these compounds. This is also true for *Mössbauer* spectroscopic investigations if the impurity in question is investigated in absorption. Considerably higher sensitivities are possible in "source-type" experiments. However, preparation of suitable sources may not be easy, and the interpretation of data may be complicated by defect creation during the radioactive decay of the source isotope.

Other nuclear spectroscopic techniques such as positron or μ^+ meson annihilation also bear promise for sensitive defect detection in III–V compounds.

The above qualitative survey underlines that the electrical and the microscopic techniques for defect assessment have their own merits and drawbacks. In many respects they are complementary, and a complete characterization thus requires application of several of these techniques. This is quite a difficult task, since electrical measurements most often are performed on inhomogeneous semiconductor junctions, whereas a neutral bulk piece of material is generally used for optical or ESR studies. Nevertheless, joint experiments should be possible and would be highly desirable.

III. Donors and Acceptors

Donors and acceptors are mostly deliberate dopants enabling *n*- or *p*-type conductivity of the crystal. In the elemental semiconductors Si and Ge this role is played by group V and group III impurities, respectively, when they have entered a substitutional site. In the binary III–V compounds more possibilities exist, since the dopant can occupy either cation or anion sites. Thus, donor states are created by group VI elements on anion (P, As) sites as well as by group IV elements on cation (Ga, In) sites. Correspondingly, acceptor states are formed by electron-deficient dopants, i.e., by group II elements on cation sites, but also by group IV elements on anion sites. There is also the possibility that donor states are formed by interstitial impurities. A well-studied example is the interstitial lithium donor that is electrically active in GaP and GaAs as well as in silicon. The scheme of doping III–V semiconductors is illustrated in Fig. 1.

The ionization energies E_D, E_A of substitutional "shallow" donors and acceptors in GaP and GaAs are quoted in Tables I and II, respectively. It is seen that values of E_D and E_A in GaAs are distinctly smaller than those found in GaP. This is mainly a consequence of the smaller effective masses of electrons and holes in GaAs as compared to GaP. It may also be noted that the values for E_D and E_A depend much less on the chemical identity of the dopant in GaAs than they do in GaP. Such an equivalence is demanded in the spirit of effective mass theory (EMT), which considers all shallow donors and acceptors as being equal, thus disregarding their different chemical natures.

In a *direct* semiconductor such as GaAs and InP the electronically excited states of shallow impurities are very precisely determined by Balmer's formula

$$E(n) = 13.6(m_{\text{eff}}/m)(1/\epsilon^2)(1/n^2) \quad [\text{eV}] \tag{1}$$

where n is the principal quantum number, ϵ the static dielectric constant, and m_{eff} the effective electron or hole mass. Optical transitions from the 1s

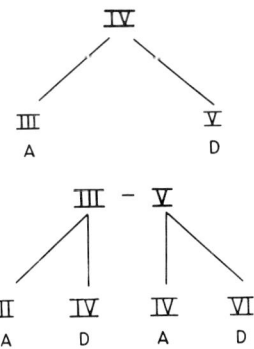

FIG. 1. Scheme of doping group IV and group III–V semiconductors with substitutional donors (D) and acceptors (A).

TABLE I

GaAs Ionization Energies (meV) of Substitutional Donors
and Acceptors at Low Temperatures[a]

As site	Ionization energy (meV)	Ga site	Ionization energy (meV)
Donors			
S	5.870	C	5.913
Se	5.789	Si	5.839
—	—	Ge	5.882
Acceptors			
C	27	Be	28
Si	34.8	Mg	28.8
Ge	40.4	Zn	30.7
Sn	167	Cd	34.7

[a] The experimental data are, in most cases, close to the theoretical EMT values, $E_D = 5.715$ meV or $E_A = 27.0$ meV [from Bimberg et al. (1981)].

ground state to the excited np states are electric-dipole allowed. They have rather large transition moments because of the greatly inflated radius of the weakly bound carrier.

The Lyman-type excitation spectrum of shallow donors in GaAs is shown in Fig. 2. Donor impurities of different chemical identity can be distinguished by a very small (<1 cm^{-1}) "chemical shift" that accounts for the splitting of the 1s → 2p transition in Fig. 2.

In InP chemical shifts of the 1s → np transitions are even more difficult to resolve, since the binding energy of all shallow donors is very close to the EMT value, 7.14 meV (Bimberg et al., 1981).

Electronic spectra of shallow donors in the *indirect* semiconductor GaP are considerably more complicated. Because of the multivalley structure of the conduction band and the resulting anisotropy of the effective electron mass, the simple Balmer relation [Eq. (1)] for the energy levels is no longer valid. For the excited states, $n \geq 2$, of a donor in GaP, the effective mass (EMT) description is similar to that in silicon (Kohn, 1957; Faulkner, 1969), since in both semiconductors the minima of the conduction band occur along the $\langle 100 \rangle$ axes of **k**-space.

Because of the anisotropy of the effective mass in each valley, the spherical degeneracy of the np, nd, ..., states is partially lifted. Thus, the 2p and 3p states are split into the sublevels $2p_\pm$, $2p_0$ and $3p_\pm$, $3p_0$, respectively. This splitting is clearly resolved in the IR excitation spectra, as seen in Fig. 3 for Te and S donors in GaP. Because donors in GaP are much deeper than

TABLE II

GaP Ionization Energies (meV) of Shallow Donors
and Acceptors at Low Temperatures[a]

P site	Ionization energy (meV)	Ga site	Ionization energy (meV)
Donors			
O	897	Si	85
S	107	Ge	204
Se	105	Sn	72
Te	93	—	—
Acceptors			
C	54	Be	57
Si	210	Mg	60
Ge	265	Zn	70
—	—	Cd	102

[a] The experimental values deviate considerably from the EMT predictions, $E_D = 59$ meV and $E_A \approx 50$ meV [from Bimberg et al. (1981)].

those in GaAs and InP, different donor species now exhibit much stronger chemical shifts of their 1s → np electronic Lyman spectra.

Effective mass theory has been very successful in silicon (and germanium), where the energetic separation of excited donor states from the conduction-band edge was predicted with remarkable precision (Kohn, 1957; Faulkner, 1969). In GaP, such attempts are rendered more difficult because of the slight "camel-back" structure of the conduction-band minima (Carter et al., 1977).

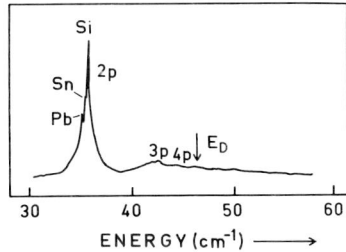

Fig. 2. Lyman-type far-IR excitation spectrum of shallow donors in n-GaAs, as measured by extrinsic photoconductive response. Donors of different chemical origin are distinguished by a small chemical shift of the 1s → 2p transition. The arrow indicates the donor ionization energy E_D. [From Cooke et al. (1978).]

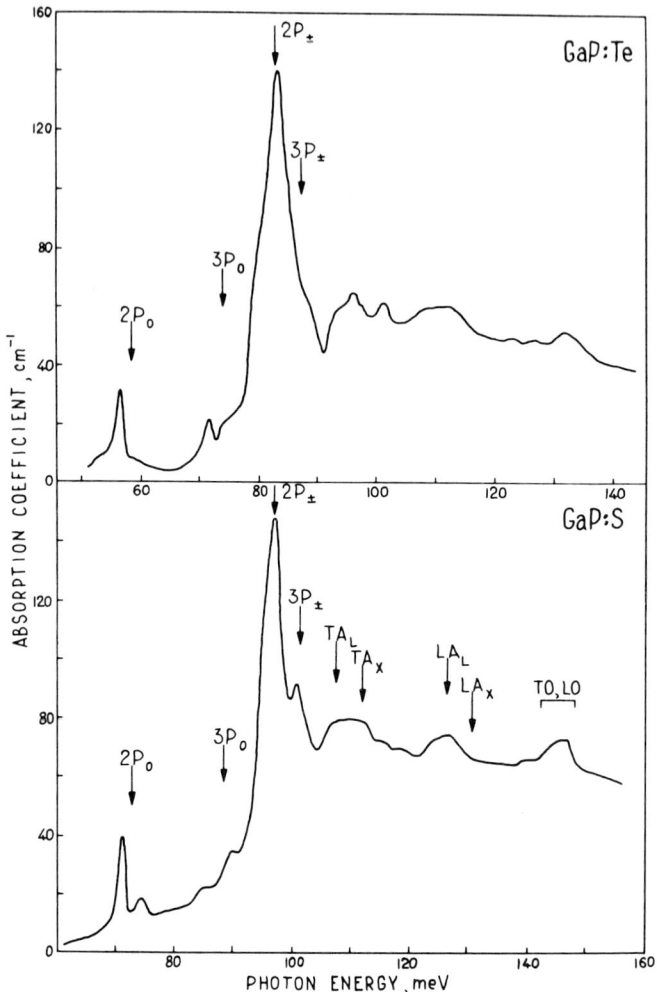

FIG. 3. Absorption coefficient of Te and S donors in GaP. Interfering lattice absorption lines have been subtracted from the actual spectrum to show only the impurity lines. [From Kopylov and Pikhtin (1978).]

1s *Ground State*

However, the 1s ground state of a shallow donor is not described correctly by EMT. One reason is that deviations from the simple Coulomb potential are here more sensitively probed by the 1s wave function. This explains the slightly differing donor ionization energies in GaAs (see Table I) that were determined by chemical shifts of the Lyman spectra. In indirect semiconductors further complications arise, since the multivalley structure of the conduction band may also cause a splitting of the 1s ground state.

In the binary compound GaP, each electronic donor state is *three*fold degenerate in **k**-space if certain complications arising from the slight camel-back structure are neglected. Whether or not this degeneracy of the 1s ground state is lifted by valley–orbit interaction depends on the site occupied by the donor. This difference stems from the fact that the conduction-band minima involved transform as X_1 or X_3 for donors substituting on the anion or cation site, respectively (Morgan, 1968).

For chalcogen donors on P sites, the three X_1 minima induce states of local symmetry $1s(A_1)$ and $1s(E)$. They are split by valley–orbit interaction such that the singlet state $1s(A_1)$ lies below the doublet $1s(E)$ (see Fig. 4a). For the deep oxygen donor in GaP, $E_D = 897$ meV, the valley–orbit depression of the $1s(A_1)$ ground state below the $1s(E)$ level is unusually large: 841 meV (Dean and Henry, 1968).

In contrast, for group IV donors on Ga sites the three X_3 minima induce a local p-like state of T_2 symmetry which cannot be split by any interaction having cubic symmetry. Inclusion of the electron spin however lifts the degeneracy through the rather weak spin–valley interaction (see Fig. 4b). For the Sn donor in GaP this splitting amounts to only 2.1 meV (Dean et al., 1970).

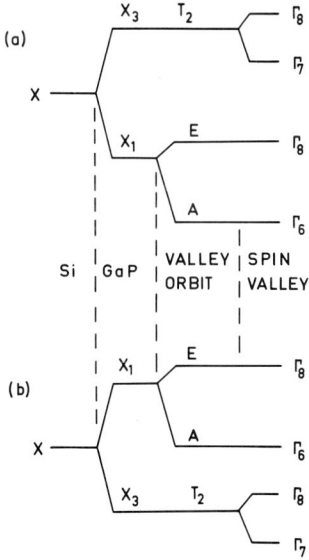

FIG. 4. Schematic configurations of electronic ground states of substitutional donors in GaP: (a) P-site donors; (b) Ga-site donors. [From Dean et al. (1974).]

The deep Ga-site germanium donor in GaP behaves anomalously, since its ground state has been found to have $1s(A_1)$ symmetry (Mehran et al., 1972), as in the case of the P-site donors. It has been argued that this level reversal is caused by a strong valley–orbit depression of the $1s(A_1)$ level deriving from the excited X_1 minimum in Fig. 4b (Dean et al., 1974; Altarelli, 1980).

Acceptors

The effective mass theory of holes bound to shallow acceptor impurities is quite similar to that outlined above for donors. Because the top of the valence band always occurs at the Γ point of the Brillouin zone, no multi-valley complications exist. However, the top of the valence band is *four*fold degenerate (Γ_8) because of the spin–orbit interaction in the p-like valence band. Thus, a hole in the upper valence band has angular momentum $j = 3/2$, rather than $j = 1/2$ for a conduction electron.

Bound Exciton Spectra

There are other possibilities for assessment of shallow donors and acceptors spectroscopically: They can also be studied by their bound exciton or by their donor–acceptor recombination spectra. For shallow impurities, these do not occur in the far IR but in a more convenient spectral range, i.e., at energies slightly below the band gap of the host.

A neutral donor–bound exciton complex, D^0X, consists of the charged-donor core D^+ and three carriers: $D^+e^-e^-e^+$. Correspondingly, an A^0X complex is $A^-e^+e^+e^-$. In either case the total angular momentum of the complex is *odd*. This multiplicity is directly confirmed by the Zeeman splitting of the sharp bound exciton lines into an *even* number of components.

The D^0X and A^0X emission occurs at energies slightly less than that of the free exciton: $h\nu = E_g - E_X - Q$. Here E_X is the binding energy of the free exciton, which is an intrinsic property of the crystal. In GaP donors (and acceptors) of different chemical identity can be distinguished by their different binding energies for the free exciton. The approximate relations $Q = 0.26 E_D - 7$ meV and $Q = 0.056 E_A + 3$ meV have been found to be valid for excitons bound to neutral shallow donors and acceptors, respectively (Dean, 1973a). In silicon simpler relations, Q/E_D or $Q/E_A \approx 0.1$, exist (Haynes, 1961). In GaAs and InP typical donors and acceptors are too shallow to be easily distinguished by a chemical shift of their normal D^0X or A^0X emission lines (Reynolds and Collins, 1981).

In GaP, exciton annihilation may also be strongly favored at "isoelectronic" defect centers, as group V nitrogen and bismuth impurities substituting on P-sites. The nitrogen centers in GaP have been extensively investigated after they were identified as being responsible for the electroluminescence of green light-emitting diodes (LEDs). In contrast to the more complex three-particle D^0X and A^0X recombination process, only *two* carriers are involved in the NX emission.

Donor–Acceptor Recombination

Another radiative channel for electron–hole annihilation is that of donor–acceptor recombination. In a partially compensated semiconductor ionized donors and acceptors may trap carriers under excitation. These subsequently recombine according to the scheme $D^0 A^0 = D^+ A^- e^- e^+ \to D^+ A^- + h\nu$. If the donors and acceptors are sufficiently separated in space, the emitted light has the energy

$$h\nu = E_g - E_D - E_A + e^2/\epsilon R \quad (2)$$

where R is the distance between donor and acceptor of a specific pair. Because R is discrete, but may aquire many values, spectra of remarkable multiplicity are expected. At low temperatures, the multiline structure is resolved in GaP, but not in GaAs or InP.

If both donors and acceptors occupy sites of the same sublattice, e.g., O and C substituting P-sites, the discrete values of the DA distance R can be numbered by $R = R(n) = (n/2)^{1/2} a$, where a is the lattice constant and n is an integer. It is remarkable that for some "magic" numbers, $n = 14, 30, 46, \ldots$, no lattice points exist (Hopfield *et al.*, 1963). The missing values of n produce gaps in the spectra labeled G in Fig. 5, which served as a key for the numbering of lines in such type I DA spectra. No such gaps exist in DA spectra of type II, where donors and acceptors occupy different sublattices. Here, $R(n) = (n/2 - 5/16)^{1/2} a$.

It should be noted that such multiline DA spectra are only observed at LHe temperatures. If the temperature is raised, the carriers are excited to the band states, from which they subsequently recombine ("free–free" transitions). An intermediate situation occurs for GaAs, where the donors have much lower ionization energies than the acceptors (Table I). Here, in a certain temperature range, free electrons and bound holes coexist. Their recombination then proceeds via "free–bound" transitions. An example for such emission is given for the deep manganese acceptor, $E_A = 113$ meV, in Section V.

As already seen in Table II the oxygen donor in GaP has an unusually high ionization energy. This causes the oxygen-related DA emission to occur in the near IR and not near the band edge.

The emission of distant donor–acceptor pairs observed in O, Zn- or O, Cd-doped GaP is superimposed with another luminescence band, which is

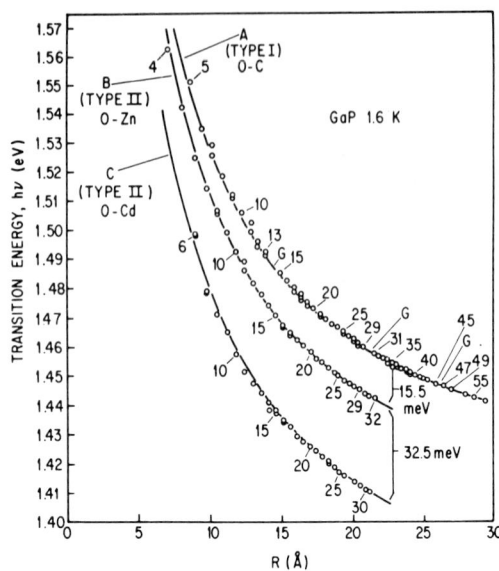

FIG. 5. Energetic positions of discrete DA recombination lines in GaP plotted versus pair separation. The curves are fits to Eq. (2). [From Dean et al. (1968).]

displaced by ~0.5 eV to higher energies. For GaP:O:Cd a sharp structure is resolved at low temperature. Zeeman studies have shown that the luminescence center in question has trigonal symmetry. It was therefore identified as a nearest neighbor O–Cd pair (Henry et al., 1968a). The presence of oxygen and cadmium in the defect was proven unambiguously by the observation of isotope shifts of the zero-phonon lines (Fig. 6). The O–Cd or O–Zn nearest neighbor pair may be viewed as an "isoelectronic molecule." The binding energy of an exciton at the O–Zn center, $Q = 0.3$ eV, is considerably larger than that at the isoelectronic trap nitrogen, $Q = 0.02$ eV. Consequently, the O–Zn emission occurs in the red, rather than in the green spectral range, as in the case of the nitrogen center.

The deep oxygen donor in GaP also causes a third type of emission ("capture luminescence"), arising from radiative capture of an electron by the ionized donor core (Dean and Henry, 1968). This emission occurs at $h\nu \leq E_D \approx 0.8$ eV. Later it was discovered by photocapacitance techniques that the deep oxygen donor in GaP can bind a *second* electron (Kukimoto et al., 1973). This finding raised the question whether the above capture luminescence is caused by the one-electron or by the two-electron state of oxygen (Morgan, 1978).

Recent Zeeman and piezospectroscopic investigations (Gal et al., 1981) of the 0.841-eV zero-phonon line associated with the capture luminescence

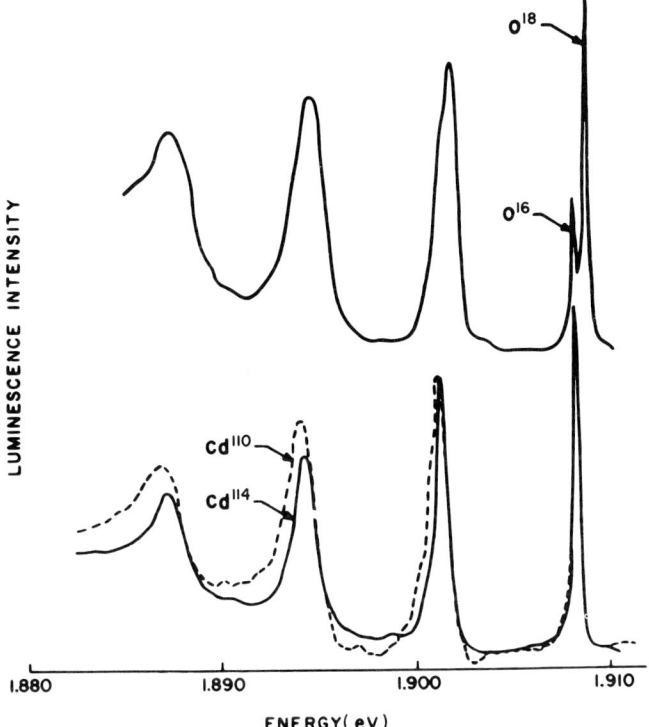

FIG. 6. Isotope shifts observed for the O–Cd pair luminescence in GaP. [From Henry et al. (1968b).]

strongly indicate that this transition arises from a transition between the valley–orbit split states, $1s(E) \rightarrow 1s(A_1)$, of the neutral donor state, as originally suggested by Dean and Henry (1968).

Localized Vibrational Modes

Defects in solids can occasionally also be studied by their "local mode" vibrational spectra. This is only possible if the corresponding energies are sufficiently separated from those of the intrinsic, much stronger lattice phonon absorption bands. Such a situation is favorable for light impurities having a smaller mass than the lattice atoms they replace.

A representative example for GaP is shown in Fig. 7. The impurities C, B, N, Si, and Al are identified by their different vibrational frequencies.

A drawback of such studies is the fact that spectra are not readily recorded on *n*-type or *p*-type samples. Here it is compulsary to eliminate the

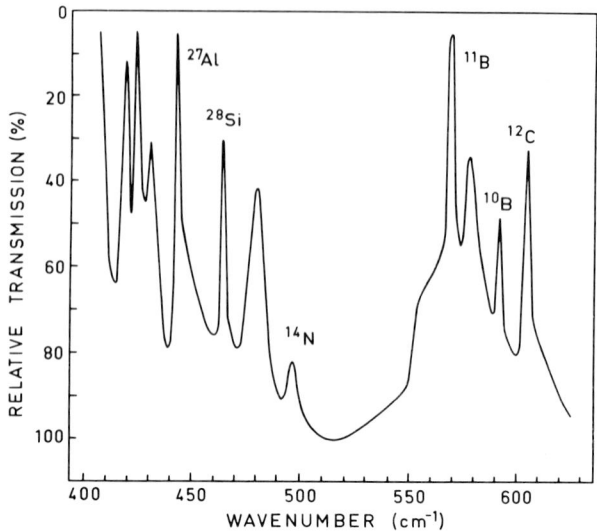

FIG. 7. Localized vibrational modes of C, B, N, Si, and Al in GaP; $T = 77$ K. [From Thompson and Newman (1971).]

strong free-carrier absorption before IR measurements can be made. This can be achieved by fast-electron irradiation of the sample, by which deep traps for the carriers are created (Newman, 1973). It should be added that impurity phonon modes can also be detected as side bands of electronic transitions of a defect.

IV. INTRINSIC DEFECTS

In strongly ionic binary compounds, such as the alkali halides, the most prominent intrinsic defects are vacancies and interstitials. Here anion or cation vacancies can trap electrons or holes, respectively, thus forming the so-called F and V centers. The trapped carriers have bound excited electronic states which are responsible for a characteristic defect-induced coloration of the otherwise transparent crystal. The corresponding optical absorption and fluorescence processes arising from these color centers have been extensively investigated for more than five decades.

Vacancy-related defects in tetrahedrally coordinated binary compounds are much less well understood. In the group II_B–VI semiconductor family, fair progress in the assessment of F- and V-type centers was achieved in the past. However, present knowledge of such defects in III–V compounds is only marginal. It is not known whether the F-center analog exists in GaP,

GaAs, or InP despite much work and speculation about anion vacancy centers, especially in GaAs. This means that it is uncertain whether an anion vacancy in these compounds is capable of binding an unpaired electron, thus forming a localized state in the forbidden zone of the semiconductor.

However, the V-center analog, i.e., a hole trapped at a cation vacancy, was recently detected by ESR in electron-irradiated GaP (Kennedy and Wilsey, 1978). By ESR it was also discovered that in III–V semiconductors a third possibility for lattice disorder exists: the "antisite" defects. They are formed by anions on cation sites (or vice versa) and were identified in GaP (Kaufmann *et al.*, 1976) and GaAs (Wagner *et al.*, 1980).

A. The Isolated Gallium Vacancy in GaP

Early ESR studies of fast-electron-irradiated silicon have revealed that isolated vacancy centers are mobile already far below room temperature (Watkins, 1965). In compound semiconductors a lower mobility of vacancies is expected, since migration now proceeds by jumps to the second nearest position, whereas in silicon a jump to the nearest site is sufficient. This was confirmed by the observation that isolated zinc vacancies in ZnS and ZnSe are frozen in at room temperature (Watkins, 1973). Here isolated cation vacancies are thermally destroyed above 300°C.

The thermal stability of isolated cation vacancies in III–V compounds should be intermediate—between that in silicon and that in the isoelectronic II_B–VI compounds. This suggests that *isolated* cation vacancies do *not* exist in noticeable concentrations in as-grown material. The reason is that during the slow cooling process following crystallization, vacancies have sufficient time to diffuse out of the crystal or to associate with other defects, as is discussed in Section IV,B.

In as-grown GaP crystals, pulled by the liquid encapsulation Czochralski (LEC) technique, isolated V_{Ga} centers could not be detected by ESR down to the 10^{15}-cm^{-3} level (Kaufmann and Kennedy, 1981).

However, the creation of isolated vacancy centers in concentrations far above the thermal equilibrium value can be enforced by fast-electron irradiation. Thus, ESR of the isolated gallium vacancy could be detected in fast-electron-irradiated GaP (Kennedy and Wilsey, 1978). A typical ESR spectrum is shown in Fig. 8 for $H \| \langle 100 \rangle$.[1] The five-line pattern arises from the hyperfine (hf) interaction of the unpaired spin with the four equivalent ^{31}P ligands around V_{Ga}. More complex patterns are observed for other

---

[1] Very recently Scheffler *et al.* (1981) questioned the assignment of this spectrum to V_{Ga} and suggested that it originates from C_{Ga}. Experiments are presently being performed to settle this controversy.

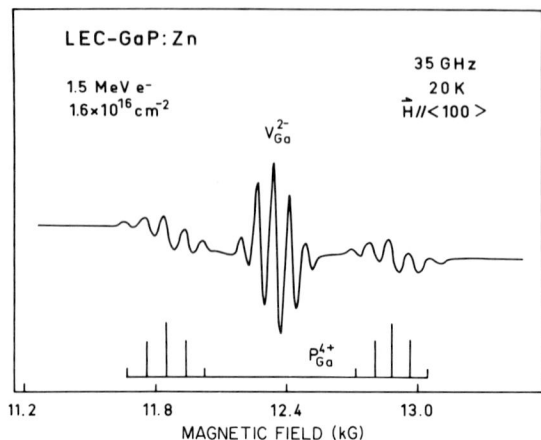

FIG. 8. ESR spectrum of the isolated gallium vacancy, V_{Ga}^{2-}, observed after 1.5 MeV irradiation of p-type GaP. The signal appears only under optical excitation. The two outer five-line patterns originate from the P_{Ga} antisite defect (Section IV,C). [From Schneider and Kaufmann (1981).]

geometries, such as $H\|\langle 111\rangle$ and $H\|\langle 110\rangle$. They are caused by the axial anisotropy of the ligand hf interaction, from which it can be inferred that 88% of the unpaired spin's wave function resides at the four nearest P-ligands around V_{Ga} (Kennedy and Wilsey, 1978, 1981a).

However, the overall symmetry of the V_{Ga} center was found to remain strictly cubic (T_d) even at 1.4 K (Kennedy and Wilsey, 1981a). This is a rather surprising observation since, in contrast, isolated vacancy centers in silicon, as well as the isolated V_{Zn} centers in ZnS and ZnSe, are known to undergo a strong static Jahn–Teller distortion (Watkins, 1965, 1973). No fully convincing explantion for the puzzling lack of a Jahn–Teller distortion of the V_{Ga} center in GaP can be offered at present (Kennedy and Wilsey, 1981a).

By electron irradiation the V_{Ga} center is preferentially introduced into p-type GaP. In situ optical excitation, $hv \geq 1.7$ eV, is required to convert the defect into its paramagnetic charge state V_{Ga}^{2-}. The occupancy level related to V_{Ga}^{2-} lies in the upper half of the band gap (Kennedy and Wilsey, 1981a).

The fate of the Ga interstitial, simultaneously created when a V_{Ga} defect is formed by the electron beam, remains uncertain. There is evidence that Ga interstitials become trapped at ubiquitous 3d impurity sites. For instance, it was observed by ESR that trigonal Fe centers are formed after electron irradiation of iron-doped LEC InP (Kennedy and Wilsey, 1981b). Similar observations have been made for GaP and GaAs (Igelmund, 1979), as well as for fast-neutron-irradiated ZnO and ZnS (Leutwein and Schneider, 1971). The trapping of primary radiation defects by transition metal background impurities may be a common event in III–V semiconductors.

Such findings are also remarkable in the sense that they provide strong evidence for fast-diffusing radiation-induced defects in compound semiconductors.

B. Vacancies in GaAs

1. *General*

Numerous attempts have been made to relate spectroscopic features to vacancies in GaAs. It appears, however, that an unambiguous identification of isolated vacancies or vacancy complexes has not been achieved to date with any spectroscopic technique.

Most studies were performed by photoluminescence on samples, either grown under conditions favoring Ga-rich (or As-rich) material or annealed in atmospheres favoring outdiffusion of As (or Ga). The correlations between specific luminescence bands with growth or annealing conditions were then taken as evidence for an assignment to V_{As} (or V_{Ga}). Even if such studies are done very carefully they do not yield a convincing identification of vacancies. There are two reasons which make the results of such studies ambiguous to interpret. *First*, growth under nonstoichiometric, say As-rich, conditions will not only favor formation of V_{Ga} but also formation of As_{Ga} antisites and arsenic interstitials As_i. Correspondingly, growth under Ga-rich conditions favors V_{As} as well as Ga_{As} and Ga_i formation. *Second*, studies on thermally annealed samples are severely hampered by effects due to impurities either in-diffusing from the surroundings or accumulating near the surface by out-diffusion from the bulk of the material.

Hwang (1969) reported unstructured luminescence bands near 1.37 eV for Zn- and Cd-doped GaAs, which he relates to V_{As} complexed with Zn and Cd acceptors, respectively (Fig. 9). This assignment has been questioned by Chiang and Pearson (1975), who suggest that Cu is responsible for this emission. Their criticism does not appear to be fully justified since the 1.35-eV Cu_{Ga} emission (Queisser and Fuller 1966) shows a distinct structure at 20 K. The corresponding complexes of V_{As} with Si_{As} and Ge_{As} have also been invoked in luminescence studies of Si- and Ge-doped GaAs (Kressel *et al.*, 1968; Williams and Elliott, 1969).

Chang *et al.* (1971) attribute luminescence bands at 1.35 and 1.40 eV to isolated V_{Ga} and isolated V_{As}, respectively, and they explicitly exclude

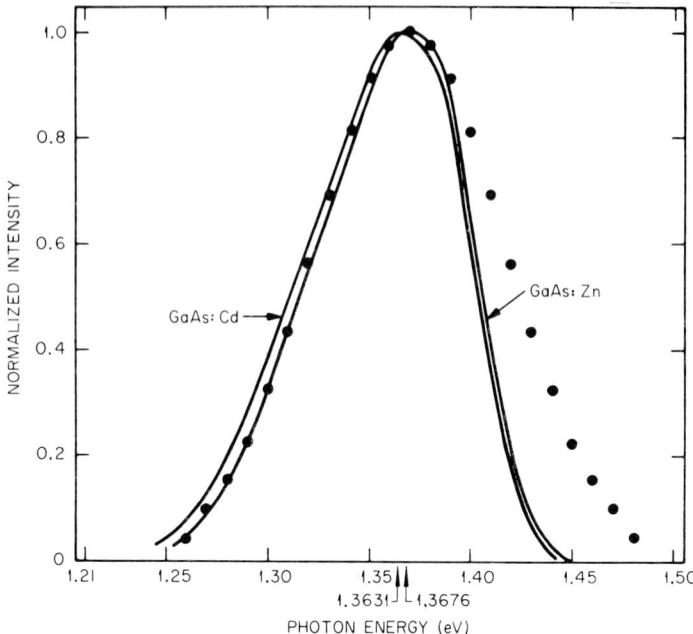

FIG. 9. The 1.37-eV emission band in GaAs:Zn and GaAs:Cd at 20 K. The solid circles are points of a Gaussian curve computed to get the best fit to the GaAs:Zn curve. [From Hwang (1969).]

copper as the source for the 1.35-eV luminescence. Chatterjee *et al.* (1975) also relate the 1.35 eV peak to V_{Ga}. Whether this 1.35-eV band is actually different from the Cu 1.35-eV emission appears to be an open question.

Koschel *et al.* (1977) assign two structured emission bands with relatively sharp zero-phonon lines at 1.409 and 1.413 eV to V_{As}–Si_{As} and V_{As}–C_{As}, respectively. Similar conclusions have been reached by Lum *et al.* (1977) and Lum and Wieder (1977, 1978). These assignments remained puzzling since other workers had concluded that a luminescence band with nearly the same spectral shape and position is due to the deep Mn acceptor (Section V,E). Recently Klein *et al.* (1980), by combining luminescence, Hall-effect, and mass spectroscopic measurements, convincingly demonstrated that the 1.41-eV emission is due to manganese. They also concluded that Mn, although being present at low concentrations in the bulk of the material, migrates during heat treatment and accumulates near the surface. The results of Klein *et al.* (1980) also cast doubt on the conclusions of Munoz *et al.* (1970) who invoked V_{Ga} and V_{As} to explain surface carrier concentration changes after annealing. The work of Klein *et al.* (1980) emphasizes the importance of impurity effects in luminescence or electrical studies of heat-treated GaAs. It should teach us not to be too ready to invoke vacancies for the interpretation of experimental data even if they appear at first sight to be the most obvious choice.

The controversies about vacancy assignments exemplified above, as well as experience with Si, the II_B–VI compounds, and in particular GaP, tell us that the samples most suitable for the study of vacancies in GaAs are particle-irradiated crystals. Some progress toward identification of vacancies and other possibly simple intrinsic defects has been made by studying such samples with DLTS.

Lang and Kimerling (1975) and Lang et al. (1977) studied 1-MeV electron-irradiated GaAs by the DLTS technique. They found five radiation-induced deep electron traps, which were labeled E1–E5. From the orientation dependence of the introduction rate for these defects they inferred that E1, E2, and E3 are simple point defects related to Ga displacements. The shift of the E3-level depth in $Ga_{1-x}Al_xAs$ alloys finally led these authors to conclude that E3 most likely corresponds to V_{Ga}. This assignment is at variance with the DLTS results of Pons et al. (1980). Based on the thermal annealing behavior of the traps E2–E5 they tentatively identify E2 rather than E3 with V_{Ga}. In view of such disputes further attempts to observe V_{Ga} by ESR are highly desirable. Even if the signal were an order of magnitude weaker than that of V_{Ga} in GaP, ESR should be detectable.

2. *Association of Vacancies with Donors*

When a semiconductor is doped with donors (acceptors), at least partial self-compensation can occur (Kröger and Vink, 1956; Mandel, 1964). This means that during cooling from the growth temperature, the crystal reacts to the doping by forming intrinsic defects, i.e., acceptors (donors), such as to minimize its total free energy. Whether this mechanism is thermodynamically favored depends on the balance of formation energy for the compensating defects on one hand, and the energy gained by trapping of free carriers by the defects on the other hand. Self-compensation effects are known to be important for II_B–VI compounds (Aven and Prener, 1967). For example, in donor-doped ZnS and ZnSe, formation of Zn vacancies acting as acceptors is known to occur. They associate with the donor impurity to form the so-called "self-activated" luminescent centers (Prener and Williams, 1956), also called A centers. ESR investigations provided an unambiguous microscopic confirmation of their existence (Watts, 1977).

For the most important III–V semiconductors, GaP, GaAs, and InP, controlled *p*- and *n*-type doping over at least three orders of magnitude of free carrier concentration is possible. Thus, if self-compensation occurs, the effect must be small and this has been demonstrated by many electrical

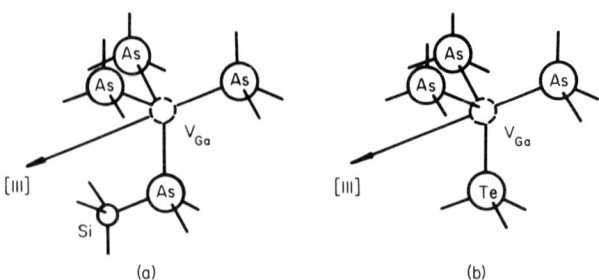

FIG. 10. V_{Ga}-donor associates in GaAs. They represent the direct counterpart of the "self-activated" luminescent centers in ZnS or ZnSe: (a) Si doped; (b) Te doped. [From Williams and Bebb (1972).]

investigations. However, this does not mean that defects produced by self-compensation might not be important. Even if only 1% of the dopant atoms were compensated by intrinsic centers this would result in 10^{14}–10^{16} cm^{-3} defects at the typical doping levels. Although the effect on free carrier concentrations would be negligibly small, other material parameters might be severly affected, since the compensating intrinsic centers are expected to act as deep traps.

Pairs V_{Ga}–D, analogous to the A centers identified for some II$_B$–VI compounds, have also been proposed to exist in GaAs (Williams and Bebb, 1972) (Fig. 10). They are thought to be responsible for an unstructured deep luminescence in donor-doped GaAs (Fig. 11). Much work has been devoted to this emission, and considerable evidence in favor of its assignment to gallium–vacancy donor pairs has been obtained (Williams and Bebb, 1972). However, it is noted that a direct microscopic confirmation is still lacking. It would be crucial to establish the symmetries of the centers in question. If the defects responsible for the luminescences in Fig. 11 form by a self-compensation mechanism, the analogy with II$_B$–VI compounds does not necessarily support an assignment to V_{Ga}–D centers. For III–V compounds formation of other compensating intrinsic acceptors might be more favorable (van Vechten, 1975a,b), e.g., Ga$_{As}$ antisite defects (Section IV,C). During crystallization such defects could tend to associate with donor impurities to form Ga$_{As}$–D pairs, similar to A-center formation in II$_B$–VI compounds.

Donor-related defects, so-called DX centers, have also been found in Ga$_{1-x}$Al$_x$As alloys. Ballistic phonon attenuation experiments (Narayanamurti *et al.*, 1979) indicate that their symmetry is trigonal and orthorhombic for Ga- and As-site donors, respectively. The model suggested for the DX defect consists of a donor paired with an As vacancy (Lang *et al.*, 1979; Lang, 1980). Further work to confirm this suggestion would be desirable.

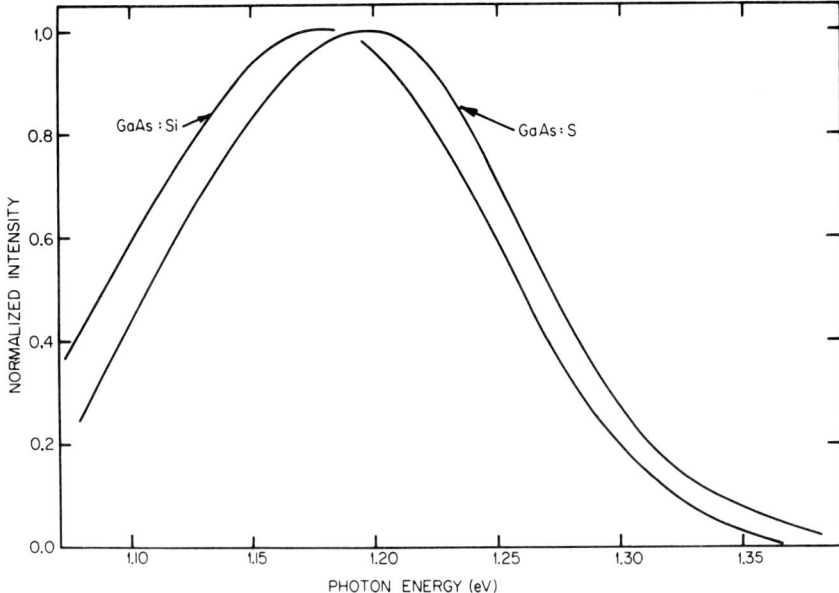

FIG. 11. The 1.2-eV luminescence band in GaAs:Si and GaAs:S at 74 K. [From Williams and Bebb (1972).]

The above examples provide strong evidence for the existence of complexes consisting of donor impurities and intrinsic lattice defects. However, the almost exclusive discussion (Logan and Hurle, 1971; Hurle 1977) of these centers as vacancy complexes does not appear to be fully justified, either from a theoretical point of view (van Vechten, 1975a,b) or from the convincing experimental data that now exist for other intrinsic defects in *as-grown* III–V compounds, namely antisite defects.

C. Antisite Defects

In a binary AB compound antisite defects B_A, A_B, and $B_A\ A_B$ pairs represent a third fundamental possibility for lattice disorder. This was pointed out by Kröger and Vink already in 1956. Although antisite defects have occasionally been invoked to explain electrical data for GaAs (Blanc et al., 1964) and GaSb (Reid et al., 1966; Van der Meulen, 1967) their study has been neglected for a long time even in careful work about stoichiometry effects in GaP (Jordan et al., 1974a,b). Recently antisites were also invoked to explain near-band gap luminescence lines in GaSb (Jakowetz et al., 1977;

Allègre and Avéruns, 1979) and GaAs (Reynolds et al., 1980; Dean and Herbert, 1980).

Thermodynamic calculations by van Vechten (1975b) suggest that antisite defects should be especially important for III–V compounds. Intuitively this is not surprising since they are among those AB compounds having the smallest electronegativity difference between the constituent atoms. Van Vechten (1976) also noted that no description of a compound semiconductor can be complete without considering antisites.

1. *Isolated P_{Ga} in GaP*

As in the case of the gallium vacancy in GaP, ESR so far is the only experimental method that provides definite, quantitative information about antisites. The first firm microscopic evidence for the existence of such centers was obtained for GaP. Here the $P_{Ga}P_4$ cluster was detected by ESR (Kaufmann et al., 1976). The identification is based on the hyperfine interactions of the unpaired electron with the central P_{Ga}^{4+} ion and the four P ligands (^{31}P, 100%, $I = 1/2$), giving rise to a very characteristic 2×5-line ESR spectrum (Fig. 12). The ESR parameters of this center are listed in Table III. From the hf data information about the spatial distribution of the unpaired electron's wave function can be obtained. Using a simple molecular orbital model Kaufmann and Schneider (1980a) found that 26% and 66% of the electron's density are localized at the central P_{Ga}^{4+} ion and at the four P ligands, respectively.

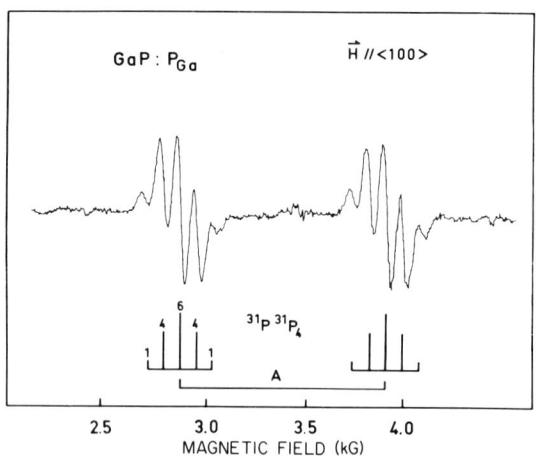

FIG. 12. ESR spectrum of the P_{Ga} antisite defect as observed in as-grown LEC GaP at 9.7 GHz and 20 K. A decomposition of the spectrum into the 2×5 hyperfine components of the $^{31}P^{31}P_4$ cluster is shown in the lower part. [From Schneider and Kaufmann (1981).]

TABLE III

ESR PARAMETERS OF ANTISITE DEFECTS IN GaP AND GaAs[a]

Defect	g	A	$T_\|$	$T_\perp$	$\Delta H(G)$
$P_{Ga}^{4+}P_4$	2.007	966	105	60	61
$P_{Ga}P_3X$	2.006	704	($T_{100}=81$)		85
			($T_{100}=109$)		
P_{Ga} ?	2.002	625	—	—	≈180
$As_{Ga}^{4+}As_4$	2.04	900	—	—	≳400

[a] Hyperfine parameters are in units of 10^4 cm^{-1}.

Until now, P_{Ga} has only been observed in acceptor-doped (Zn, Cr, V) GaP where concentrations up to 4×10^{16} cm^{-3} were found (Kaufmann and Kennedy, 1981). This raises the question whether P_{Ga} enters primarily by a self-compensation mechanism in acceptor-doped material or from stoichiometry deficiencies.

Van Vechten (1975b) noted that formally one would expect P_{Ga} in GaP to be a donor, probably multiply ionizeable, and with a first ionization energy comparable to the effective mass value (59 meV). In the ionic limit the neutral donor state D^0 can be denoted as P_{Ga}^{3+} since three of the five outer electrons of the P atom are needed to complete bonds. No experimental data for the first ionization energy (P_{Ga}^{3+}/P_{Ga}^{4+}) of P_{Ga} is available. However, by comparison with the chalcogen double donors S, Se, Te in Si (Grimmeiss et al., 1981), which have first ionization energies roughly an order of magnitude larger than the effective mass value (31 meV for Si), one would expect that the P_{Ga}^{3+}/P_{Ga}^{4+} level in GaP could be as deep as ~ 0.5 eV. It is possible that the P_{Ga}^{3+}/P_{Ga}^{4+} level has been observed in electrical studies of GaP (Fabre and Bhargava, 1974; Brunwin et al., 1980) but could not be identified because of lacking microscopic evidence. Some experimental results have been obtained for the second ionization energy (P_{Ga}^{4+}/P_{Ga}^{5+}) using ESR (Kaufmann et al., 1981a). While the P_{Ga}^{4+} ESR in semi-insulating GaP:Cr is observed in the dark, in p-type GaP:Zn it appears only after optical excitation with $hv \geq (1.25 + 0.10)$ eV (Fig. 13). These observations provide strong evidence that the doubly ionized state P_{Ga}^{5+} is stable if the Fermi level is close to the valence band and that P_{Ga} actually is a deep double donor with a second optical ionization energy of 1.1 eV. This value is at variance with the theoretical estimate 0.2–0.5 eV of Jaros (1978).

The internal levels of the singly ionized P_{Ga} donor, P_{Ga}^{4+}, must be similar to those of the isoelectronic neutral gallium-site donor Ge (Mehran et al., 1972), but different from those of Si and Sn (Section III). For the D^+ state of

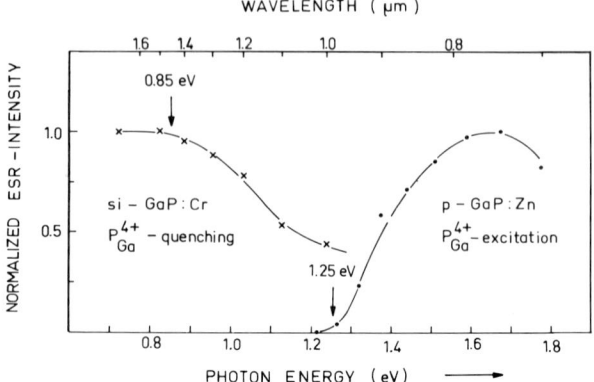

FIG. 13. Spectral dependence of light sensitivity of the P_{Ga}^{4+} antisite ESR signal in semi-insulating and p-type GaP, respectively. Note that under dark conditions the signal is present in the semi-insulating material but unobservable in the p-type sample. [From Kaufmann et al. (1981).]

P_{Ga} and the D^0 state of Ge, ESR is observed with a g value close to the free spin value *without* application of uniaxial stress. In both cases this indicates that the donor ground state is an orbital singlet (Dean et al., 1974), rather than the p-like T_2 state that one expects (Morgan, 1968) on the basis of effective mass theory after taking the multivalley structure of the conduction band near the X minimum into account (Section III).

Optical spectra associated with $P_{Ga}P_4$ have not yet been identified. One can expect that the neutral P_{Ga}^{3+} antisite is able to bind an exciton. However, since P_{Ga} is a double donor, recombination presumably will occur preferentially via a nonradiative Auger-type mechanism rather than radiatively (Dean and Herbert, 1980). It has also been argued (van Vechten, 1975c) that associates of P_{Ga} with other defects could be efficient nonradiative recombination centers acting through a configuration coordinate mechanism. Further work is required to confirm these suggestions. An internal capture luminescence analogous to that of oxygen in GaP (Dean and Henry, 1968; Gal et al., 1981) could also occur for P_{Ga}.

P_{Ga} antisites were found to be destroyed by high-temperature treatment (Kaufmann and Schneider, 1978), and there is evidence that the gallium vacancies thus formed become filled with in-diffusing transition metal acceptors, especially iron (Kaufmann and Kennedy, 1981). Whether P diffuses out or is trapped at phosphorus vacancies is not known. In any case an intrinsic double donor is replaced by an acceptor impurity. It therefore seems very likely that P_{Ga} antisite annihilation contributes to the observed thermal p-type conversion of semi-insulating GaP:Cr. In GaAs, As_{Ga} antisites may play a corresponding role.

2. Isolated As_{Ga} in GaAs

Recently Wagner *et al.* (1980), using a far-IR 0.3 THz ESR spectrometer, observed an isotropic four-line ESR spectrum in semiinsulating GaAs:Cr (Fig. 14). They assigned this signal to the As_{Ga} antisite. The spectrum is analogous to that of P_{Ga} in GaP but now consists of four hf lines arising from hf interaction of the unpaired spin with a central ^{75}As (100%, $I = 3/2$) nucleus. In principle, each of these four lines should be split into $(2 \times 4 \times 3/2) + 1 = 13$ ligand hf components if the magnetic field is along a $\langle 100 \rangle$ direction. This splitting has not been resolved, presumably because the width of a ligand hf component is larger than the ligand hf splitting. The ESR parameters of As_{Ga} in GaAs are quoted in Table III. They are consistent with what one expects by scaling the corresponding values of P_{Ga} in GaP.

Definite results for the energy levels of As_{Ga} in GaAs are not available, although tentative assignments of a DLTS peak observed in electron-irradiated material to As_{Ga} have been made (Guillot *et al.*, 1981). As noted before, P_{Ga} in GaP has so far only been observed in acceptor-doped material, and it is worth noting that the first observation of As_{Ga} in GaAs was also for acceptor (Cr)-doped material.

Another ESR spectrum, observed by Goldstein and Almeleh (1963) in heat-treated GaAs, has been repeatedly cited as evidence for antisites (Blanc

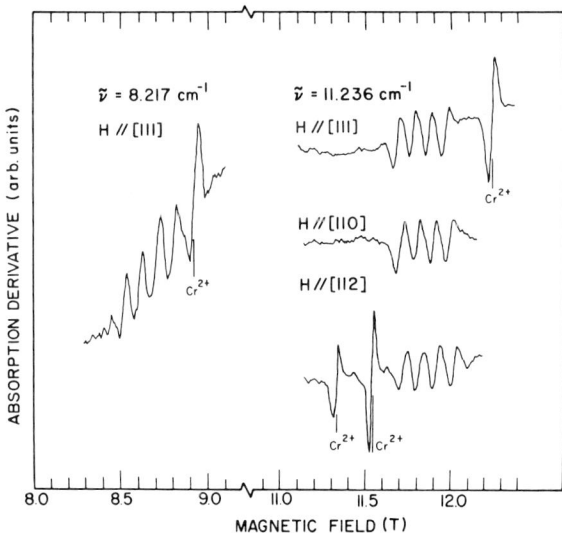

FIG. 14. Far-IR ESR spectra of semi-insulating GaAs:Cr measured for two different IR frequencies at $T \approx 15$ K. The isotropic four-line pattern arises from the As_{Ga} antisite defect. The remaining anisotropic lines are due to Cr^{2+}. [From Wagner *et al.* (1980).]

et al., 1964; van Vechten, 1975b; Reynolds *et al.*, 1980). It is emphasized that this assignment is erroneous, the ESR spectrum in question being due to isolated Fe_{Ga}^{3+} (Section V,D).

3. *Antisite-Related Defects in GaP*

Two further P-antisite-related defects in GaP have been discovered by ESR. The first center has only been observed following electron irradiation of *n*-type material (Kennedy and Wilsey, 1979). As for $P_{Ga}P_4$, the spectrum exhibits a large hf splitting due to interaction with a central P atom. However, the multiplicity of ligand hf lines implies interaction with only three equivalent P neighbors. The identity of the fourth nearest neighbor has not been established, but the impurities C, Si, N, O, or S are considered the most likely candidates (Kennedy and Wilsey, 1979). The possibility of a nearest neighbor antistructure pair P_{Ga}–Ga_P also cannot be ruled out completely. Other, sofar unidentified, radiation-induced defects were observed by Kennedy and Wilsey (1981a).

The second center has been observed in as-grown Zn-doped GaP overcompensated with sulfur. An ESR spectrum before and after illumination is shown in Fig. 15 (Kaufmann and Wörner, 1981). Again there is a large hf

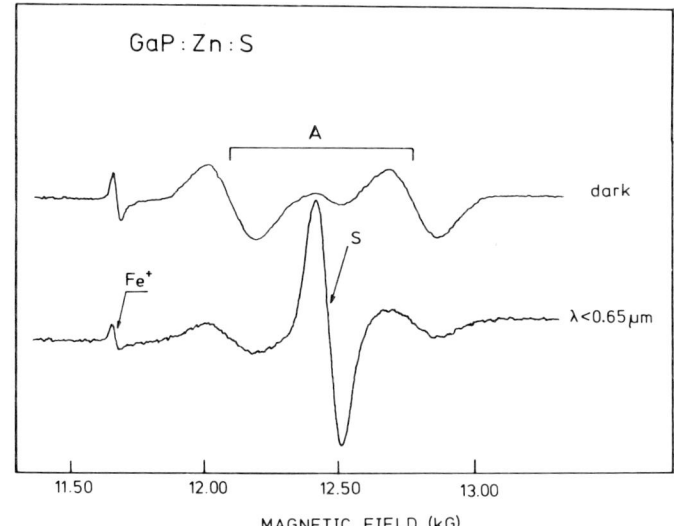

FIG. 15. ESR spectra of Zn-doped GaP slightly overcompensated with sulfur at 35 GHz and 20 K. The dark spectrum displays a P-antisite related signal A and the Fe^+ signal. Optical excitation transfers electrons from the center labeled A to ionized sulfur donors S. If the Fermi level is driven toward midgap the A signal disappears and the $P_{Ga}^{4+}P_4$ ESR pattern shows up in the dark. [From Kaufmann and Wörner (1981).]

POINT DEFECTS IN GaP, GaAs, AND InP   109

interaction with a central P atom but no ligand hf splitting is resolved. This prevents definite identification of the neighbors surrounding the P defect. The light-induced intensity changes in Fig. 15 indicate an electron transfer from the P defect to ionized sulfur donors. The center is therefore a majority carrier trap in n-type GaP. The ESR parameters of antisite-related defects are summarized in Table III.

V. 3d Transition Metals

In III–V semiconductors—and devices—some 3d elements occur as omnipresent and practically unavoidable background impurities. Their influence on minority carrier lifetime may be considerable and deleterious to device performance. The maximum solubility of 3d ions in III–V hosts is typically of the order of 10^{17} cm^{-3}, with the exception of manganese and possibly copper. For Mn solubilities in excess of 10^{19} cm^{-3} have been reported without precipitation of a second phase.

It was recognized already in the early 1960s that the 3d transition metals form deep states in GaAs (Haisty and Cronin 1964). Later work confirmed that this is also true for GaP and InP. Optical and ESR spectroscopy have shown that in most cases the 3d impurity preferably occupies a substitutional metal site. This contrasts to the situation encountered for silicon, where mainly interstitial 3d impurities have been detected by ESR (Ludwig and Woodbury, 1962).

Most 3d impurities in GaP, GaAs, and InP act as deep acceptors. Consequently, association with donors in n-type material can be expected to occur (Section V,H). Ionization energies for neutral substitutional 3d acceptors in GaP and GaAs are listed in Table IV. They are seen to exceed greatly the values of shallow acceptors quoted in Tables I and II. Obviously

TABLE IV

Ionization Energies (eV) of $3d^n$ Neutral Acceptors in GaP and GaAs[a]

Acceptor	GaP	GaAs	Acceptor	GaP	GaAs
V($3d^2$)	1.2	0.75	Co($3d^6$)	0.41	0.16
Cr($3d^3$)	1.13	0.80	Ni($3d^7$)	0.50	0.20
Mn($3d^4$)	0.40	0.11	Cu($3d^8$)	0.5	0.14
Fe($3d^5$)	0.70	0.52			

[a] The experimental accuracy is generally less than that typical for shallow defects [from Bimberg et al. (1981)].

the electronic ground state of a deep 3d acceptor cannot be described by effective mass theory.

Several authors have recently attempted to establish an adequate theoretical description of the $3d^n$ acceptor ground states (Masterov and Samorukov, 1978; Fleurov and Kikoin, 1979, Partin *et al.*, 1979a, Hemstreet, 1980). However, the ionization energies predicted often still differ considerably from the experimental values.

The electrically neutral charge state of a substitutional $3d^n$ acceptor in III–V hosts corresponds, in the ionic limit, to a trivalent ion. We shall use this nomenclature in the following. If the neutral acceptor state A^0 has the configuration $3d^n$, the first ionized acceptor state A^- has the configuration $3d^{n+1}$, corresponding to a divalent transition metal ion. Some 3d impurities can also trap a second electron, thus forming a double acceptor state A^{2-}, as found for Ni, Fe, and Cr in GaP. A tetravalent charge state, $3d^{n-1}$, can result by hole trapping at a neutral 3d impurity. The only example so far is the stable Cr^{4+} ion recently identified by ESR in GaP, GaAs, and InP, (Section V,F).

The occupation of the multiple levels of a deep impurity is determined by the position of the Fermi level. Thus, the A^{2-} states detected for Ni, Fe, and Cr in GaP are only stable in *n*-type material. However, metastable valency states can be enforced by optical excitation at low temperatures.

A survey of the multiple $3d^n$ acceptor charge states in a III–V semiconductor is given in Table V. Most of them have been assessed by ESR and optical spectrocopy. The effective spin S of the ground state of each $3d^n$ configuration is quoted in the lowest row. It is obtained after crystal-field splitting and spin–orbit coupling have been taken into account. The resulting ground-state degeneracies of $2S + 1$ are seen to be in most cases quite different from those of shallow acceptors, where $2S + 1 = 4$.

1. *Excited States of* 3d *Acceptors*

For a $3d^n$ acceptor two rather different types of electronic transitions may be induced optically:

(a) ionization to the valence band $(3d^n) \rightarrow (3d^{n+1})e^+$; with less likelihood also to the conduction band $(3d^n) \rightarrow (3d^{n-1})e^-$.

(b) intra-3d-shell (crystal-field) transitions $(3d^n) \rightarrow (3d^n)^*$; these do not change the charge state of the impurity.

Optical spectra associated with photoionization of deep $3d^n$ impurities unfortunately do not exhibit a characteristic sharp line structure resulting

TABLE V

MULTIPLE CHARGE STATES OF TRANSITION METAL ACCEPTORS IN III–V SEMICONDUCTORS

	$3d^2$	$3d^3$	$3d^4$	$3d^5$	$3d^6$	$3d^7$	$3d^8$	$3d^9$
A^0	V^{3+}	Cr^{3+}	Mn^{3+}	Fe^{3+}	Co^{3+}	Ni^{3+}	Cu^{3+}	
A^-		V^{2+}	Cr^{2+}	Mn^{2+}	Fe^{2+}	Co^{2+}	Ni^{2+}	Cu^{2+}
A^{2-}				Cr^+		Fe^+		Ni^+
S	1	3/2	2	5/2	0, 1/2, 1	3/2	0	1/2

from transitions into bound shallow hydrogenic states that are precisely described by EMT. Thus, an exact determination of the ionization energy by optical spectroscopy is rendered rather difficult.

The spectral dependence of the photoionization cross section $\sigma(hv)$ of deep levels is frequently described by the relation

$$\sigma(hv) \propto (hv - E_0)^{3/2}/(hv)^3 \quad (3)$$

which is valid for photoionization from a δ-function potential (Lucovsky, 1965). Here $\sigma(hv)$ has its maximum at twice the binding energy, but disappears at the ionization threshold E_0. This contrasts with photoionization from a Coulomb potential, where threshold and maximum of $\sigma(hv)$ coincide at $hv = E_0$.

Intra-3d-shell excitations of transition metal impurities in solids can readily be analyzed by optical spectroscopy. In general, very sharp spectral features may be resolved, typically occuring in the 1–3.5-μm spectral range. Interpretation of experimental data here proceeds in the framework of crystal-field theory, often suitably modified to take into account electron–phonon coupling [Jahn–Teller effect (JTE)]. Typical examples for individual $3d^n$ ions will be discussed in the following sections. As a result of these studies it became apparent that $(3d^n) \leftrightarrow (3d^n)^*$ spectra of transition metals in III–V compounds do not, in general, differ drastically from those observed in more ionic hosts, such as the II_B–VI compounds.

In III–V compounds, sharp $(3d^n)$–$(3d^n)^*$ line spectra are observed for A^--state divalent transition metal ions. For nickel in GaP, such spectra were also detected for the A^{2-} state Ni^+. No sharp $(3d^n)$–$(3d^n)^*$ spectra have been found for A^0-state trivalent ions, presumably because of their overlap with the much stronger photoionization bands.

We shall now discuss specific features of the individual $3d^n$ acceptors, with special emphasis on chromium.

A. Copper

In the III–V as well as in the elemental and II_B–VI semiconductors, copper diffuses interstitially very rapidly. Several Cu-related defects have been detected by their IR and visible luminescence in GaP (Dean, 1973b; Monemar, 1972) and in GaAs (Gross et al., 1969; Willmann et al., 1973). Zeeman and piezospectroscopic studies reveal that several of these copper centers have lower than cubic symmetry, indicating association with other defects. Further examples will be given in Section V,H. So far, the chemical identity of the associated defects has not been definitely established. This seems to be analogous to the unsatisfactory situation encountered for copper-related defects and luminescent centers in II_B–VI phosphors (Holton et al., 1969).

B. Nickel

Nickel has been reported to be a persistent inadvertent impurity in LEC GaP (Kaufmann and Schneider, 1978) and even in device-grade VPE epitaxial layers (Dean et al., 1977). Since it diffuses quite rapidly in GaP (Hamilton and Peaker, 1978; Ennen and Kaufmann, 1980) as well as in GaAs (Partin et al., 1979b), further Ni contaminations are expected to occur during thermal processing steps. In addition Ni plays an essential role in GaAs contact technology, but it is also known that Ni traces disadvantageously affect minority carrier properties in n-type GaAs (Partin et al., 1978, 1979b).

A^0 State ($3d^7$). In GaP, the neutral Ni acceptor has an ionization energy of about 0.5 eV (Abagyan et al., 1976). Values reported in the literature for Ni in GaAs scatter considerably, ranging from 0.20 to 0.42 eV. The value quoted in Table IV, 0.20 eV, is assumed to be that of isolated Ni_{Ga}. It was recently discovered (Ennen et al., 1981) that Ni in GaP and GaAs has a strong tendency to form associates with donors (Section V,H). Ionization energies of such complexes have not been positively identified.

The electronic ground state of Ni^{3+} ($3d^7$) is orbitally nondegenerate, but has fourfold spin degeneracy. ESR has been observed for Ni^{3+} in GaP, GaAs, and InP (Kaufmann and Schneider, 1978).

A^- State ($3d^8$). This charge state of Ni, presumably being diamagnetic, has not been detected by ESR. However, in n-type GaP samples heavily compensated by Ni indiffusion a sharply structured impurity absorption band near 1 μm is observed (Baranowski et al., 1968; Ennen and Kaufmann, 1980). Although its assignment to a $^3T_1(F) \rightarrow {}^3T_1(P)$ crystal field transition may be debated, it certainly arises from an electronic excitation of the A^- state of isolated nickel, Ni^{2+}.

A^{2-} *State* ($3d^9$). Detailed optical and ESR studies have shown that in n-type GaP also the two-electron trap state Ni^+ is stable (Kaufmann *et al.*, 1979). In contrast to Ni^{2+}, the Ni^+ charge state is strongly fluorescent in the IR region. A typical spectrum is shown in Fig. 16. The emission can also be observed in nominally pure, device-grade VPE epitaxial layers (Dean *et al.*, 1977). This indicates that nickel contamination may lower the quantum efficiency of green-emitting LEDs.

A Zeeman analysis of the sharp zero-phonon line A_0 at 5355 cm^{-1} in Fig. 16 confirmed that this emission results from the $\Gamma_8(^2E) \rightarrow \Gamma_7(^2T_2)$ crystal field transition of Ni^+ on cubic lattice sites (Kaufmann *et al.*, 1979). This identification was further supported by uniaxial stress measurements (Hayes *et al.*, 1979) and luminescence excitation spectroscopy (Bishop *et al.*, 1980).

The occupation of the three possible charge states of Ni in GaP was found to depend strongly on the position of the Fermi level (Ennen and Kaufmann, 1980). It was also observed that nonequilibrium charge states could be created by optical excitation at low temperatures. The data suggest that the A^- state disproportionates according to the photoreaction

$$2Ni^{2+} + h\nu \rightarrow Ni^+ + Ni^{3+}$$

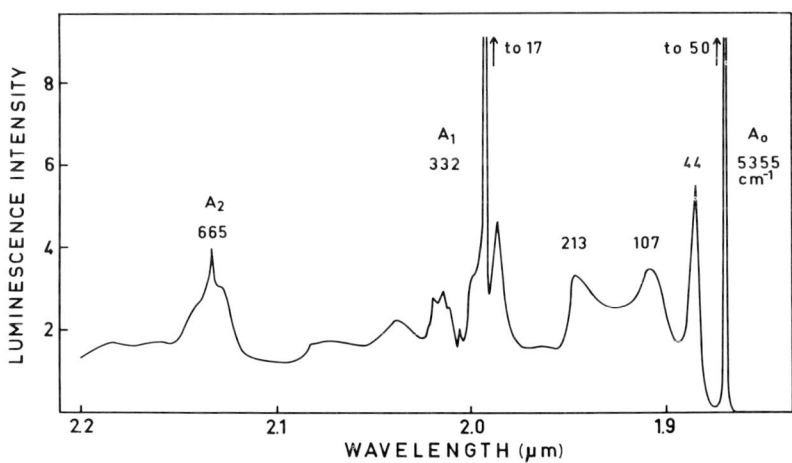

FIG. 16. Infrared luminescence spectrum of Ni^+ in GaP at 4.2 K. The numbers associated with the phonon side bands represent their energetic separation (in cm^{-1}) from the 5355 cm^{-1} zero-phonon line. [From Kaufmann and Schneider (1980a).]

C. Cobalt

A^0 *State* ($3d^6$). The ionization energy of the neutral Co acceptor in GaP and GaAs amounts to 0.41 (Loescher *et al.*, 1966) and 0.16 eV (Brown and Blakemore, 1972), respectively. The electronic configuration of the A^0 ground state is assumed to be $3d^6$, ^5E, corresponding to a Co^{3+} valency. The tenfold-degenerate ^5E state is split, by second-order spin–orbit interaction, into five close-lying sublevels. The same situation occurs for the A^- state, $3d^6$, of iron (Section V,D).

A^- *State* ($3d^7$). The electronic ground state of $Co^{2+}(3d^7)$ is the spin-only quartet $^4A_2$, in analogy to Ni^{3+} (Section V,B). ESR has been observed for Co^{2+} in GaP (Kaufmann and Schneider, 1978) and GaAs (Godlewski and Hennel, 1978).

Intra-d-shell excitations of Co^{2+} occur in the near IR. The one at lowest energy, $^4A_2 \rightarrow {^4T_2}(F)$, is dominated by a sharp zero-phonon line at 4506 cm^{-1} in GaP (Weber *et al.*, 1980) and at 4039 cm^{-1} in GaAs (Ennen *et al.*, 1980). In GaP:Co this transition is also strongly luminescent. Here, the above assignment was further confirmed by the details of the Zeeman and stress splitting of the zero-phonon line (Weber *et al.*, 1980; Hayes *et al.*, 1980).

In GaAs:Co no such luminescence was found. Instead, a strong near-band-gap emission at 1.30 eV is observed that may result from electron capture: $A^0(3d^6) + e^- \rightarrow A^-(3d^7) + h\nu$ (Ennen *et al.*, 1980). This radiative channel is also the dominant one in GaAs:Mn (Section V,E).

D. Iron

Apart from copper and nickel, iron is one of the most frequent 3d contaminants in III–V compounds. Iron introduces rather deep acceptor states in GaP and GaAs (Table IV). In InP, the Fe acceptor level lies almost exactly at midgap (Mizuno and Watanabe, 1975). Iron is therefore the preferred dopant to grow high-resistivity LEC InP material. For GaAs this is achieved by chromium doping (Section V,F).

A^0 *State* ($3d^5$). The ground state of Fe^{3+} is the spin-only sextet $^6A_1$, all other crystal field states being separated by more than 1 eV. ESR is readily detected and has been reported for Fe^{3+} in GaP, GaAs, InP, and also InAs. ESR parameters have been compiled by Bimberg *et al.* (1981).

A^- *State* ($3d^6$). The free-ion state ^5D of Fe^{2+} is split by the tetrahedral crystal field into an excited $^5T_2$ and a ^5E ground state. The tenfold-degenerate ^5E state is further split by second-order spin–orbit interaction into five close-lying sublevels. Their energetic separation is typically about 2 meV, depending on the host. ESR of Fe^{2+} has not been reported, but transitions between these low-lying levels should occur in the far IR. They can also be enforced by THz phonons. This explains the strong reduction of thermal conductivity induced by Fe^{2+} ions on tetrahedral lattice sites (Slack et al., 1966).

The crystal-field transition $^5E \leftrightarrow {}^5T_2$ occur near 3 μm. They have been observed, in absorption or emission, in GaP (Baranowski et al., 1967; Vasil'ev et al., 1976b; Andrianov et al., 1976; West et al., 1980), GaAs (Baranowski et al., 1967; Ippolitova and Omel' yanovskii, 1975), and InP (Koschel et al., 1977; Ippolitova et al., 1977). The spectra are very similar to those observed for Fe^{2+} in the more ionic II_B–VI compounds (Slack et al., 1966; Baranowski et al., 1967).

A^{2-} State $(3d^7)$. In GaP the two-electron trap state of iron is also stable and has been identified by ESR (Kaufmann and Schneider, 1977) in n-type material, see Fig. 17. The chemical origin of the Fe^+ ESR signal was unambiguously confirmed by electron nuclear double resonance (ENDOR) on ^{57}Fe-enriched GaP (Kirillov and Teslenko, 1979). The level corresponding to the process $Fe^+ \rightarrow Fe^{2+} + e^-$ is only about 0.1–0.2 eV below the conduction band (Suto and Nishizawa, 1972; Ennen and Kaufmann, 1980).

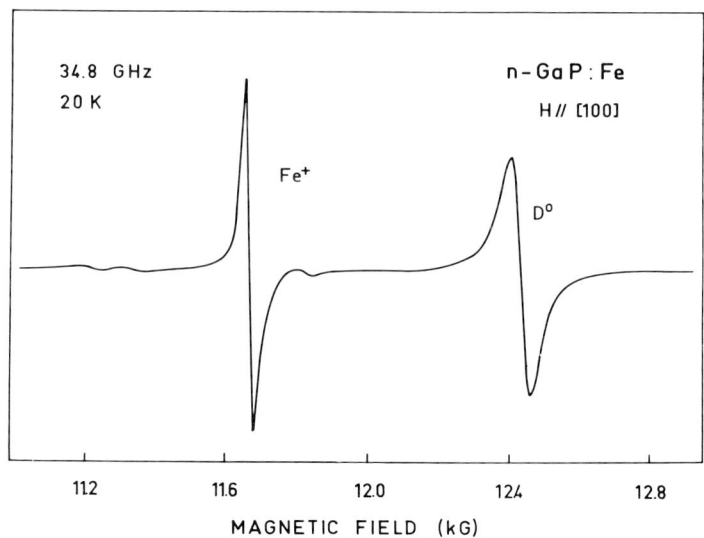

FIG. 17. ESR spectrum of n-type iron-doped GaP showing the signal of the two-electron trap state of iron and the signal of neutral donors. [From Kaufmann and Schneider (1977).]

E. Manganese

Among the deep $3d^n$ acceptors in III–V semiconductors, Mn is the one with the lowest ionization energy (Table IV). According to a general rule of experience, a low ionization energy is accompanied by a high impurity solubility (Queisser, 1971; Milnes, 1973). Thus, for Mn in GaP concentrations up to 3×10^{19} cm^{-3} have been reported without precipitation of a second phase (Abagyan et al., 1975).

It has been pointed out that the high solubility of the deep Mn acceptor in GaP may enable hopping conductivity at temperatures as high as 200 K (Abagyan et al., 1975). This is unique for that type of conduction in crystalline semiconductors.

A^0 *State.* In the tight-binding limit, the acceptor hole would enter the 3d core shell, resulting in a $Mn^{3+}(3d^4)$ configuration. However, in view of the rather low ionization energy, and the great stability of the $3d^5$ configuration, the hole may be energetically favored in a more delocalized orbit. In this case, the A^0 ground state would be $Mn^{2+}(3d^5)$ + hole. Experimental data so far do not permit a clear distinction between these two possibilities.

The spectral response of photoionization of the neutral Mn acceptor, $A^0 + h\nu \rightarrow A^- +$ hole, has been investigated in GaAs (Chapman and Hutchinson, 1967) and GaP (Brown and Blakemore, 1972; Abagyan et al., 1975; Evwaraye and Woodbury, 1976). In GaAs:Mn the onset of the associated broad absorption (Fig. 18) is at $E_A \approx 0.1$ eV, with its peak at about twice that energy. The Mn acceptor has been found to be a good example for the application of Lucovsky's formula [Eq. (3)] describing the

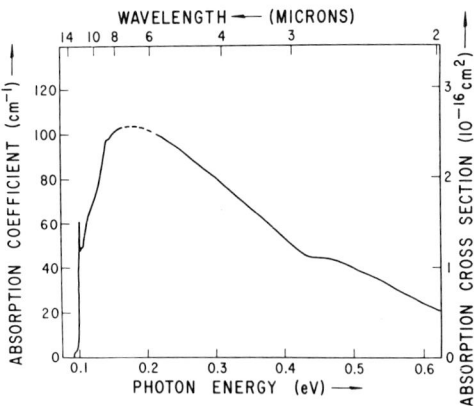

FIG. 18. IR absorption band associated with photoionization of the neutral manganese acceptor in GaAs. [From Chapman and Hutchinson (1967).]

photoionization cross section (Queisser, 1971; Abagyan *et al.*, 1975; Evwaraye and Woodbury, 1976).

The sharp line structure observed at the ionization threshold of the Mn acceptor in GaAs (Fig. 18) results from transitions into bound excited states. These are quite shallow and are accurately described by effective mass theory. From their position E_A can be determined very precisely: 113 meV at LHe temperatures. Lucovsky's relation does not account for such transitions, since in its derivation no long-range Coulomb potential is taken into account.

A^- *State* ($3d^5$). Since the ground state of Mn^{2+} is a ^6S state, ESR is readily detected, as in the case of the isoelectronic ion Fe^{3+}. The ESR spectra bear a characteristic six-line hyperfine label, arising from the nuclear spin $I = 5/2$ of the 100% isotope ^{55}Mn.

In GaP, co-doped with Mn and S, nearest neighbor S–Mn donor–acceptor associates were detected by ESR (van Gorkom and Vink, 1972). Association of Li^+ ions, presumably on interstitial sites, with substitutional Mn^{2+} has also been investigated (Title, 1969). A strong tendency of Mn to associate with other defects was further reported to occur in liquid-phase epitaxial GaAs layers (Segsa and Spenke, 1975).

In GaP the Mn^{2+} charge state gives rise to an IR emission, characterized by a zero-phonon line at 1.534 eV (Vink and van Gorkom, 1972). It was assigned to the El-forbidden crystal-field transition $^4T_1(^4G) \rightarrow {}^6A$. Its low oscillator strength is reflected in a rather long lifetime, 1 msec, of the luminescent level. This type of emission is very prominent in many inorganic manganese-activated phosphors where, however, it occurs in the green-to-orange spectral part. The unusual low energy of this emission in GaP indicates a reduction of the 3d electron–electron interaction and an increase of the crystal-field splitting, as compared to Mn^{2+} in the more ionic II_B–VI hosts. Similar trends have been observed for Co^{2+} (Weber *et al.*, 1980).

Also in GaAs, manganese impurities are readily detected by photoluminescence (Lee and Anderson, 1964; Schairer and Schmidt, 1974; Klein *et al.*, 1980)—often also in device-grade material. At low temperatures, the emission is peaked near 1.405 eV. In contrast to the situation encountered for Mn in GaP, and for most other 3d ions so far investigated, it does *not* arise from an intracenter 3d–3d crystal-field transition. Instead, we are confronted with the first well-documented example of a donor–acceptor recombination involving a 3d ion. The emission is of the type $A^0 + e^- \rightarrow A^- + hv$. The nonoccurrence of the $^4T_1(^4G) \rightarrow {}^6A$ crystal-field transition of Mn in GaAs may indicate that the $^4T_1$ level lies above the conduction-band edge. The temperature dependence of the Mn emission zero-phonon lines

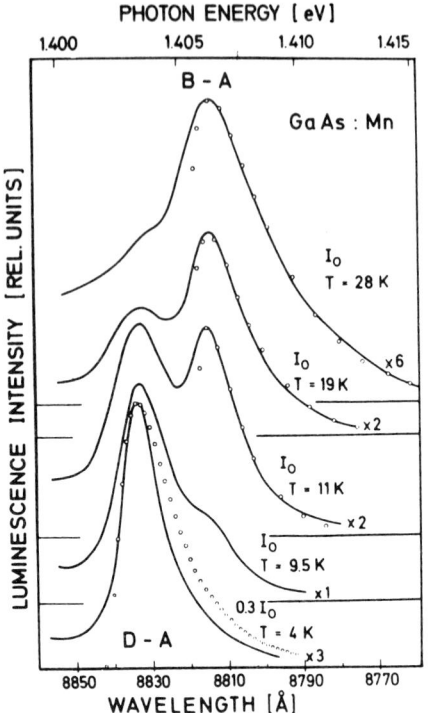

FIG. 19. Manganese luminescence in GaAs. The peak at lower energy is due to donor–acceptor transitions (D–A). At elevated temperatures band–acceptor (B–A) recombination is dominant. [From Schairer and Schmidt (1974).]

in GaAs is shown in Fig. 19. It is seen that up to 10 K the recombining electron is trapped at distant shallow donor (D) sites. At higher temperature, however, the donors become ionized, and the emission then arises from conduction band (B)–acceptor recombination, $hv = E_g - E_A = 1.406$ eV. This process may be viewed as the complement of the previously discussed photoionization of the neutral Mn acceptor in the far IR, which starts at energies above $E_A = 113$ meV (Fig. 19). We have already pointed out in Section IV,B that the Mn emission may also be observed after thermal treatment of accidentally Mn-doped GaAs (Klein et al., 1980).

F. Chromium

1. *GaAs*

Chromium is one of the few deep impurities used as deliberate dopants for III–V compounds. Doping of GaAs with Cr is the cheapest and most reliable method for reproducible growth of large semi-insulating crystals by the LEC technique. Such material is of very great technological importance as a substrate for epitaxial layer growth. It is a prerequisite for the fabrication of GaAs field effect transistors (FETs), which are the basic circuitry elements for the rapidly developing GaAs IC technology (Zucca et al., 1980).

So-called "undoped" bulk GaAs, as well as GaP and InP, is most often n-type with carrier concentrations in the range $10^{15}-10^{16}$ cm^{-3}. It contains significant amounts of electrically active shallow impurities like C, Si, and S.

The role of Cr as a dopant is to trap free electrons so deeply that thermal emission becomes negligible and the material acquires a room-temperature resistivity greater than 10^4 Ω cm. This conventional view of the compensation process is possibly too simple. Several authors have concluded (Lindquist, 1977; Zucca, 1977; Martin et al., 1980a) that a deep donor also plays an important role. The chemical origin of this donor is uncertain. Oxygen is most often proposed, but an intrinsic donor like As$_{Ga}$ could also be invoked.

Electrical and Photoelectric Studies. Since the early work of Haisty and Cronin (1964) it is known that Cr introduces a deep acceptor level in GaAs, slightly above midgap. More recent detailed photoconductivity (Lin and Bube, 1976; Plesiewicz, 1977; Stocker, 1977), photocapacitance (Szawelska and Allen, 1979), and DLTS (Martin et al., 1980b) studies fully support this view. It appears that most electrical data as well as many light-induced effects are consistent with a Cr acceptor-level position at $E_V + (0.80 \pm 0.03)$ eV. The correlation with ESR results (see the next section) convincingly shows that this level refers to the process $Cr^{3+} \rightarrow Cr^{2+} + e^+$ of isolated Cr. Recent studies of the near-band-gap emission of high-purity LPE semiinsulating GaAs:Cr excited far below E_g have demonstrated that the 0.8-eV level acts as an efficient intermediate state for two-step generation of electron–hole pairs (Engemann and Hornung, 1981). This result shows very clearly that photoinduced effects in semiinsulating GaAs:Cr above a threshold of $hv = 0.82$ eV in general involve both electrons and holes.

A second Cr acceptor level at $E_c - 0.5$ eV has been deduced from Hall effect (Brozel et al., 1978) and photocapacitance (Szawelska and Allen, 1979) measurements. Recent ESR results no longer support the view that this level is connected with isolated Cr$_{Ga}$. Resistivity and optical absorption measurements (Hennel et al., 1980; Hennel and Martinez, 1980) under hydrostatic pressure provide clear evidence that the second acceptor level (Cr^{2+}/Cr^+) of isolated Cr lies above the conduction-band edge. Under hydrostatic pressure it moves into the gap and thus becomes electrically active. The microscopic nature of the $E_c - 0.5$-eV level is not known.

Photocapacitance on bulk GaAs:Cr (White et al., 1976) as well as DLTS on GaAs FET structures show the existence of hole traps about

0.5 eV above E_V (Houng and Pearson, 1978; Zylberstein et al., 1979; Meignant et al., 1979). Hole trapping effects near the GaAs:Cr substrate–epilayer interface were also observed by Queisser and Theodorou (1979). Although the origin of these hole traps once was obscure, it now appears almost certain that they correspond to a hole trap level of isolated Cr, according to $Cr^{3+} + e^+ \rightarrow Cr^{4+}$ (Fig. 23). This identification relies on ESR and is discussed below.

ESR Studies. The microscopic understanding of Cr centers in GaAs, as well as in GaP and InP, has advanced significantly during the last few years. It has been derived primarily from ESR results which rather convincingly show that isolated Cr_{Ga} is an amphoteric impurity. In other words, neutral Cr traps electrons in *n*-type material (acceptor behavior) but holes in *p*-type material (donor-like behavior).

The most likely charge states that an isolated Cr_{Ga} atom might acquire are the one-hole trap state Cr^{4+}, the neutral state Cr^{3+}, the one-electron trap state Cr^{2+}, and the two-electron trap state Cr^+. Present theories fail to predict which charge states are actually stable in a particular III–V compound, so this question must be solved experimentally.

Based on the predictions of crystal-field theory and experience with the behavior of 3d transition metals in crystals—and in semiconductors in particular—one expects "simple" ESR spectra for $Cr^{4+}(3d^2)$ and $Cr^+(3d^5)$. This is because these ions have an orbitally nondegenerate ground state with spin $S = 1$ and $S = 5/2$, respectively. For that reason these ions are insensitive to symmetry-reducing Jahn–Teller (JT) distortions (Ham, 1971). Any Cr-related ESR spectrum showing a cubic anisotropy or none must, therefore, arise from one of these two charge states. For sufficiently narrow ESR lines the spectra of Cr^{4+} and Cr^+ can be distinguished without any ambiguity. However, if the fine-structure splitting of Cr^+ is masked by very broad ESR lines, a situation to be expected for III–V compounds, it is difficult or even impossible to discriminate between Cr^{4+} and Cr^+ on the basis of the characteristics of ESR spectra alone. Nevertheless, a clear-cut distinction is possible. Considering *n*- and *p*-type material, Cr^+ (Cr^{4+}) should be stable only in the former (latter) case. One anticipates that the ESR spectra of $Cr^{3+}(3d^3)$ and $Cr^{2+}(3d^4)$ might be more complicated. Since these ions have an orbitally degenerate ground state in addition to the fourfold and fivefold spin degeneracy, they might suffer a dynamic or a static JT distortion. Whether or not this effect is operative and which type of distortion occurs must be inferred from experiment.

Three different Cr-related ESR spectra have been identified. They all arise from the same isolated Cr center in three different charge states. In semiinsulating GaAs:Cr all three spectra are observed simultaneously (at $T \leq 6$ K) after *in situ* optical excitation. The Cr^{3+} spectrum exhibits a complicated angular dependence, similar to the one shown in Fig. 27. It reflects a superposition of ESR transitions from six differently oriented Cr^{3+} centers having orthorhombic symmetry (Krebs and Strauss, 1977b). Uniaxial stress measurements confirm that the reduced symmetry is due to a quasi-static JT distortion (Stauss and Krebs, 1980). The ESR angular dependence of Cr^{2+} on the other hand shows that in this charge state the center has tetragonal symmetry (Krebs and Stauss, 1977a). Again, the symmetry reduction is enforced by a static JT distortion (Krebs and Stauss, 1979). It is worth noting that the X-band (≈ 9 GHz) Cr^{2+} ESR data were found to be in excellent agreement with a far IR (300 GHz) ESR analysis of the same defect (Wagner and White, 1979). The simplest Cr-related spectrum is a broad isotropic line with $g = 1.993$, see Fig. 20. In semi-insulating material it is observed only after optical excitation. At the time when the amphoteric nature of Cr had not been recognized, this spectrum was thought to arise from the two-electron trap state Cr^+ (Kaufmann and Schneider, 1976; Stauss and Krebs, 1977). This assignment was erroneous, as more recent work has shown. Several attempts to observe this line in n-GaAs:Cr without optical excitation remained unsuccessful. However, in as-grown p-GaAs:Zn:Cr, as well as in Cr-diffused GaAs:Zn and in p-type converted semiinsulating GaAs:Cr the line is observed already in the dark and is insensitive to *in situ* illumination of the sample. It was, therefore, reassigned to the Cr_{Ga}^{4+} charge state (Kaufmann and Schneider, 1980b; Stauss *et al.*, 1980). Goswami *et al.* (1980) also considered an alternative explanation, namely an interstitial $Cr^+(3d^5)$. In view of the quantitative concentration determination of Stauss *et al.* (1979) this possibility is considered very unlikely. It would imply that practically all the Cr introduced into the crystal

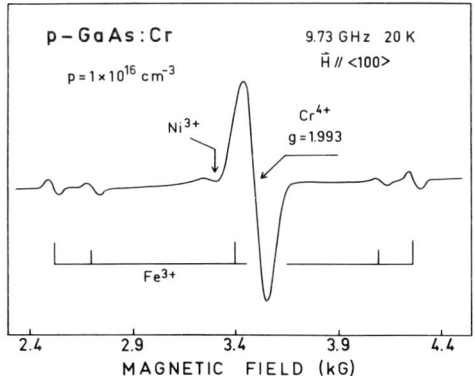

FIG. 20. ESR spectrum of p-type GaAs:Cr showing the signals of the neutral Fe and Ni acceptor states and of the Cr one-hole trap state. In contrast to semi-insulating GaAs:Cr, neither signal is light-sensitive. [From Schneider and Kaufmann (1981).]

enters a cubic interstitial site. It therefore appears that the amphoteric character of Cr in GaAs is now well established.

The quantitative ESR results of Stauss et al. (1979, 1980) show that isolated Cr_{Ga} is the dominant Cr-related defect in GaAs:Cr. Acoustic paramagnetic resonance data (see Bury et al., 1980), however, indicate additional Cr-related defects. One could invoke Cr_{Ga}–X associates, interstitial Cr complexed with an acceptor, or even Cr_{As}. No direct ESR evidence for such centers exists. In any case, if they should exist, their concentration is small compared to that of isolated Cr_{Ga}.

Effect of Light on Chromium ESR Spectra in GaAs. If high-resistivity samples are illuminated with near- or below-band-gap light, drastic ESR signal intensity changes can occur. This is because the dark near-midgap Fermi level splits into quasi-Fermi levels, which for electrons and holes can be close to the conduction and valence bands, respectively. Thus nonequilibrium charge states of a defect may show up. This effect is very pronounced in semi-insulating GaAs:Cr. The ESR intensity changes for the three Cr_{Ga} charge states are plotted in Fig. 21 as a function of monochromatic illumination. Little or no effect occurs for an initial low- to high-energy scan at energies below 0.8 eV. For $hv \gtrsim 0.8$-eV Cr^{4+} appears, Cr^{2+} increases, and Cr^{3+} decreases in concentration. This behavior implies the following light-induced reaction:

$$2Cr^{3+} + hv \rightarrow Cr^{4+} + Cr^{2+}$$

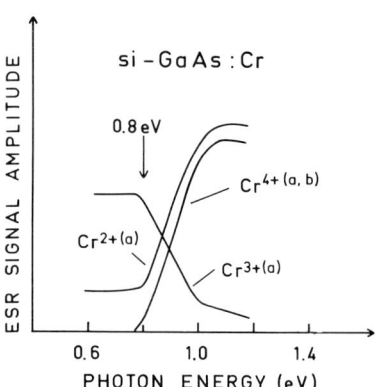

FIG. 21. ESR signal intensity changes in semi-insulating GaAs:Cr as a function of monochromatic *in situ* illumination (schematic) for an initial low- to high-energy scan. Note that the thresholds for strong intensity changes are the same for all three charge states and that Cr^{2+} and Cr^{3+} are already present in the dark. [After (a) Krebs and Stauss (1977); (b) Kaufmann and Schneider (1976).]

POINT DEFECTS IN GaP, GaAs, AND InP   123

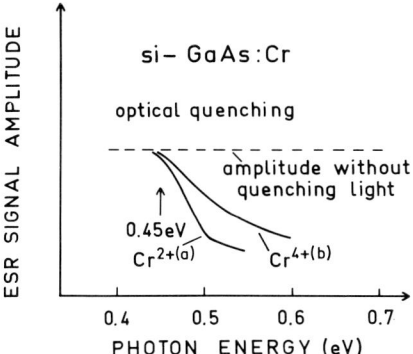

FIG. 22. Optical quenching of nonequilibrium Cr^{2+} and Cr^{4+} ESR in semi-insulating GaAs:Cr (schematic). Note that both charge states have the same quenching threshold (≈ 0.45 eV). [After (a) White *et al.* (1980); (b) Kaufmann and Schneider (1976).]

When the excitation light is switched off, the nonequilibrium Cr^{4+} and Cr^{2+} concentrations decay with similar decay characteristics, even at 20 K or below, whereas the Cr^{3+} intensity recovers. This (thermal) decay and recovery can be optically enhanced as schematically shown in Fig. 22. All these observations allow a consistent and simple explanation (see Fig. 23). An electron from the valence band is excited into a Cr^{3+} center thus converting it to Cr^{2+} (transition 1). The hole left behind in the valence band is trapped by

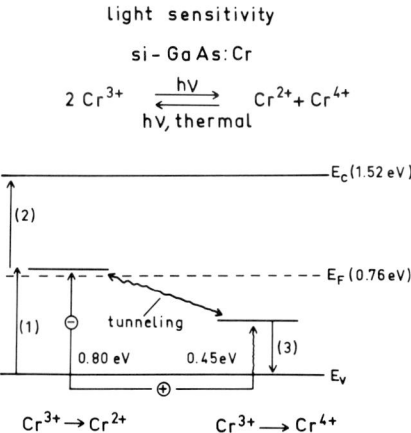

FIG. 23. Level scheme for isolated Cr_{Ga} in GaAs. The strong light sensitivity of semi-insulating material results from the $Cr^{3+} \rightarrow Cr^{2+}$, Cr^{4+} conversion. The nonequilibrium Cr^{2+}, Cr^{4+} concentrations decay via a tunneling mechanism. This decay may be optically enhanced by hole ionization from Cr^{4+}: transition (3).

another Cr^{3+} center, whereby it is converted to Cr^{4+}. The subsequent thermal decay of Cr^{4+} and Cr^{2+} most probably involves the tunneling of electrons and/or holes between these ions, as suggested by White *et al.* (1980), and corresponds to the reverse process in the above reaction. The optical quenching of Cr^{2+} and Cr^{4+} results from hole ionization of Cr^{4+} and subsequent trapping at Cr^{2+}.

It is remarkable that the energy of the Cr^{3+}/Cr^{4+} hole trap, $E_V + 0.45$ eV, inferred from ESR, is nearly identical with that of hole traps observed by DLTS near the substrate–epilayer interface in FET structures (Houng and Pearson 1978; Meignant *et al.*, 1979; Zylberstein *et al.*, 1979). This could be accidental but probably is not. Actually it seems rather likely that the sometimes pronounced light sensitivity, as well as hole-trapping effects near the interface, of FETs are directly related to the processes described above. The improved performance (Crossley *et al.*, 1977; Butlin *et al.*, 1977; Yokoyama *et al.*, 1977) of FETs grown on *lightly* Cr-doped or non-Cr-doped buffer layers supports this view. The fact that the ≈ 0.5-eV hole traps are detected with DLTS only at or on the substrate side of the interface is also indicative of a $Cr^{3+} + e^+ \rightarrow Cr^{4+}$ trapping process. The implications for high-quality semiinsulating GaAs substrate growth are evident. Despite being the most convenient dopant for semiinsulating material growth, chromium doping is not ideal with respect to FET performance.

Among the many-level schemes proposed for chromium in GaAs, the one in Fig. 23 is a relatively simple one. It involves only isolated Cr_{Ga}. It consistently explains the available ESR results, the near-IR absorption (ignoring internal excitations), and most photoelectric measurements. The detailed effects in the latter case sensitively depend on the Fermi-level position in the dark, the presence of other defects, and the intensity of the exciting light. These parameters primarily govern the balance $Cr^{4+} \rightleftarrows Cr^{3+} \rightleftarrows Cr^{2+}$ (Blakemore, 1980).

Near-IR Absorption. The spectral shape of the near-IR absorption in GaAs:Cr depends critically on the degree of compensation, i.e., on the $Cr^{3+}:Cr^{2+}$ concentration ratio. It is complicated since three different absorption mechanisms of isolated Cr_{Ga} can contribute to the absorption, all of which have thresholds near 0.8 eV. A further complication is due to the fact that another band, arising from an unidentified deep donor, is apparently superimposed on the Cr absorptions (Martin *et al.*, 1979).

The spectrum of conducting *n*-GaAs:Cr, (Fig. 24), displays a well-defined band peaked at 0.9 eV. It is superimposed on a monotonically rising absorption (Bois and Pinard, 1974; Ippolitova *et al.*, 1976; Martin *et al.*, 1981; Hennel *et al.*, 1981). Clerjaud *et al.* (1980) have recently resolved the characteristic three-line zero-phonon structure at 0.82 eV of the 0.9-eV band. This data conclusively proves that the band arises from an internal crystal-

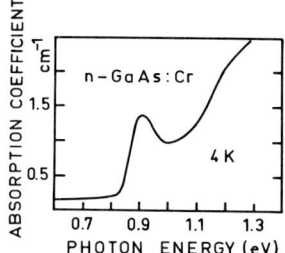

FIG. 24. Near-IR absorption spectrum of *n*-type GaAs:Cr. The band peaked at 0.9 eV arises from the Cr^{2+} crystal-field absorption. [After Bois and Pinard (1974).]

field transition of isolated Cr^{2+} (cf. Fig. 25). One should note that the notorious 0.84-eV luminescence of GaAs:Cr is *not* the counterpart of the 0.9-eV absorption (see the following). At least part of the background absorption in Fig. 24 is due to electron excitation from the Cr acceptor level to the conduction-band transition (2) in Fig. 23.

Semi-insulating GaAs:Cr crystals with $[Cr^{3+}] \gg [Cr^{2+}]$ do not exhibit the 0.9-eV band. However, a featureless, monotonically rising absorption with a threshold near 0.8 eV is again present (Ippolitova *et al.*, 1976; Martin

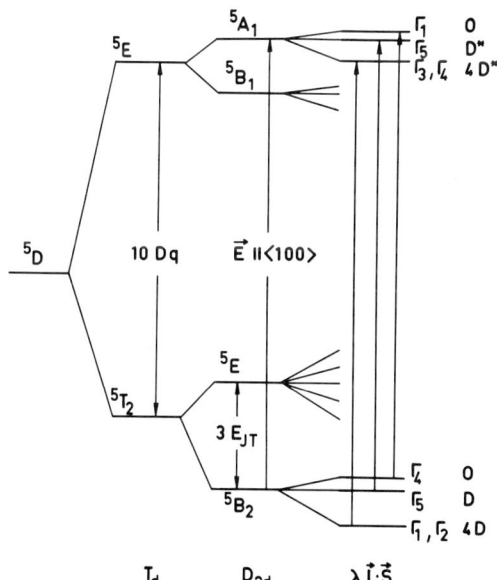

FIG. 25. Crystal-field level scheme of $Cr^{2+}(3d^4)$ in a tetragonally distorted, tetrahedral environment. The symmetry reduction from T_d to D_{2d} is due to a static Jahn–Teller distortion. First-order electric dipole transitions are indicated. Splittings are not to scale.

et al., 1979; 1981). It arises from transition (1) in Fig. 23. A detailed analysis of this $Cr^{3+} \rightarrow Cr^{2+}$ charge transfer has been performed by Martinez *et al.* (1981). A very weak Cr-related absorption with many sharp zero-phonon-lines (ZPLs) near 0.84 eV has been detected by Lightowlers *et al.* (1979).

Luminescence Bands of GaAs:Cr. The dominant luminescence feature of semiinsulating GaAs:Cr is the notorious 0.84-eV band (cf. Fig. 26). It is absent in *p*- and *n*-type Cr-doped material (Wight *et al.*, 1980). Its complicated zero-phonon structure, consisting of more than 10 lines (Lightowlers and Penchina, 1978; Lightowlers *et al.*, 1979), was also observed in absorption (see the preceding). Zeeman measurements (Killoran *et al.*, 1980; Eaves *et al.*, 1980a, 1981) indicate that the luminescent center has approximate or even perfect trigonal symmetry. It is not obvious how this can be reconciled with the uniaxial stress data and the conclusions of Schmidt and Stocker (1978). The interpretations of the band are highly controversial and comprise such opposing ideas as excitonic recombination at a Cr_{Ga}–donor complex on one hand (White, 1979b) and, on the other hand, a crystal-

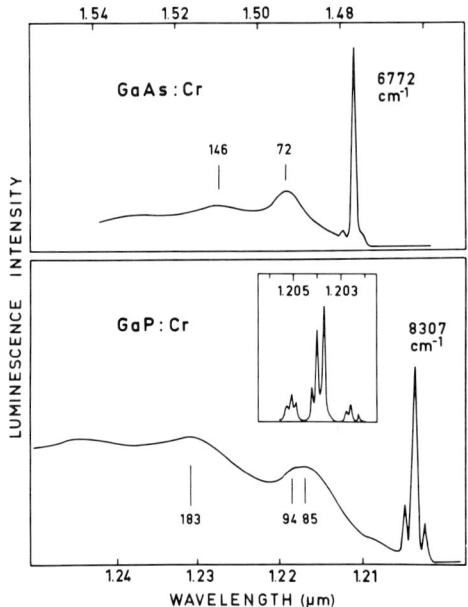

FIG. 26. Near-IR emission bands of semi-insulating GaAs:Cr (upper spectrum, 0.84-eV band) and semi-insulating GaP:Cr (lower trace, 1.03-eV band) recorded with medium spectral resolution at 2 K. The insert for GaP:Cr shows the zero-phonon structure under high resolution. Note the differing wavelength scales for GaAs and GaP. [After Kaufmann and Schneider (1980a).]

field transition of a ^5D ion in trigonal symmetry, e.g., an interstitial Cr^{2+} complexed with an acceptor (Picoli et al., 1981). What is evident is that the band is not due to isolated Cr^{2+} and it even appears unlikely that it is related to the isolated Cr_{Ga} system.

Other luminescence bands dominate the IR emission of p- and n-type GaAs:Cr. There is growing evidence that two bands peaked at 0.57 and 0.61 eV arise from the processes $Cr^{2+} + e^+ \rightarrow Cr^{3+}$ and $Cr^{3+} + e^- \rightarrow Cr^{2+}$, respectively (Wight et al., 1980; Deveaud et al., 1980; Leyral et al., 1981).

2. GaP and InP

Relatively few electrical studies have been reported on Cr in GaP and InP. Abagyan et al. (1974) found that the conductivity of semi-insulating GaP:Cr is thermally activated with an energy of 0.84 eV. Photocapacitance data (Gloriozova and Kolesnik, 1978; Brunwin et al., 1981) for n-GaP:Cr also indicate an electron trap at $E_C - 0.85$ eV. A Cr-related electron trap with an activation energy of 1.0 eV has been observed by DLTS. It is thought to be identical with the 0.85-eV level mentioned above (Brunwin et al., 1981). Another Cr-related level is responsible for a sharp photoconductivity threshold at 1.2 eV in semi-insulating GaP:Cr (Eaves et al., 1980a). As in the case of GaAs:Cr, this level acts as an intermediate state for efficient two-step electron–hole generation (Engemann and Hornung, 1981; Clerjaud et al., 1981), the threshold being $hv = 1.22$ eV. There are many analogies between this level and the 0.8-eV Cr acceptor level in GaAs, and it is tempting to assign it to Cr^{2+}/Cr^+. The $E_C - 0.85$-eV level could correspond to the second acceptor level, Cr^{3+}/Cr^{2+}, since ESR suggests the existence of stable Cr^+ in n-GaP:Cr. A Cr-related hole trap at $E_V + 0.5$ eV was inferred from photocapacitance data (Gloriozova and Kolesnik, 1978).

In semi-insulating InP:Cr n-type conductivity is thermally activated with $E_A = 0.39$ eV (Iseler, 1979). Photoconductivity studies suggest that the optical threshold for electron emission from this level is at 0.47 eV (Fung and Nicholas, 1980). It is probable that the Cr^{3+}/Cr^{2+} acceptor level is involved.

ESR Studies. The ESR spectra of Cr which have been reported for GaP and InP are similar to those for GaAs. Fig. 27 shows the angular dependence of a spectrum which has been observed in lightly p-type converted GaP:Cr and has been attributed to an orthorhombic Cr^{3+} center (Kaufmann and Ennen, 1979, 1981). It is not known whether the low symmetry is JT induced or results from association with another defect. In InP, ESR of Cr^{3+} has not been reported.

The ESR spectrum of Cr^{2+} in InP shows that the center is again tetragonally distorted (Stauss et al., 1977) as in GaAs. Despite several attempts,

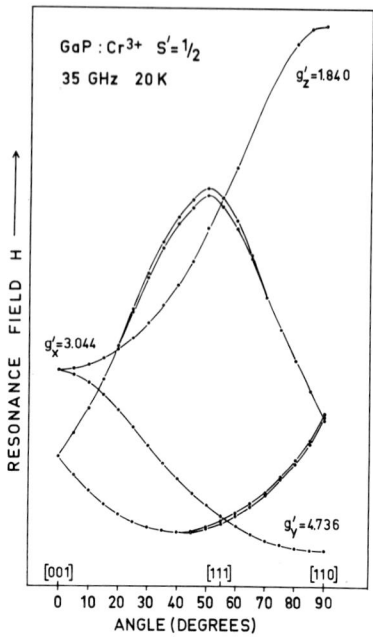

FIG. 27. ESR angular dependence of an orthorhombic Cr^{3+} center on rotating the magnetic field H in the (110) plane. Note that two of the four branches count twice, as evidenced by the small splitting arising from a slight misorientation. [Kaufmann and Ennen (1979, 1981).]

no analogous spectrum has yet been identified in GaP (Clerjaud and Kaufmann, 1980; Krebs and Stauss, 1980). This fact is puzzling since optical absorption measurements on the same material convincingly demonstrate the presence of tetragonally distorted Cr^{2+} centers.

After optical excitation semi-insulating GaP:Cr shows an isotropic 130-G broad line with $g = 1.999$ that has been assigned to the two-electron trap state Cr^+ of isolated Cr (Kaufmann and Koschel, 1978). In n-type-Cr-diffused material this signal is already observed in the dark (Kirillov et al., 1978), thus supporting its assignment[2] to Cr^+_{Ga}. Excitation and quenching curves for the Cr^+ signal are presented in Fig. 28. The threshold for Cr^+ generation is very close to that seen in photoconductivity and two-step luminescence excitation spectroscopy. Presumably one and the same Cr level (Cr^{3+}/Cr^{2+}) is involved.

[2] One should mention that Kirillov et al. (1978) consider the signal as most likely due to $Cr^{4+}(3d^2)$ on a P-site. This alternative cannot be ruled out definitely. Although electronegativity arguments and the general behavior of transition metal ions are at variance with such an interpretation a small amount of Cr_P cannot be excluded.

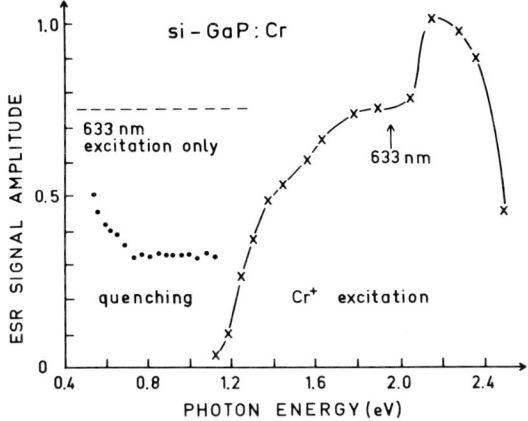

FIG. 28. Excitation and quenching curves for the Cr$^+$ ESR signal in semi-insulating GaP:Cr. The quenching data points were obtained under permanent excitation with a second light source of 633-nm wavelength. [After Kaufmann and Koschel (1978).]

In p-type converted semi-insulating GaP:Cr, as well as in p-type-Cr-diffused GaP:Zn and as-grown p-GaP:Zn:Cr, another isotropic 86-G broad line with $g = 1.986$ has been reported (Kaufmann and Schneider, 1978, 1980b; Goswami et al., 1980). This signal is insensitive to light only in p-type material. Thus its assignment to Cr_{Ga}^{4+} is most likely, in full analogy with the behavior of that charge state in GaAs. The corresponding signal (linewidth ≈ 350 G) was also reported for p-InP:Zn:Cr (Goswami et al., 1980).

Near-IR Absorption and Emission. Semi-insulating GaP:Cr displays an absorption band peaked at 0.95 eV (Abagyan et al., 1974). Its characteristic three-line zero-phonon structure (Fig. 29) and the behavior of the ZPLs under uniaxial stress demonstrate that the band is due to the $^5T_2 \rightarrow {}^5E$ crystal field transition of a tetragonally JT distorted Cr^{2+} ion (Kaufmann and Ennen, 1979, 1981). As in the case of GaAs:Cr, this crystal-field transition has not been observed in emission. Instead, the dominant IR luminescence of semi-insulating GaP:Cr is the band shown in Fig. 26 (Dean, 1972; Kaufmann and Koschel, 1978). Its complicated zero-phonon structure and the shape of the vibronic sidebands suggest that it is the counterpart of the 0.84-eV emission of GaAs:Cr. However, Zeeman measurements indicate that the defect in GaP has tetragonal symmetry (Eaves et al., 1980a, 1981), rather than trigonal as in GaAs. Several interpretations have been invoked (see, e.g., White, 1979b), but a fully convincing model is still lacking. A slightly structured Cr-related luminescence peaked at 0.85 eV was observed in semi-insulating InP:Cr by Koschel et al. (1977).

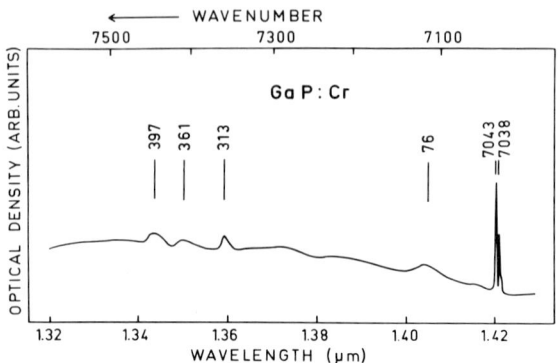

FIG. 29. Crystal-field absorption band of Cr^{2+} as observed in semi-insulating GaP:Cr at 5 K. The zero-phonon lines arise from the transitions shown in Fig. 25. [After Kaufmann and Ennen (1979, 1981).]

G. Vanadium

The electrical behavior of V in GaAs is not well understood. Early Hall-effect measurements of Haisty and Cronin (1964) indicate a medium-deep donorlike level, 0.22 eV below the conduction band. A V-related level deeper than 0.2 eV could not be detected in the DLTS spectrum of GaAs:V (Mircea-Roussel et al., 1980). On the other hand, Vasil'ev et al. (1976a) found that V doping produces semiinsulating material. From their electrical studies they concluded that V gives rise to a near midgap acceptor level. If this observation can be substantiated, V could be a promising dopant for semi-insulating GaAs growth. Recent attempts to grow semi-insulating GaAs:V by the LEC technique (Köhl, 1980) were also successful. The material can be semiinsulating, but problems with reproducibility remain.

An ESR spectrum due to a cubic V center has been reported for semi-insulating GaAs:V (Kaufmann and Schneider, 1980a). This spectrum is almost insensitive to illumination of the sample and most likely arises from $V^{3+}(3d^2)$, i.e., from the neutral acceptor state. ESR attributed to V^{3+} was also reported for GaP (Masterov and Samorukov, 1978).

Vanadium induces a characteristic, sharply structured, near IR emission band in both GaAs and GaP (Fig. 30) (Kaufmann et al., 1981b). The Zeeman splitting (Weber et al., 1981) of the ZPLs indicates that a cubic V center with an odd number of electrons is involved. Therefore, the bands appear to arise from a crystal-field transition of isolated $V^{2+}(3d^3)$, i.e., from the ionized V acceptor.

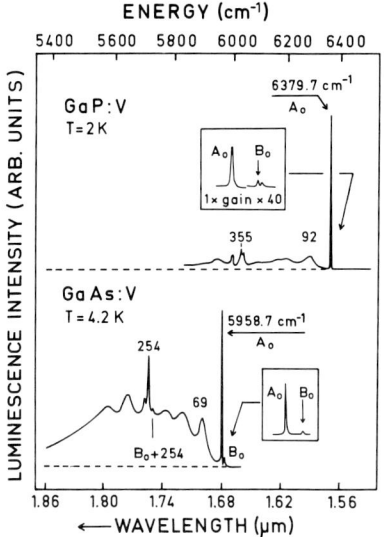

FIG. 30. Luminescence bands of vanadium-doped GaP and GaAs. The inserts display the zero-phonon structure under high spectral resolution. [After Kaufmann et al. (1981b).]

H. Ni–Donor Pairs in GaP and GaAs

Donor–acceptor pairing is a well-established phenomenon for shallow acceptors (Zn, Cd) and the deep donor oxygen in GaP (Section III). Until recently only one example for a near-neighbor pair consisting of a deep acceptor and a shallow substitutional donor was known, namely the trigonal Mn–S pair in GaP (Section V,E). Although copper–donor pairs are frequently invoked in the literature, and their existence in GaAs (Nakashima, 1971) and GaP (Dean, 1973b) is rather likely, a definite microscopic identification is still lacking.

Convincing evidence for the existence of Ni–X (X means group IV or group VI donors) near-neighbor pairs has been reported for GaP and GaAs (Ennen et al., 1981). These pairs were inferred from sharp line crystal-field absorptions and emissions in the 2-μm range. The luminescence spectra attributed to Ni–S and Ni–Ge in GaP are shown in Fig. 31. The dominant zero-phonon lines B_0 and F_0 and their vibronic replicas arise from Ni–S and Ni–Ge pairs, respectively. The crystal-field band of isolated Ni^+ impurities A_0 with its phonon progressions (cf. Fig. 16), is only weakly present. The expected trigonal symmetry for the nearest-neighbor Ni–S pair was confirmed by uniaxial stress measurements. Fig. 32 shows absorption spectra

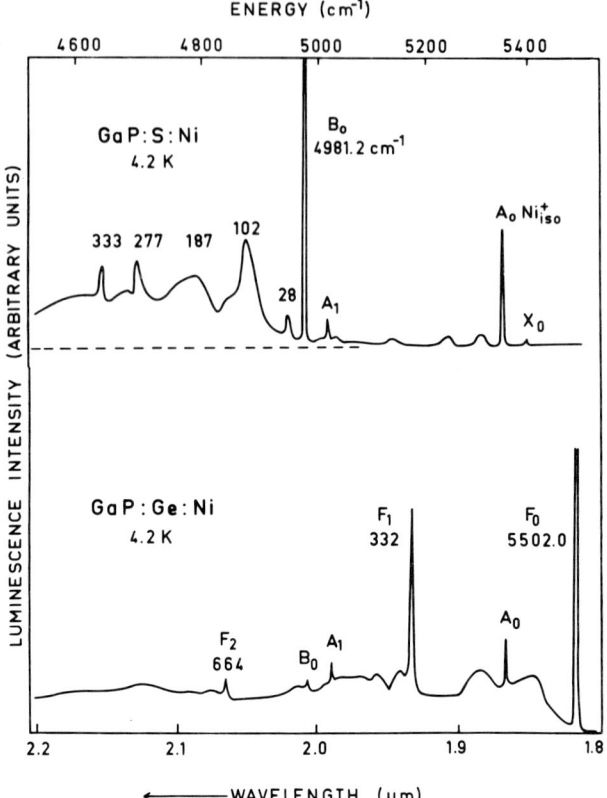

FIG. 31. Luminescence spectra of n-type GaP:S and GaP:Ge diffused with Ni. The dominant zero-phonon lines B_0 and F_0 with their phonon replicas have been attributed to Ni–S and Ni–Ge near-neighbor pairs, respectively, The line A_0 with its side bands is due to isolated Ni^+. [After Ennen et al. (1981).]

of Ni-diffused GaAs which, in addition to a crystal-field band of isolated Ni—presumably Ni^+, exhibit sharp ZPLs attributed to Ni–S, Ni–Se, and Ni–Te nearest neighbor pairs. Table VI summarizes the positions of the dominant Ni–X ZPLs identified. Ennen et al. (1981) concluded that nickel has a strong tendency to form such pairs. It is especially pronounced for Ni and S in both GaP and GaAs. Since S as well as Ni is an important trace impurity the Ni–S complex may be present also in nominally pure GaP and GaAs.

DLTS studies on vapor-phase epitaxial GaP have revealed an ubiquitous hole trap with a thermal activation energy of 0.95 eV (Hamilton and Peaker, 1978; Tell and Kuijpers, 1978; Hamilton et al., 1979). In epitaxial GaAs an electron trap at $E_C - 0.39$ eV was observed Partin et al. (1979c). Diffusion

POINT DEFECTS IN GaP, GaAs, AND InP   133

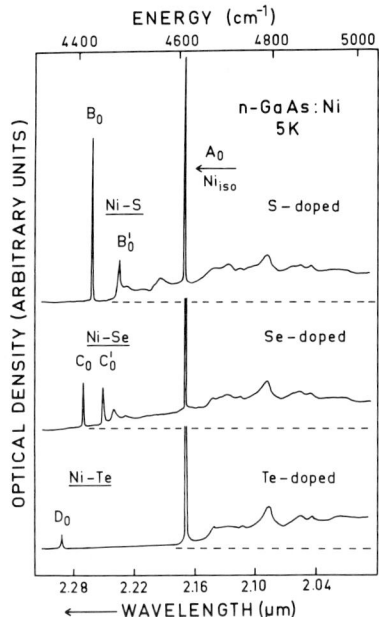

FIG. 32. Absorption spectra of n-type GaAs:S, GaAs:Se, and GaAs:Te diffused with Ni. The zero-phonon line A_0 and its side bands are due to a crystal-field transition of isolated Ni. The zero-phonon lines B_0, B_0', C_0, C_0', and D_0 have been assigned to Ni–S, Ni–Se, and Ni–Te nearest neighbor pairs. [After Ennen et al. (1981).]

studies showed both levels to be Ni-related, and it was concluded that a complex, rather than isolated, Ni is involved. In view of the above optical results it is therefore tempting to assign the 0.95-eV hole trap in GaP to the Ni–S pair. The $E_C - 0.39$ eV level in n-GaAs:Te might correspond to a Ni–Te pair.

TABLE VI

ZERO-PHONON LINES OF ISOLATED NICKEL (Ni_{iso}) AND Ni-DONOR PAIRS IN GaP AND GaAs[a]

	Ni_{iso} A_0	Ni–S B_0	Ni–Se C_0	Ni–Te D_0	Ni–Si E_0	Ni–Ge F_0	Ni–Sn G_0
GaP	5355.5 (A, E)	4981.2 (A, E)		4712.9 (A, E)		5502.0 (A, E)	
GaAs	4617.8 (A)	4427.0 (A, E)	4409.8 (A, E)	4369.0 (A, E)	4699.2 (A, E)	4739.6 (E)	4626.8 (A, E)

[a] A, absorption; E, emission.

I. 4d and 5d Transition Metals

Very little is known about the behavior of the elements of the $4d^n$ and $5d^n$ transition metal series in III–V semiconductors. This is rather unsatisfactory since some 4d and 5d metals, such as Pd, Ta, and Pt, find increasing application in contact technology of III–V devices (Partin et al., 1978). The first spectroscopic data were recently reported by Ushakov and Gippius (1980) and Vavilov et al. (1980). After ion implantation of Zr, Nb, Mo, Pd, and W into GaAs and GaP and subsequent thermal annealing, sharp-line intra-d-shell luminescence in the 1.5–1.9-μm range was observed. The emissions presumably arise from the divalent A^- state of the transition metals. Early work on the acceptors Ag and Au in GaAs has been reviewed by Queisser (1971).

J. 4f-Rare Earth Metals

So far, no sharp-line optical spectra, characteristic for the 4f–4f transitions of rare earth ions, have been reported for III–V compounds. An emission at 3.15 eV, observed for dysprosium-activated GaN (Pankove et al., 1973) appears to be of different origin. Also, no ESR data are as yet available.

On the other hand, rare earth ions are well known to be strongly luminescent in many groups of different hosts, including the family of II_B–VI semiconductors (Brown et al., 1974). However, their large ionic radius and their preference for higher than fourfold ligand coordination will limit the solubility of 4f elements in III–V compounds. Nevertheless, by radioactive ^{170}Tm tracer analysis, solubilities of 4×10^{17} cm^{-3} were determined for thulium in GaAs at 1150°C (Casey and Pearson, 1964). Higher solubilities may be expected in InP because of the larger size of the indium cation. It has already been speculated, in the early period of semiconductor laser research, that stimulated emission of 4f ions in III–V hosts may be excited by electrical dc pumping (Bell, 1963). Experimental confirmation is lacking. Sharp line intracenter 4f–4f emissions have been recently observed for Yb and Pr ions in GaP (Kasatkin et al., 1980, 1981).

ACKNOWLEDGMENTS

We thank P. J. Dean and M. S. Skolnick for a critical reading of the manuscript.

References

Abagyan, S. A., Ivanov, G. A., Kuznetsov, Yu. N., Okunev, Yu. A., and Shanurin, Yu. E. (1974). *Sov. Phys. Semicond. (Engl. Transl.)* **7**, 989.
Abagyan, S. A., Ivanov, G. A., Koroleva, G. A., Kuznetsov, Yu. A., and Okunev, Yu. A. (1975). *Sov. Phys. Semicond. (Engl. Transl.)* **9**, 243.
Abagyan, S. A., Ivanov, G. A., and Koroleva, K. A. (1976). *Sov. Phys. Semicond. (Engl. Transl.)* **10**, 1956.
Abakumov, V. N., Perel, V. I., and Yassievich (1978). *Sov. Phys. Semicond. (Engl. Transl.)* **12**, 1.
Allègre, J., and Avérous, M. (1979). *Conf. Ser.–Inst. Phys.* **46**, 379.
Altarelli, M. (1980). *J. Phys. Soc. Jpn.* **49**, Suppl. A, 169.
Andrianov, D. G., Grinshtein, P. M., Ippolitova, G. K., Omel'yanovskii, E. M., Suchkova, N. I., and Fistul, V. I. (1976). *Sov. Phys. Semicond.* **10**, 696.
Aven, M., and Prener, J. S., eds. (1967). "Physics and Chemistry of II–VI Compounds." North-Holland Publ., Amsterdam.
Baranowski, J. M., Allen, J. W., and Pearson, G. L. (1967). *Phys. Rev.* **160**, 627.
Baranowski, J. M., Allen, J. W., and Pearson, G. L. (1968). *Phys. Rev.* **167**, 758.
Bassani, F., Iadonisi, G., and Preziosi, B. (1974). *Rep. Prog. Phys.* **37**, 1099.
Bell, R. L. (1963). *J. Appl. Phys.* **34**, 1563.
Bergh, A. A., and Dean, P. J. (1976). "Light-Emitting Diodes," Oxford Univ. Press (Clarendon), London and New York.
Bimberg, D., Dean, P. J., Kaufmann, U., and Schneider, J. (1981). Landolt-Börnstein New Series *17a*, "Defects in III–V Semiconductors," Springer-Verlag, (in press).
Bishop, S. G., Dean, P. J., Porteous, P., and Robbins, D. J. (1980). *J. Phys. C* **13**, 1331.
Blakemore, J. S. (1980). In "Semi-Insulating III–V Materials" (G. J. Rees, ed.), p. 29. Shiva Publishing Ltd.
Blanc, J., Bube, R. H., and Weisberg, L. R. (1964). *J. Phys. Chem. Solids* **25**, 225.
Bois, D., and Pinard, P. (1974). *Phys. Rev. B* **9**, 4171.
Brown, M. R., Cox, A. F. J., Shand, W. A., and Williams, J. M. (1974). In "Advances in Quantum Electronics" (D. W. Goodwin, ed.) Vol. 2, p. 69. Academic Press, N.Y.
Brown, W. J., and Blakemore, J. S. (1972). *J. Appl. Phys.* **43**, 2242.
Brozel, M. R., Butler, J., Newman, R. C., Ritson, A., Stirland, D. J., and Whitehead, C. (1978). *J. Phys. C* **11**, 1857.
Brunwin, R. F., Hamilton, B., Hodgkinson, J., Peaker, A. R., and Dean, P. J. (1981). *Solid State Electron.* **24**, 249.
Bube, R. H. (1960). "Photoconductivity of Solids." Wiley.
Bury, P., Challis, L. J., King, P. J., Monk, D. J., Ramdane, A., Rampton, V. W., and Wiscombe, P. (1980). In "Semi-Insulating III–V Materials" (G. J. Rees, ed.), p. 214. Shiva Publ. Ltd.
Butlin, R. S., Parker, D., Crossley, I., and Turner, J. (1977). *Conf. Ser.–Inst. Phys.* **33a**, 237.
Carter, A. C., Dean, P. J., Skolnick, M. S., and Stradling, R. A. (1977). *J. Phys. C* **10**, 5111.
Casey, H. C., and Pearson, G. L. (1964). *J. Appl. Phys.* **35**, 3401.
Casey, H. C., and Panish, M. B. (1978). "Heterostructure Lasers." Academic Press, New York.
Chang, L. L., Esaki, L., and Tsu, R. (1971). *Appl. Phys. Lett.* **19**, 143.
Chapman, R. A., and Hutchinson, W. G. (1967). *Phys. Rev. Lett.* **18**, 443.
Chatterjee, P. K., Vaidyanathan, K. V., Durschlag, M. S., and Streetman, B. G. (1975). *Solid State Commun.* **17**, 1421.
Chiang, S. Y., and Pearson, G. L. (1975). *J. Lumin.* **10**, 313.

Clerjaud, B., and Kaufmann, U. (1980), unpublished.
Clerjaud, B., Hennel, A. M., and Martinez, G. (1980). *Solid State Commun.* **33**, 983.
Clerjaud, B., Gendron, F., and Porte, C. (1981). *J. Appl. Phys. Lett.* **38**, 212.
Cooke, R. A., Hoult, R. A., Kirkman, R. F., and Stradling, R. A. (1978). *J. Phys.* D **11**, 945.
Crossley, I., Goodridge, I. H., Cardwell, M. J., and Butlin, R. S. (1977). *Conf. Ser.–Inst. Phys.* **33b**, 289.
Dean, P. J., Henry, C. A., and Frosch, C. J. (1968). *Phys. Rev.* **168**, 812.
Dean, P. J. (1972). In "Progress in Solid State Chemistry" (J. O. McCaldin and G. Somorjai, eds.). Vol. 8, p. 1. Pergamon, Oxford.
Dean, P. J. (1973a). In "Luminescence of Crystals, Molecules and Solutions" (F. Williams, ed.), p. 523. Plenum, New York.
Dean, P. J. (1973b). *J. Lumin.* **7**, 51.
Dean, P. J., and Henry, C. H. (1968). *Phys. Rev.* **176**, 928.
Dean, P. J., and Herbert, D. C. (1979). In "Excitons" (K. Cho, ed.), p. 55. Springer-Verlag, Berlin and New York.
Dean, P. J., and Herbert, D. C. (1980). *J. Appl. Phys.* **51**, 2297.
Dean, P. J., Faulkner, R. A., and Kimura, S. (1970). *Phys. Rev.* B **2**, 4062.
Dean, P. J., Schairer, W., Lorenz, M., and Morgan, T. N. (1974). *J. Lumin.* **9**, 343.
Dean, P. J., White, A. M., Hamilton, B., Peaker, A. R., and Gibb, R. M. (1977). *J. Phys.* D **10**, 2545.
Deveaud, B., Hennel, A. M., Szuszkiewicz, W., Picoli, G., and Martinez, G. (1980). *Rev. Phys. Appl.* **15**, 671.
Eaves, L., Englert, T., Instone, T., Uihlein, C., Williams, P. J., and Wright, H. C. (1980a). In "Semi-Insulating III–V Materials" (G. J. Rees, ed.), p. 145. Shiva Publ. Ltd.
Eaves, L., Englert, T., Uihlein, C., and Williams, P. J. (1980b). *J. Phys. Soc. Jpn.* **49**, Suppl. A, 279.
Eaves, L., Englert, T., and Uihlein, C. (1981). In "Physics in High Magnetic Fields" (S. Chikazumi and N. Miura, eds.). Springer-Verlag, Berlin and New York, in press.
Engemann, D., and Hornung, T. (1981). *Phys. Rev.* B, in press.
Ennen, H., and Kaufmann, U. (1980). *J. Appl. Phys.* **51**, 1615.
Ennen, H., Kaufmann, U., and Schneider, J. (1980). *Solid State Commun.* **34**, 603.
Ennen, H., Kaufmann, U., and Schneider, J. (1981). *Appl. Phys. Lett.* **38**, 355.
Evwaraye, A. O., and Woodbury, H. H. (1976). *J. Appl. Phys.* **47**, 1595.
Fabre, E., and Bhargava, R. N. (1974). *Appl. Phys. Lett.* **24**, 322.
Faulkner, R. A. (1969). *Phys. Rev.* **184**, 713.
Fleurov, V. N., and Kikoin, K. A. (1979). *J. Phys.* C **12**, 61.
Fung, S., and Nicholas, R. J. (1980). In "Semi-Insulating III–V Materials" (G. J. Rees, ed.), Shiva Publ., Ltd.
Gal, M., Cavenett, B. C., and Dean, P. J. (1981). *J. Phys.* C, **14**, 1507.
Gloriozova, R. I., and Kolesnik, L. I. (1978). *Sov. Phys. Semicond.* (*Engl. Transl.*) **12**, 66.
Godlewski, M., and Hennel, A. M. (1978). *Phys. Status Solidi* B **88**, K11.
Goldstein, B., and Almeleh, N. (1963). *Appl. Phys. Lett.* **2**, 130.
Goswami, N. K., Newman, R. C., and Whitehouse, J. E. (1980). *Solid State Commun.* **36**, 897.
Grimmeiss, H. G. (1977). *Annu. Rev. Mater. Sci.* **7**, 341.
Grimmeiss, H. G., Janzen, E., Ennen, H., Schirmer, O., Schneider, J., Wörner, R., Holm, C., Sirtl, E., and Wagner, P. (1981). *Phys. Rev.* B, submitted.
Gross, E. F., Safarov, V. I., Sedov, V. E., and Marushchack, V. A. (1969). *Soc. Phys. Solid State* (*Engl. Transl.*) **11**, 277.
Guillot, G., Loualiche, S., Nouailhat, A., and Martin, G. M. (1981). *Conf. Ser.–Inst. Phys.*, **59**, 323.
Ham, F. S. (1972). In "Electron Paramagnetic Resonance" (S. Geschwind, ed.), p. 1. Plenum, New York.

Hamilton, B., and Peaker, A. R. (1978). *Solid State Electron.* **21,** 1513.
Hamilton, B., Peaker, A. R., and Wight, D. R. (1979). *J. Appl. Phys.* **50,** 6373.
Haisty, R. W., and Cronin, G. R. (1964). *Proc. Int. Conf. Phys. Semicond.*, p. 1161.
Hayes, W., Ryan, J. F., West, C. L., and Dean, P. J. (1979). *J. Phys. C* **12,** L815.
Hayes, W., Ryan, J. F., West, C. L., Dean, P. J. (1980). *J. Phys. C* **13,** L149.
Haynes, J. R. (1961). *Phys. Rev. Lett.* **4,** 361.
Hemstreet, L. A. (1980). *Phys. Rev. B* **22,** 4590.
Hennel, A. M., and Martinez, G. (1980). *J. Phys. Soc. Jpn.*, *Suppl. A* **49,** 283.
Hennel, A. M., Szuszkiewicz, W., Martinez, G., Clerjaud, B., Huber, S. A. M., Morillot, G., and Merenda, P. (1980). *In* "Semi-Insulating III–V Materials" (G. J. Rees, ed.), p. 228. Shiva Publ., Ltd.
Hennel, A. M., Szuszkiewicz, W., Balkanski, M., Martinez, G., and Clerjaud, B. (1981). *Phys. Rev. B* **23,** 3933.
Henry, C. H., Dean, P. J., and Cuthbert, J. D. (1968a). *Phys. Rev.* **166,** 754.
Henry, C. H., Dean, P. J., Thomas, D. G., and Hopfield, J. J. (1968b). *Localized Excitations in Solids, Proc. Int. Conf., 1st*, p. 267. (G. J. Wallis, ed.), Plenum, N.Y.
Hjalmarson, H. P., Vogl, P., Wolford, D. J., and Dow, J. D. (1980). *Phys. Rev. Lett.* **44,** 810.
Holton, W. C., De Wit, M., Watts, R. K., Estle, T. L., and Schneider, J. (1969). *J. Phys. Chem. Solids* **30,** 963.
Hopfield, J. J., Thomas, D. G., and Gershenzon, M. (1963). *Phys. Rev. Lett.* **10,** 612.
Houng, Y. M., and Pearson, G. L. (1978). *J. Appl. Phys.* **49,** 3348.
Hurle, D. T. J. (1977). *Conf. Ser.–Inst. Phys.* **33a,** 113.
Hwang, C. J. (1969). *Phys. Rev.* **180,** 827.
Igelmund, A. (1979). Thesis, Technische Hochschule Aachen, unpublished.
Ippolitova, G. K., and Omel'yanovskii, E. M. (1975). *Sov. Phys. Semicond.* (*Engl. Transl.*) **9,** 156.
Ippolitova, G. K., Omel'yanovskii, E. M., and Pervova, L. Y. (1976). *Sov. Phys. Semicond.* (*Engl. Transl.*) **9,** 864.
Ippolitova, G. K., Omel'yanovskii, E. M., Pavlov, N. M., Nashel'skii, A. Ya, and Yakobson, S. V. (1977). *Sov. Phys. Semicond.* (*Engl. Transl.*) **11,** 773.
Iseler, G. W. (1979). *Conf. Ser.–Inst. Phys.* **45,** 144.
Jakowetz, W., Barthruff, D., and Benz, K. W. (1977). *Conf. Ser.–Inst. Phys.* **33a,** 41.
Jaros, M. (1978). *J. Phys. C* **11,** L213.
Jaros, M. (1980). *Adv. Phys.* **29,** 409.
Jordan, A. S., Caruso, R., von Neida, A. R., and Weiner, M. E. (1974a). *J. Appl. Phys.* **45,** 3472.
Jordan, A. S., von Neida, A. R., Caruso, R., and Kim, C. K. (1974b). *J. Electrochem. Soc.* **121,** 153.
Kasatkin, V. A., Kesamanly, F. P., Makarenko, V. G., Masterov, V. F., and Samorukov, B. E. (1980). *Sov. Phys. Semicond.* (*Engl. Transl.*) **14,** 1092.
Kasatkin, V. A., Kesamanly, Г. P., and Samorukov, B. E. (1981). *Sov. Phys. Semicond.* (*Engl. Transl.*) **15,** 352.
Kaufmann, U., and Ennen, H. (1979). Poster paper presented at 2nd "Lund" Conf. Deep Level Impurities Semicond., St. Maxime, unpublished.
Kaufmann, U., and Ennen, H. (1981). To be published.
Kaufmann, U., and Kennedy, T. A. (1981). *J. Electron. Mater.* **10,** 347.
Kaufmann, U., and Koschel, W. H. (1978). *Phys. Rev. B* **17,** 2081.
Kaufmann, U., and Schneider, J. (1976). *Solid State Commun.* **20,** 143.
Kaufmann, U., and Schneider, J. (1977). *Solid State Commun.* **21,** 1073.
Kaufmann, U., and Schneider, J. (1978). *Solid State Commun.* **25,** 1113.
Kaufmann, U., and Schneider, J. (1980a). *Festkoerperprobleme* **20,** 87.
Kaufmann, U., and Schneider, J. (1980b). *Appl. Phys. Lett.* **36,** 74.
Kaufmann, U., and Wörner, R. (1981). To be published.

Kaufmann, U., Schneider, J., and Räuber, A. (1976). *Appl. Phys. Lett.* **29**, 312.
Kaufmann, U., Koschel, W. H., Schneider, J., and Weber, J. (1979). *Phys. Rev.* B **19**, 3343.
Kaufmann, U., Schneider, J., Wörner, R., Kennedy, T. A., and Wilsey, N. D. (1981a). *J. Phys.* C **14**, L951.
Kaufmann, U., Ennen, H., Schneider, J., Wörner, R., Weber, J., and Köhl, F. (1981b). Submitted to *Phys. Rev. B*.
Kennedy, T. A., and Wilsey, N. D. (1978). *Phys. Rev. Lett.* **41**, 977.
Kennedy, T. A., and Wilsey, N. D. (1979). *Conf. Ser.–Inst. Phys.* **46**, 375.
Kennedy, T. A., and Wilsey, N. D. (1981a). *Phys. Rev.* B, **23**, 6585.
Kennedy, T. A., and Wilsey, N. D. (1981b). *Conf. Ser.–Inst. Phys.* **59**, 257.
Killoran, N., Cavenett, B. C., and Hagston, W. E. (1980). In "Semi-Insulating III–V Materials" (G. J. Rees, ed.), p. 190. Shiva Publ., Ltd.
Kirillov, V. I., and Teslenko, V. V. (1979). *Sov. Phys. Solid State (Engl. Transl.)* **21**, 1852.
Kirillov, V. I., Pribylov, N. N., Rembeza, S. I., and Spirin, A. I. (1978). *Sov. Phys. Semicond. (Engl. Transl.)* **12**, 1342.
Klein, P. B., Nordquist, P. E. R., and Siebenmann, P. G. (1980). *J. Appl. Phys.* **51**, 4861.
Köhl, F. (1980). Private communication.
Kohn, W. (1957). *Solid State Phys.*, **5**, 257.
Koschel, W. H., Bishop, S. G., and McCombe, B. D. (1976). *Phys. Semicond., Proc. Int. Conf., 13th, 1976*, p. 1065.
Koschel, W. H., Kaufmann, U., and Bishop, S. G. (1977a). *Solid State Commun.* **21**, 1069.
Koschel, W. H., Bishop, S. G., McCombe, B. D., Lum, W. Y., and Wieder, H. H. (1977b), *Conf. Ser.–Inst. Phys.* **33a**, 98.
Kopylov, A. A., and Pikhtin, A. N. (1978). *Solid State Commun.* **26**, 735.
Krebs, J. J., and Stauss, G. H. (1977a). *Phys. Rev.* B **16**, 971.
Krebs, J. J., and Stauss, G. H. (1977b). *Phys. Rev.* B **15**, 17.
Krebs, J. J., and Stauss, G. H. (1979). *Phys. Rev.* B **20**, 795.
Krebs, J. J., and Stauss, G. H. (1980). Private communication.
Kressel, H., and Butler, J. K. (1977). "Semiconductor Lasers and Heterojunction LEDs." Academic Press, New York.
Kressel, H., Dunse, J. U., Nelson, H., and Hawrylo, F. Z. (1968). *J. Appl. Phys.* **39**, 2006.
Kröger, F. A., and Vink, H. J. (1956). *Solid State Phys.* **3**, 307.
Kukimoto, H., Henry, C. H., and Merrit, F. R. (1973). *Phys. Rev.* B **7**, 2486.
Lang, D. V. (1974). *J. Appl. Phys.* **45**, 3023.
Lang, D. V. (1980). *J. Phys. Soc. Jpn., Suppl. A* **49**, 215.
Lang, D. V., and Kimerling, L. C. (1975). *Conf. Ser.–Inst. Phys.* **23**, 581.
Lang, D. V., Logan, R. A., and Kimerling, L. C. (1977). *Phys. Rev.* B **15**, 4874.
Lang, D. V., Logan, R. A., and Jaros, M. (1979). *Phys. Rev.* B **19**, 1015.
Lee, T. C., and Anderson, W. W. (1964). *Solid State Commun.* **2**, 265.
Leutwein, K., and Schneider, J. (1971). *Z. Naturforsch.* **26a**, 137.
Leyral, P., Litty, F., Loualiche, S., Nouailhat, A., and Guillot, G. (1981). *Solid State Commun.* **38**, 333.
Liechti, C. A. (1976). *IEEE Trans. Microwave Theory Tech.* **MTT-24**, 279.
Liechti, C. A. (1977). *Conf. Ser.–Inst. Phys.* **33a**, 227.
Lightowlers, E. C., and Penchina, C. M. (1978). *J. Phys.* C **11**, L405.
Lightowlers, E. C., Henry, M. O., and Penchina, C. M. (1979). *Conf. Ser.–Inst. Phys.* **43**, 307.
Lin, A. L., and Bube, R. H. (1976). *J. Appl. Phys.* **47**, 1859.
Lindquist, P. F. (1977). *J. Appl. Phys.* **48**, 1262.
Loescher, D. H., Allen, J. W., and Pearson, G. L. (1966). *J. Phys. Soc. Jpn.* **21**, 239.
Logan, R. M., and Hurle, D. T. J. (1971). *J. Phys. Chem. Solids* **32**, 1739.

Lucovsky, G. (1965). *Solid State Commun.* **3**, 299.
Ludwig, G. W., and Woodbury, H. H. (1962). *Solid State Phys.* **13**, 223.
Lum, W. Y., and Wieder, H. H. (1977). *Appl. Phys. Lett.* **31**, 213.
Lum, W. Y., and Wieder. H. H. (1978). *J. Appl. Phys.* **49**, 6187.
Lum, W. Y., Wieder, H. H., Koschel, W. H., Bishop, S. G., and McCombe, B. D. (1977). *Appl. Phys. Lett.* **30**, 1.
Madelung, O. (1964). "Physics of III–V Compounds." Wiley, New York.
Mandel, G. (1964). *Phys. Rev. A* **134**, 1073.
Martin, G. M. (1980). In "Semi-Insulating III–V Materials" (G. J. Rees, ed.), p. 13. Shiva Publ., Ltd.
Martin, G. M., Verheijke, M. L., Jansen, J. A. J., and Poiblaud, G. (1979). *J. Appl. Phys.* **50**, 467.
Martin, G. M., Farges, J. P., Hallais, J. P., and Poiblaud, G. (1980a). *J. Appl. Phys.* **51**, 2840.
Martin, G. M., Mitonneau, A., Pons, D., Mircea, A., Woodard, D. W. (1980b). *J. Phys. C* **13**, 3855.
Martin, G. M., Jacob, G., Poiblaud, G., Goltzene, A., and Schwab, C. (1981). *Conf. Ser.–Inst. Phys.*, **59**, 281.
Martinez, G., Hennel, A. M., Szuszkiewicz, W., Balkanski, and Clerjaud, B. (1981). *Phys. Rev. B* **23**, 3920.
Masterov, V. F., and Samorukov, B. E. (1978). *Sov. Phys. Semicond.* (*Engl. Transl.*) **12**, 363.
Mehran, F., Morgan, T. N., Title, R. S., and Blum, S. E. (1972). *Solid State Commun.* **11**, 661.
Meignant, D., Boccon-Gibod, D., and Bourgeois, J. M. (1979). *Electron. Lett.* **15**, 781.
Miller, G. L., Lang, D. V., and Kimerling, L. C. (1977). *Annu. Rev. Mater. Sci.*, **7**, 377.
Milnes, A. G. (1973). "Deep Impurities in Semiconductors." Wiley, New York.
Mircea, A., and Bois, D. (1979). *Conf. Ser.–Inst. Phys.* **46**, 82.
Mircea-Roussel, A., Martin, G. M., and Lowther, J. E. (1980). *Solid State Commun.* **36**, 171.
Mizuno, O., and Watanabe, H. (1975). *Electron. Lett.* **11**, 118.
Monemar, B. (1972). *J. Lumin.* **5**, 239.
Morgan, T. N. (1968). *Phys. Rev. Lett.* **21**, 819.
Morgan, T. N. (1978). *Phys. Rev. Lett.* **40**, 190.
Mott, N. F. (1978). *Solid-State Electron.* **21**, 1275.
Munoz, E., Snyder, W. L., and Moll, J. L. (1970). *Appl. Phys. Lett.* **16**, 262.
Nakashima, H. (1971). *Jpn. J. Appl. Phys.* **10**, 1737.
Narayanamurti, V., Logan, R. A., and Chin, M. A. (1979). *Phys. Rev. Lett.* **43**, 1536.
Newman, R. C. (1973). "Infrared Studies of Crystal Defects." Taylor and Francis, London.
Pankove, J. I., Duffy, M. T., Miller, E. A., Berkeyheiser, J. E. (1973). *J. Lumin.* **8**, 89.
Pantelides, S. T. (1978). *Rev. Mod. Phys.* **50**, 787.
Partin, D. L., Milnes, A. G., and Vassamillet, L. F. (1978). *J. Electron. Mater.* **7**, 279.
Partin, D. L., Chen, J. W., Milnes, A. G., and Vassamillet, L. F. (1979a). *Solid State Electron.* **22**, 455.
Partin, D. L., Milnes, A. G., and Vassamillet, L. F. (1979b). *J. Electrochem. Soc.* **126**, 1584.
Partin, D. L., Chen, J. W., Milnes, A. G., and Vassamillet, L. F. (1979c). *J. Appl. Phys.* **50**, 6845.
Picoli, G., Deveaud, B., and Galland, D. (1981). *J. Phys.* (*Orsay, Fr.*) **42**, 133.
Pilkuhn, M. (1981). In "Handbook on Semiconductors, Vol. 4: Device Physics" (C. Hilsum, ed.), North-Holland Publ. Amsterdam.
Plesiewicz, W. (1977). *J. Phys. Chem. Solids* **38**, 1079.
Pons, D., Mircea, A., and Bourgoin, J. (1980). *J. Appl. Phys.* **51**, 4150.
Prener, J. S., and Williams, F. E. (1956). *J. Chem. Phys.* **25**, 261.
Queisser, H. J. (1971). *Festkoerperprobleme* **11**, 45.
Queisser, H. J. (1976). *Appl. Phys.* **10**, 275.
Queisser, H. J. (1978). *Solid State Electron.* **21**, 1495.

Queisser, H. J., and Fuller, C. S. (1966). *J. Appl. Phys.* **37**, 4895.
Queisser, H. J., and Theodorou, D. E. (1979). *Phys. Rev. Lett.* **43**, 401.
Reid, F. J., Baxter, R. D., and Miller, S. E. (1966). *J. Electrochem. Soc.* **113**, 713.
Reynolds, D. C., Litton, C. W., Almassy, R. J., McCoy, G. L., and Nam, S. B. (1980). *J. Appl. Phys.* **51**, 4842.
Reynolds, D. C., and Collins, T. C. (1981). "Excitons." Academic Press, N.Y.
Roitsin, A. B. (1974). *Sov. Phys. Semicond.* (*Engl. Transl.*) **8**, 1.
Sah, C. T., Forbes, L., Rosier, L. L., and Tasch, Jr., A. F. (1970). *Solid State Electron.* **13**, 759.
Schairer, W., and Schmidt, M. (1974). *Phys. Rev. B* **10**, 2501.
Scheffler, M., Pantelides, S. T., Lipari, N. O., and Bernholc, J. (1981). *Phys. Rev. Lett.* **47**, 413.
Schmidt, M., and Stocker, H. J. (1978). *J. Appl. Phys.* **49**, 4438.
Schneider, J., and Kaufmann, U. (1981). *Conf. Ser.-Inst. Phys.*, **59**, 55.
Segsa, K. H., and Spenke, S. (1975). *Phys. Status Solidi A* **27**, 129.
Slack, G. A., Ham, F. S., and Chrenko, R. M. (1966). *Phys. Rev.* **152**, 376; see also references therein.
Stauss, G. H., and Krebs, J. J. (1977). *Conf. Ser.-Inst. Phys.* **33a**, 84.
Stauss, G. H., and Krebs, J. J. (1980). *Phys. Rev. B* **22**, 2050.
Stauss, G. H., Krebs, J. J., and Henry, R. L. (1977). *Phys. Rev. B* **16**, 974.
Stauss, G. H., Krebs, J. J., Lee, S. H., and Swiggard, E. M. (1979). *J. Appl. Phys.* **50**, 6251.
Stauss, G. H., Krebs, J. J., Lee, S. H., and Swiggard, E. M. (1980). *Phys. Rev. B* **22**, 3141.
Stocker, H. J. (1977). *J. Appl. Phys.* **48**, 4583.
Suto, K., and Nishizawa, J. (1972). *J. Appl. Phys.* **43**, 2247.
Szawelska, H. R., and Allen, J. W. (1979). *J. Phys. C* **12**, 3359.
Tell, B., and Kuijpers, F. P. (1978). *J. Appl. Phys.* **49**, 5938.
Title, R. S. (1969). *J. Appl. Phys.* **40**, 4902.
Thompson, F., and Newman, R. C. (1971). *J. Phys. C* **4**, 3249.
Ushakov, V. V., and Gippius, A. A. (1980). *Sov. Phys. Semicond.* (*Engl. Transl.*) **14**, 333.
van der Meulen, Y. J. (1967). *J. Phys. Chem. Solids* **28**, 25.
van Gorkom, G. G. P., and Vink, A. T. (1972). *Solid State Commun.* **11**, 767.
van Vechten, J. A. (1975a). *J. Electrochem. Soc.* **122**, 419.
van Vechten, J. A. (1975b). *J. Electrochem. Soc.* **122**, 423.
van Vechten, J. A. (1975c). *J. Electron. Mater.* **4**, 1159.
van Vechten, J. A. (1976). *Phys. Semicond., Proc. Int. Conf., 13th, 1976*, p. 577.
Vasil'ev, A. V., Ippolitova, G. K., Omel'yanovskii, E. M., and Ryskin, A. I. (1976a). *Sov. Phys. Semicond.* (*Engl. Transl.*) **10**, 341.
Vasil'ev, A. V., Ippolitova, G. K., Omel'yanovskii, E. M., and Ryskin, A. I. (1976b). *Sov. Phys. Semicond.* (*Engl. Transl.*) **10**, 713.
Vavilov, P. N., Ushakov, V. V., and Gippius, A. A. (1980). *J. Phys. Soc. Jpn.* **45**, Suppl. A, 267.
Vink, A. T., and van Gorkom, G. G. P. (1972). *J. Lumin.* **5**, 379.
Wagner, R. J., and White, A. M. (1979). *Solid State Commun.* **32**, 39.
Wagner, R. J., Krebs, J. J., Stauss, G. H., and White, A. M. (1980). *Solid State Commun.* **36**, 15.
Watkins, G. D. (1965). "Radiation Damage in Semiconductors," p. 97. Dunod, Paris.
Watkins, G. D. (1973). *Conf. Ser.-Inst. Phys.* **16**, 228.
Watkins, G. D. (1975). "Point Defects in Solids" (Crawford, J. H., Jr, and Slifkin, L. M., eds.), Vol. 2, p. 333. Plenum, N.Y.
Watts, R. K. (1977). "Point Defects in Crystals". Wiley, N.Y.
Weber, J., Ennen, H., Kaufmann, U., and Schneider, J. (1980). *Phys. Rev. B* **21**, 2394.
Weber, J., *et al.* (1981). To be published.
West, C. L., Hayes, W., Ryan, J. F., Dean, P. J. (1980). *J. Phys. C.* **13**, 5631.

White, A. M. (1979a). *Conf. Ser.–Inst. Phys.* **43**, 123.
White, A. M. (1979b). *Solid State Commun.* **32**, 205.
White, A. M., Dean, P. J., and Porteous, P. (1976). *J. Appl. Phys.* **47**, 3230.
White, A. M., Krebs, J. J., and Stauss, G. H. (1980). *J. Appl. Phys.* **51**, 419.
Wight, D. R., Blenkinsop, I. D., and Bass, S. J. (1980). *In* "Semi-Insulating III–V Materials" (G. J. Rees, ed.), p. 174. Shiva Publ., Ltd.
Willardson, R. K., and Beer, A. C. (1965–1968). *Semicond. Semimet.* **1–4**.
Williams, E. W., and Bebb, H. B. (1972). *Semicond. Semimet.* **8**, 321.
Williams, E. W., and Elliott, C. T. (1969). *J. Phys. D* **2**, 1657.
Willmann, F., Bimberg, D., and Blätte, M. (1973). *Phys. Rev. B* **7**, 2473.
Yokoyama, N., Shibatomi, Ohkawa, S., Fukuta, M., and Ishikawa, H. (1977). *Conf. Ser.–Inst. Phys.* **33b**, 201.
Zucca, R. (1977). *J. Appl. Phys.* **48**, 1987.
Zucca, R., Welch, B. M., Asbeck, P. M., Eden, R. C., and Long, S. I. (1980). *In* "Semi-Insulating III–V Materials" (G. J. Rees, ed.), p. 335. Shiva Publ., Ltd.
Zylbersztejn, A., Bert, G., and Nuzillat, G. (1979). *Conf. Ser.–Inst. Phys.* **45**, 315.

Collisional Detachment of Negative Ions

R. L. CHAMPION

Department of Physics
The College of William and Mary
Williamsburg, Virginia

I. Introduction	143
II. Nomenclature and Experimental Techniques	146
III. Atomic Reactants	148
A. $H^-(D^-)$–Rare Gases	148
B. H^-–H	161
C. Halogen Negative Ions–Rare Gases	164
D. Halogen Negative Ions–Alkali Atoms	169
E. $H^-(D^-)$–Alkali Atoms	172
F. O^-–Rare Gases	174
G. Detachment Rate Constants	176
IV. Molecular Reactants	177
A. Atomic Negative Ions–Molecular Targets	178
B. Molecular Negative Ions	184
V. Summary	186
References	187

I. Introduction

Collisions of negative ions with neutral atoms and molecules in the gas phase are important in many areas of physics and chemistry. One of the most fundamental processes in such collisions is referred to as collisional detachment: in this process a negative ion collides with a neutral atom or molecule, producing a free electron

$$X^- + B \rightarrow X + B + e \qquad (1)$$

This process is a principal mechanism for the destruction of negative ions and is frequently the dominant inelastic collision channel for negative ions. Moreover, the theoretical problem of collisional detachment, which may involve bound state–continuum interactions, is an interesting and broadly applicable problem in the field of atomic and molecular physics.

The purpose of this review is to provide a discussion of the process of collisional detachment and to survey the activity in this subfield of atomic and molecular physics. The reader who is interested in the general field of negative ions and various other collisional processes associated with negative ions should refer to the thorough and excellent text by Massey (1976).

The collisional detachment of negative ions has been studied for about three decades (Hasted, 1952; Dukel'skii and Zandberg, 1951); but a recent awareness of the importance of negative ions in many physical and chemical processes has led to an increased level of both experimental and theoretical activity in this field. A few examples of areas of applied interest where the collisional detachment of negative ions (or the reverse process of three-body attachment) is important include (1) plasma physics, where intense beams of H^- or D^-, after being accelerated, are neutralized and serve to "heat" magnetic containment fusion devices such as the Tokamaks; (2) ionospheric chemistry, where molecular negative ions may play an important role in determining the free electron density; (3) flame chemistry, in which a variety of negative ions have been identified in mass spectrometric studies; (4) magnetohydrodynamic convertors, where the conductivity of the discharge depends on the concentration of negative ions. This small list is far from exhaustive simply because the majority of elements in the periodic table and a large number of molecules can form stable atomic or molecular negative ions: i.e., they have positive electron affinities. A complete review of the subject of atomic electron affinities is given by Hotop and Lineberger (1975). To illustrate the frequency of stable atomic negative ions, a periodic table of "recommended electron affinities" is taken from their review (Fig. 1). There are also several recent reviews concerned with molecular electron affinities (Janousek and Brauman, 1979; Franklin and Harland, 1974). The techniques associated with the most precise method of determining electron affinities of free atoms and molecules (photodetachment) were surveyed most recently by Corderman and Lineberger (1979).

Collisions of negative ions with neutral atoms or molecules can lead to the detachment of the loosely bound electron of the negative ion via several distinct mechanisms; these may be enumerated as follows.

(1) Detachment in which there are no excited intermediate or final states of the reactants: direct detachment,

$$X^- + B \rightarrow X + B + e \qquad (2a)$$

(2) Detachment in which excited states of either target or negative ion (or its neutral parent) are involved:
  (a) excitation to autodetaching levels,

$$X^- + B \rightarrow (X^-)^* + B \rightarrow X + B + e \qquad (2b)$$

  (b) excitation of target or negative ion parent,

$$X^- + B \rightarrow X + B^* + e \quad \text{or} \quad X^* + B + e \qquad (2c)$$

1 H 0.7542																	2 He <0
3 Li 0.620	4 Be <0											5 B 0.28	6 C 1.268	7 N ≤0	8 O 1.462	9 F 3.399	10 Ne <0
11 Na 0.546	12 Mg <0											13 Al 0.46	14 Si 1.385	15 P 0.743	16 S 2.0772	17 Cl 3.615	18 Ar <0
19 K 0.5012	20 Ca <0											31 Ga 0.3	32 Ge 1.2	33 As 0.80	34 Se 2.0206	35 Br 3.364	36 Kr <0
37 Rb 0.4860	38 Sr <0											49 In 0.3	50 Sn 1.25	51 Sb 1.05	52 Te 1.9708	53 I 3.061	54 Xe <0
55 Cs 0.4715	56 Ba <0											81 Tl 0.3	82 Pb 1.1	83 Bi 1.1	84 Po 1.9	85 At 2.8	86 Rn <0

21 Sc <0	22 Ti 0.2	23 V 0.5	24 Cr 0.66	25 Mn <0	26 Fe 0.25	27 Co 0.7	28 Ni 1.15	29 Cu 1.226	30 Zn <0
39 Y ≈0	40 Zr 0.5	41 Nb 1.0	42 Mo 1.0	43 Tc 0.7	44 Ru 1.1	45 Rh 1.2	46 Pd 0.6	47 Ag 1.303	48 Cd <0
57 La 0.5	72 Hf <0	73 Ta 0.6	74 W 0.6	75 Re 0.15	76 Os 1.1	77 Ir 1.6	78 Pt 2.128	79 Au 2.3086	80 Hg <0

FIG. 1. Electron affinities of the elements. The width of the solid bar below each element indicates the percentage uncertainty in the number. [From Hotop and Lineberger (1975).]

(c) charge transfer to temporary negative ion state of target,

$$X^- + B \to X + (B^-)^* \to X + B + e \quad (2d)$$

(d) detachment with ionization,

$$X^- + B \to X^+ + B + 2e \quad (2e)$$

(3) Detachment in which the neutral reactants can form bound molecules:
 (a) associative detachment,

$$X^- + B \to XB + e \quad (2f)$$

 (b) ion molecule reaction with detachment,

$$X^- + BC \to XB + C + e \quad (2g)$$

These last two processes [Eqs. (2f) and (2g)] are important only for relative collision energies less than several tens of electron volts and for selected reactants. Direct detachment is important (and often the dominant detachment channel) at all energies above the threshold, whereas the remaining mechanisms [with the possible exception of Eq. (2d)] are important only at higher collision energies, say $E \simeq 1$ keV. This review has as its principal concern the experimental observations associated with processes (2a)–(2e) and will cover the energy range from the threshold for detachment up to several hundred keV.

In an attempt to understand fully the dynamics of collisional detachment, experiments have been performed in which the absolute total detachment cross section as well as the angular differential and doubly differential cross sections (i.e., both the angular and product energy spectra are obtained) have been measured. There have also been experiments in which energy, and sometimes angular, spectra of the detached electrons have been measured, and others in which the light emitted from the products of collisional detachment has been monitored.

In what follows, no attempt will be made to compile all of the various cross-section measurements since, for the most part, such compilations exist (Risley, 1980). Rather, the approach will be to examine the various phenomena observed in collisional detachment for a variety of reactants.

II. NOMENCLATURE AND EXPERIMENTAL TECHNIQUES

The notation for the total electron detachment cross section is usually $\sigma_{-10}(E)$, or more generally $\sigma_{ij}(E)$, where i and j refer to the incident and final charge states of the projectile. Thus, $\sigma_{-11}(E)$ refers to collisional ionization of a negative ion, which was given as (2e) earlier. The collisional energy is denoted by E and in some papers it is unclear whether the "energy" is the relative collision energy of the reactants or the laboratory energy of the projectile. In most cases this distinction is obvious and we clarify only those cases where the coordinate system may appear ambiguous.

The differential cross section for a given collision energy is usually denoted by $\sigma(\theta)$ [rather than $\sigma(\theta, E)$], where

$$\sigma_{\text{Total}}(E) = 2\pi \int \sigma(\theta) \sin \theta \, d\theta$$

for cases of azimuthal symmetry. In addition, some further symbol must be assigned to $\sigma(\theta)$ to identify the product channel. For example, $\sigma_e(\theta)$ refers to elastic differential cross sections and $\sigma_n(\theta)$ refers to the differential cross section for neutrals that are the products of collisional detachment.

It has become common practice for experimentalists to report their results for differential cross-section measurements with "reduced" coordinates:

$$\tau \equiv E \cdot \theta, \qquad \rho \equiv \theta \cdot \sin\theta \cdot \sigma(\theta)$$

This representation is attractive since it allows results at various energies to be compared conveniently with each other. Structure in $\sigma(\theta)$ which is due to some particular feature of the intermolecular potential (e.g., curve crossing) occurs at an approximately fixed value of τ. Moreover, the magnitude of $E \cdot \theta$ is approximately independent of the coordinate system, i.e., $E \cdot \theta \simeq E_{\rm rel}\theta_{\rm rel}$. The usefulness and efficacy of these reduced coordinates are discussed by Smith et al. (1967).

Several methods have been used to measure total detachment cross sections. For lower collision energies, it is convenient to detect the detached electrons using weak electrostatic fields, often combined with magnetostatic fields, to steer the free electrons to some collector plate. This can be done without seriously disturbing the trajectories of the heavier negative ions, and examples of this technique can be found in the work of Hasted (1952), Roche and Goodyear (1969), Hummer et al. (1960), Smith et al. (1978), and Bydin and Dukel'skii (1957). Bailey and Mahadevan (1970) used an rf electron trap to distinguish detached electrons from slow heavy ions.

For collision energies for which large-angle elastic scattering is not too important, the total detachment cross section can be determined by an attenuation technique in which the intensity of the primary negative ion beam is observed as the target gas pressure is varied. Such an approach has been used by Risley and Geballe (1974) and Bennett et al. (1975). Meyer (1980) has employed a slight variation of this method in which the collision path length is varied (rather than the pressure), in an attenuation experiment in which the neutralization cross sections for negative ion–alkali reactants were determined.

For both the attenuation and electron-trap techniques, it is generally not possible to separate $\sigma_{-10}(E)$ from $\sigma_{-11}(E)$. For example, in an electron trap that detects all electrons which are collision products, the experiment measures

$$\sigma(E) = \sigma_{-10}(E) + 2\sigma_{-11}(E)$$

At low collision energies, however, the second term is usually negligible. At higher collision energies the fast neutral collision products can be measured directly to determine $\sigma_{-10}(E)$ and $\sigma_{-11}(E)$, as has been done, for example, by Geddes et al. (1980).

Differential cross sections have been measured with the time-of-flight method [e.g., Fayeton et al. (1978), Cheung and Datz (1980)], by using electrostatic energy analysis [e.g., Lam et al. (1974)], and with position-sensitive channel plate detectors [e.g., de Vreugd et al. (1979a)]. In order to investigate the role of excited states in collisional detachment, it is necessary to measure doubly differential cross sections, and the time-of-flight apparatus is especially well suited to this purpose. All of these methods have been used previously to measure differential cross sections for various positive ion–neutral target experiments.

III. ATOMIC REACTANTS

A. $H^-(D^-)$–Rare Gases

These reactants have been investigated extensively, and because of its relative simplicity, the H^- + He system has been the subject of numerous theoretical studies. Recent compilations of the experimental values of $\sigma_{-10}(E)$ and $\sigma_{-11}(E)$ for 10 eV $< E <$ 10 keV have been published by Tawara (1978) and Risley (1980). In order to examine the experimental observations and discuss the dynamics of collisional detachment, it is instructive to begin with the reactants H^- + He for collision energies near the threshold for detachment. For collision energies below about 100 eV, direct detachment is by far the dominant process among the detachment mechanisms listed in Eqs. (2a)–(2g). For these low relative collision energies, there exist both experimental and theoretical results for H^- and D^- + He; they are exhibited in Fig. 2.

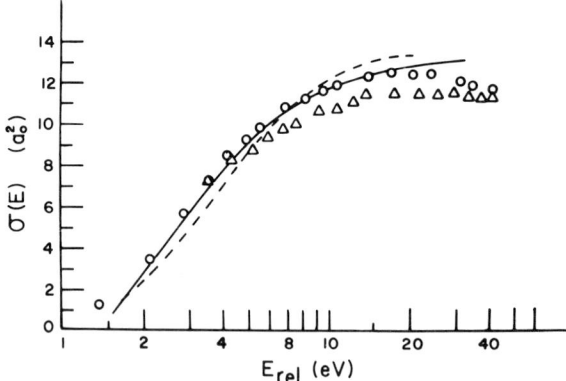

FIG. 2. Total detachment cross sections for $H^-(D^-)$ + He. The experimental points are the results of recent near-threshold measurements (Huq et al., 1982): solid line derives from a calculation by Gauyacq (1980a) for D^- + He; dashed line derives from a calculation by Taylor and Delos (1982) for D^- + He; $\bigcirc$, D^- + He; $\triangle$, H^- + He.

The qualitative behavior of the total detachment cross section at low collision energies has been discussed by Mason and Vanderslice (1958) and is easily understood if one refers to the diagram of the intermolecular potentials for the molecular systems HeH⁻ and HeH. The results of recent calculations for these potentials by Olson and Liu (1980a) are given in Fig. 3. At large internuclear separations, the HeH⁻ electronic energy lies 0.75 eV (i.e., the electron affinity of H) below the HeH curve. In a collision, as the negative ion approaches the target atom, the electronic energy of the HeH⁻ system rises until it crosses or merges with the continuum of states representing neutral atoms and a free electron. If the curves cross at $R = R_x$, as in Fig. 3, then for $R < R_x$, HeH⁻ can no longer be regarded as a stable state, but perhaps as a quasi-bound resonance. It is anticipated that detachment may occur during a collision when the separation of the negative ion and the atom is less than (or close to) the crossing radius R_x.

For these collision energies the de Broglie wavelength associated with the nuclear motion is considerably smaller than the molecular size, and the motion of the nuclei may be discussed within the framework of classical mechanics. Moreover, for low collision energies the nuclear velocities are considerably smaller than electron velocities and molecular calculations based on a near-adiabatic approximation should be appropriate for calculating the intermolecular potential.

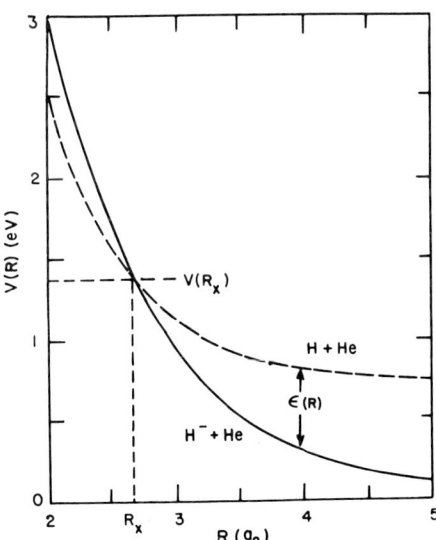

FIG. 3. Intermolecular potentials for the ground states of HeH and HeH⁻. [From a calculation by Olson and Liu (1980a).]

The cross section for collisional detachment may be expressed

$$\sigma_{-10}(E) = \int_0^\infty 2\pi b P_d(b, E)\, db \qquad (3)$$

where b is the classical impact parameter and $P_d(b, E)$ is the detachment probability depending on b and E. Several different theoretical approaches have been employed to describe $P_d(b, E)$; these will be discussed later. It may be useful, however, to first consider the simplest (but not very realistic) assumption for $P_d(b, E)$, which is

$$P_d(b, E) = 0, \quad b > b_x; \qquad P_d(b, E) = 1, \quad b \le b_x \qquad (4)$$

where b_x is the impact parameter for which the classical turning point of the nuclear motion is R_x, i.e., satisfying

$$(b_x^2/R_x^2) + [V(R_x)/E] = 1 \qquad (5)$$

In this case

$$\sigma_{-10}(E) = \pi b_x^2 = \pi R_x^2[1 - V(R_x)/E] \qquad (6)$$

Thus the threshold energy will be $V(R_x)$ as indicated in Fig. 3 and the high-energy limit for $\sigma_{-10}(E)$ will be approximately πR_x^2. As can be seen from Figs. 2 and 3, the asymptotic value actually only reaches about half of this limit, indicating that Eq. (4), as might be expected, overestimates the detachment probability.

Figure 2 shows that there is a substantial isotope effect in the total detachment cross sections for H^- and D^- + He which will require a description of $P_d(b, E)$ that is considerably more detailed than Eq. (4). One type of experimental measurement which should offer some insight into the nature of $P_d(b, E)$ is a measurement of the differential scattering cross section. Unfortunately, at such low energies, the experimental problem of detecting neutral hydrogen atoms from, e.g.,

$$H^- + He \to H + He + e \qquad (7)$$

is very difficult and there exists no low-energy ($E \lesssim 100$ eV) data concerning the differential cross section for Eq. (7). On the other hand, differential cross-section measurements for *elastic* scattering have been made, and these, as the complement to Eq. (7), can contribute to an understanding of the dynamics of direct detachment. The elastic differential cross section is

$$\sigma_{el}(\theta) = (b/\sin\theta)|db/d\theta|[1 - P_d(b)] \qquad (8)$$

where θ is the scattering angle and

$$\theta(b) = \pi - 2b \int_{R_0}^{\infty} \frac{dR}{R^2 \{1 - [V(R)/E - b^2/R^2]\}^{1/2}} \qquad (9)$$

The expression for $\sigma_{el}(\theta)$ [Eq. (8)] is particularly simple for these reactants since the potential curves are purely repulsive, $\theta(b)$ is a unique function of b, and there are no semiclassical interference effects (Bernstein, 1966) to be expected in $\sigma_{el}(\theta)$. From Eqs. (8) and (9) the elastic differential scattering cross section (for a system with potential curves similar to those of Fig. 3) would have the form given in Fig. 4. For scattering angles larger than $\theta(b_x)$, a "droop" in the elastic cross section should occur due to the onset of detachment for $b \lesssim b_x$. The magnitude of this droop depends directly on $P_d(b)$. The fraction $\sigma_{el}^D(\theta)/\sigma_{el}^H(\theta)$ should be less than or equal to unity (in order to agree with the observations of Fig. 2), since $\theta(b)$ is independent of the reduced mass of the reactants at a given relative collision energy E. Such an isotope effect for $H^-(D^-) + $ He has been observed (Lam et al., 1974) and it is in accord with the total cross-section measurements: $\sigma_{el}(\theta)$ for the D^- projectile decreases more rapidly than that for the H^- projectile. On the other hand, measurements of $\sigma_{-10}(E)$, for $H^-(D^-) + $ Ne, Ar (Champion et al., 1976) yield an isotope effect that is just the reverse of that found for $H^-(D^-) + $ He, as can be seen in Fig. 5. An immediate question arises: what is the source of the two different isotope effects exhibited for low-energy direct detachment of H^- by the rare gases?

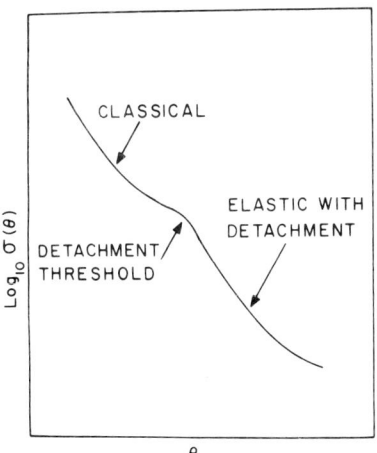

FIG. 4. Qualitative behavior of differential elastic scattering cross section when accompanied by electron detachment. [From Lam et al. (1974).]

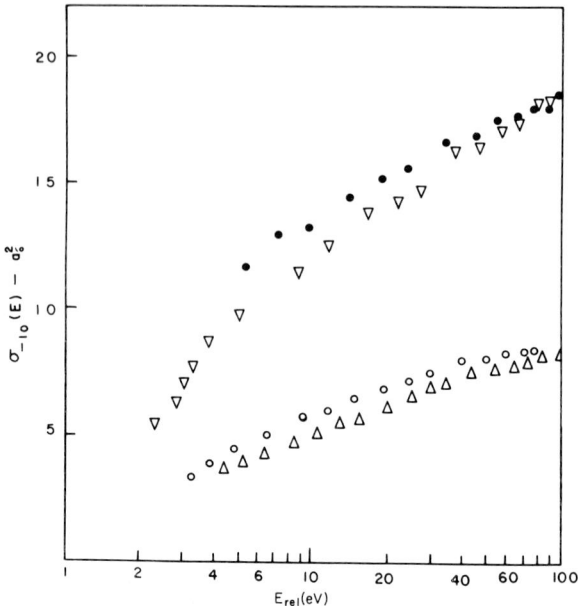

FIG. 5. Total detachment cross sections for $H^-(D^-)$ + Ne and Ar: $\triangle$, D^- + Ne; $\bigcirc$, H^- + Ne; $\triangledown$, D^- + Ar; $\bullet$, H^- + Ar. [From Champion et al. (1976).]

In addition to the HeH⁻ potential, Olson and Liu (1980a) have calculated the potential for NeH⁻ and the result is given in Fig. 6. The ionic and neutral curves do not clearly cross as for HeH⁻, but rather merge in the vicinity of $R = 2.2a_0$. Olson and Liu argue that this basic difference in the intermolecular potentials accounts for the different isotope effects observed for the He and Ne (and presumably Ar) targets in the following manner. For He, collisional detachment should be reasonably described by the use of a local complex potential model (see Lam et al., 1974, for example) in which the decay of the negative ion state is governed simply by a width, $\Gamma(R)$, which is zero for $R > R_x$ and generally increases for $R < R_x$. The lifetime of the unstable HeH⁻ state is proportional to $1/\Gamma(R)$, and the detachment probability in this model is given by

$$P_d(b, E) = \int \Gamma[R(t)]\, dt = \int [\Gamma(R)/v(R)]\, dR \tag{10}$$

where $v(R)$ is the relative nuclear velocity. Equation (10) predicts the isotope effect observed for the He target since, for a given relative collision energy, the heavy projectile spends more time in the region $R < R_x$ and is hence more likely to detach. This simple model does predict the correct size of the iso-

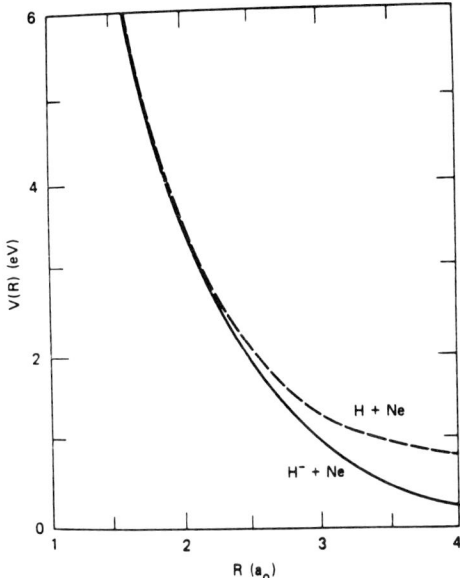

FIG. 6. Intermolecular potentials for the ground states of NeH and NeH$^-$. From a calculation by Olson and Liu (1980a).

tope effect for H$^-$(D$^-$) + He (Champion et al., 1976; Lam et al., 1974). Such a model will, however, fail to predict the increase in $\sigma_{-10}(E)$ observed for energies greater than several hundred electron volts (Risley and Geballe, 1974).

For H$^-$(D$^-$) + Ne, a local complex potential approach is inappropriate because no crossing into the continuum occurs; it should be replaced with a collision dynamics more descriptive of "charge transfer to a continuum." For cases similar to that for NeH$^-$, wherein the potential curves for charge transfer states may be close together (see, e.g., Melius and Goddard, 1974, or Wijnaendts van Resandt et al., 1978, for LiNa$^+$), the probability for a charge transfer transition increases with an increase in collision velocity. Thus, any model based on such an assumption will result in isotope effects for $\sigma_{-10}(E)$ as observed in Fig. 5.

A different approach has been taken by Gauyacq (1979, 1980a), who calculated $P_d(b, E)$ for the HeH$^-$ system by using an extension of a formalism known as the "zero-radius model," which was previously discussed by Devdarianni (1973) and Demkov (1980, 1964). In this model, detachment can occur for R near R_x, which is in the region where the binding energy $\epsilon(R)$ of the extra electron is very small. [$\epsilon(R)$ is the difference between the neutral and ionic potentials for $R > R_x$.] For small values of $\epsilon(R)$, the wavelength of the outer electron becomes much larger than the size of the HeH molecular core (which is on the order of R_x) and the probability that the electron is outside this core becomes large. The detachment problem is then addressed by dividing the space into two regions: (1) the core and (2) the outer region where the electron is essentially in a field-free environment with energy $\epsilon(R)$. In this outer region, the electron wave function (assumed to be a S state) satisfies the free-particle Schrödinger equation

$$[\nabla^2 - 2\epsilon(R)]\psi = 0 \qquad (11)$$

which gives $\psi \sim e^{-kr}/r$, where $k = [2\epsilon(R)]^{1/2}$. The logarithmic derivative of the wave function is specified on the boundary between the two regions, which is allowed to be at arbitrarily small r:

$$(r\Psi)^{-1}[\partial(r\Psi)/\partial r]|_{r\to 0} = f[R(t)] \qquad (12)$$

and where the time dependence of the boundary condition is due to the motion of the nuclei. For $R > R_x$, $f R(t)$ is negative and equal to $-[2\epsilon(R)]^{1/2}$. For $R < R_x$, $f[R(t)]$ changes sign and is not so easily obtained as for $R > R_x$. Consequently, these calculations for $H^- + He$ based on the zero-radius model use a linear extrapolation to approximate $f[R(t)]$ for $R < R_x$. The intermolecular potentials (such as those of Fig. 3) can be used to find $R(t)$ and $\epsilon[R(t)]$ for a given impact parameter and Eq. (12) can be numerically integrated to find $\Psi(r, t)$. For a large value of t, $t = T$, the collision partners will have separated and the survival probability (i.e., the complement of the detachment probability) can be found by projecting the electron wave function $\Psi(r, T)$ onto the bound eigenfunction. The result of such a procedure for $D^- + He$ at $E = 16$ eV is seen in Fig. 7. [The results of Fig. 7 are based on configuration interaction calculations for $V(R)$ by

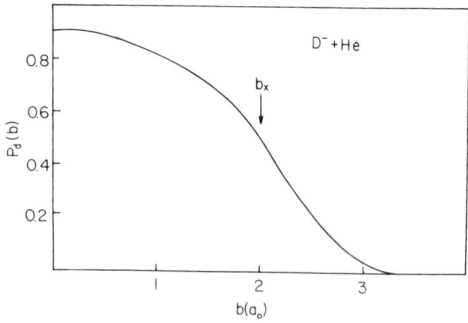

FIG. 7. Detachment probability as a function of impact parameter b for $E_{rel} = 16$ eV. The value b_x is the impact parameter for which the distance of closest approach is R_x. [From Gauyacq (1980a).]

Gauyacq (1980a) and give slightly different results from a similar calculation of Olson and Liu (1980a) shown in Fig. 3. This difference is presumably due to the fact that slightly different configurations were employed in the two calculations.] The calculation for $P_d(b)$ indicates that a substantial fraction of $\sigma_{-10}(E)$ comes from impact parameters larger than b_x [see Eq. (6)], a feature not present in a complex potential model. The calculated detachment cross section for $D^- + $ He based on the above zero-radius approximation is given in Fig. 2, and the agreement between experiment and calculation is seen to be quite good.

The isotope effects observed in the $H^- + $ He experiment are not completely accounted for by this calculation: The calculation predicts about 50% of the observed difference in $\sigma_{-10}(E)$ for $H^-(D^-) + $ He. Gauyacq (1980a) suggests that the zero-radius model inherently contains two types of isotope effects in opposition, and as a consequence it may result in $\sigma_{-10}(E)$ exhibiting the behavior of either the $H^-(D^-) + $ He *or* the $H^-(D^-) + $ Ne system. Alternatively, there may be no discernable isotope effect. These two opposing effects at a given relative energy are due to: (*i*) the shortening of time spent in the region $R < R_x$ for the lighter isotope (the same effect as for the local complex potential model) and (*ii*) the velocity dependence of "dynamical transitions" for $R > R_x$.

Similar zero-range potential calculations have been reported for $H^-(D^-) + $ Ne (Gauyacq, 1980b). The isotope effect observed for these reactants is very nicely reproduced by the calculation, thus illustrating the importance of dynamical effects (*ii*) in the detachment process.

In order to illustrate the features of the zero-range model, it is interesting to follow the temporal behavior of the electron wave function as the collision proceeds. Figure 8 illustrates this behavior for the $H^- + $ He system for a relative collision energy of 200 eV. Prior to the collision (negative time on Fig. 8) the wave function is a straight line on the semilog plot, which is indicative of the bound state. At $t = 0$ (the point where $f[R(t)]$ changes sign), the wave packet has begun to spread appreciably. For large positive times the wave function appears to be split into two distinct parts: that which remains near the origin and represents the survival probability and an outgoing wave packet representing the outgoing (detached) electron.

A different approach to the theory of these processes has been developed by Taylor and Delos (1982). They expand the electronic wave function as

$$\Psi = C_0(t)\phi_0 + \int C_E(t)\phi_E \rho(E) \, dE \qquad (13)$$

where ϕ_0 is the state in which the electron is bound to the molecule, ϕ_E is a state in which the electron is free with kinetic energy E, and $\rho(E)$ is the density of states in this continuum. Thus $C_0(t)$ is the probability amplitude for

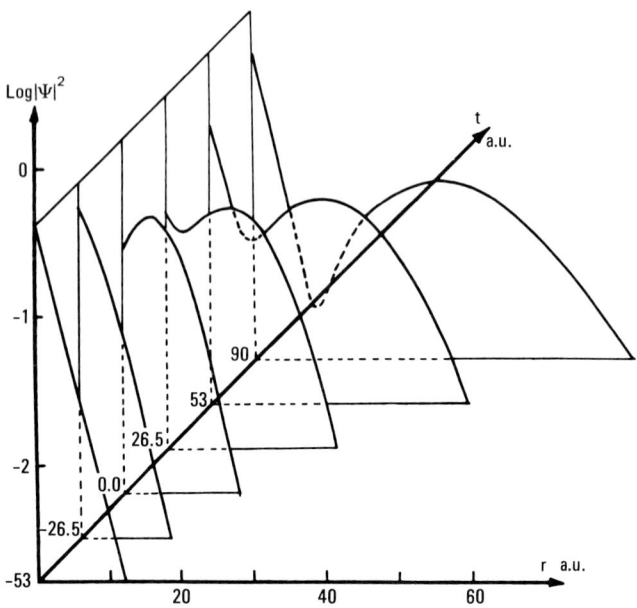

FIG. 8. Logarithm of the probability of finding the "extra" electron at a distance r is plotted as a function of time. Negative (positive) times are pre (post) collision. The system is $H^- + He$ and the relative collision energy is 200 eV. The impact parameter is $1a_0$. [From Gauyacq (1980).]

finding the electron bound; $|C_0(-\infty)|^2 = 1$, and $1 - |C_0(\infty)|^2$ is the probability of electron detachment.

Taylor and Delos assume that the states ϕ_0, ϕ_E can be constructed in such a way that nonadiabatic couplings among them are negligible, and detachment occurs because of electrostatic couplings (i.e., matrix elements V_{0E} of the electronic Hamiltonian). Furthermore, they assume that transitions only occur between the bound state and the continuum (and vice versa), and that direct continuum–continuum transitions are insignificant.

From these assumptions, they derive coupled equations that must be satisfied by the coefficients $C_0(t)$, $C_E(t)$:

$$i\hbar\{d/dt\}[C_0(t)] = V_{\text{ion}}(t)C_0(t) + \int V_{0E}(t)C_E(t)\rho(E)\,dE$$

$$i\hbar\{d/dt\}[C_E(t)] = [V_{\text{neutral}}(t) + E]C_E(t) + V_{E0}C_0(t)$$

(14)

The problem then reduces to solving this nondenumerably infinite set of coupled equations.

Neglecting the time dependence of $V_{E0}(t)$, and approximating $V_{\text{ion}}(t) - V_{\text{neutral}}(t)$ by a quadratic function of time [which for the H^-–He system is

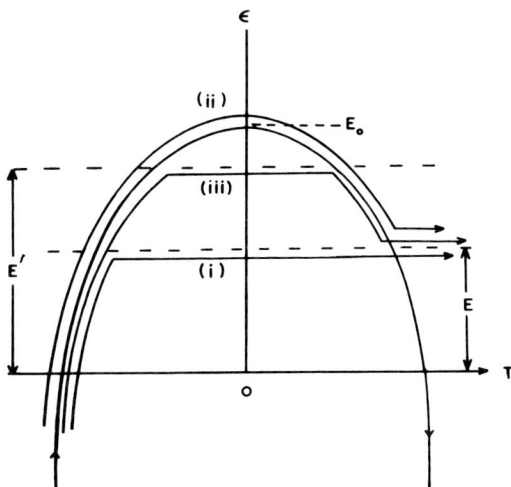

FIG. 9. Schematic representation of various mechanisms that may lead to detachment. The ordinate represents the energy available to a detaching electron as a function of the collision time T. [From Taylor and Delos (1982).]

fitted to the calculation of Olson and Liu (1980a)], they show that these equations can be solved, and proceed to derive a rather complicated formula for the survival probability. Their results can be interpreted using Fig. 9. Transitions from the bound state to the free state with electron energy E occur primarily at the crossing between these states, and can occur either on the incoming or the outgoing part of the trajectory. These two contributions add coherently, leading to a possible interference pattern in the kinetic energy spectrum of the detached electrons. In addition, transitions into the final state (of energy E) can also occur through detachment to an intermediate state (of energy E' in Fig. 9) on the incoming part of the trajectory, then reattachment into the bound state, followed by detachment into the final state on the outgoing part of the trajectory.

Results of their calculation of the total detachment cross section for $H^-(D^-)$ on He are also shown in Fig. 2. Good agreement with experimental results is obtained. It is also interesting that the calculations of Gauyacq and of Taylor and Delos agree quite well with each other, even though they are based on what appear to be very different assumptions about the dynamics of these processes.

Herzenberg and Ojha (1979) have investigated the H^- + He problem at slightly higher energies using an approach closely related to the zero-range approximation already discussed. They argue that detachment occurs as the particles approach, when the binding energy of the extra electron becomes sufficiently small so that the Born–Oppenheimer approximation is no longer valid and the electron "forgets" that it started off in a bound state of H$^-$. This "forgetting-point" R_f is defined as that value of R for which

$$\sum_{i \neq 0} |a_i[R_f(t)]|^2 \simeq 0.5 \qquad (15)$$

where the a_i are the expansion coefficients for the electron wave function:

$$\Psi(r, t) = \sum_i a_i[R(t)]\Psi_i[r, R(t)] \exp(-i/h) \int_{-\infty}^{t} E_i[R(\tau)]\, d\tau \qquad (16)$$

which satisfy the boundary conditions $a_0(-\infty) = 1$ and $a_i(-\infty) = 0$, with $i \neq 0$. In Eq. (16), Ψ_i and E_i are eigenfunctions and eigenvalues of the electronic Hamiltonian at fixed R. As in the results of Gauyacq (1980a) discussed earlier, there is substantial detachment due to "undercrossing transitions," since R_f is found to be considerably larger than the crossing radius. For a given impact parameter, R_f increases with increasing energy because the B–O breakdown occurs earlier at higher energies. Thus this model predicts that $\sigma_{-10}(E)$ will rise for $E \gtrsim 200$ eV, as is found to be the case by Risley and Geballe (1974).

For collision energies which exceed several keV, the projectile (H$^-$) velocity is comparable to that of the electron, and the adiabatic picture of the collision is not appropriate. For $E \gtrsim 5$ keV, $\sigma_{-10}(E)$ has been observed to decrease with increasing energy [see data compilation in Risley (1980), for example] and this energy range has been treated theoretically in an impulse approximation in which the target scatters the essentially free electron associated with the H$^-$ projectile [Lopantseva and Firsov (1966); Dewangan and Walters (1978)]. This high-energy theory also predicts a decreasing detachment cross section and is in reasonable agreement with the experimental observations.

Collision mechanisms other than direct detachment begin to become important for energies larger than several hundred electron volts. These mechanisms have been studied in some detail by Esaulov et al. (1978) over the energy range 80–2000 eV and by Risley et al. (1978) for the range 1–6 keV. The former experiments measured the doubly differential cross sections for the neutral hydrogen atoms that resulted from detachment. These measurements were able to separate direct detachment [Eq. (2a)] from the detachment–excitation channels [Eqs. (2b)–(2e)]. An example of an energy-loss spectrum for the H coming from collisions of H$^-$ with He is given in Fig. 10. It is clear that there is considerable production of excited-state hydrogen at this particular value of the scattering angle, as indicated by the relative size of the peak labeled "B" in the figure. The peak labeled "A" represents direct detachment, in that no excited states of the reactants are involved. In prin-

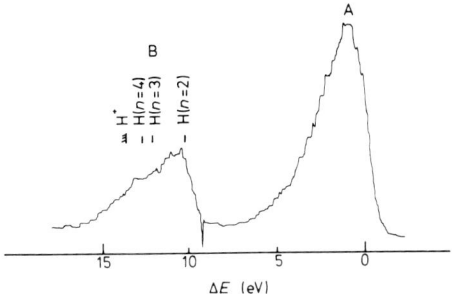

FIG. 10. Energy-loss spectrum of neutral hydrogen atoms scattered in H⁻–He collisions at $E = 730$ eV, $\theta_{lab} = 0.6°$. The zero of the energy-loss scale refers to a hydrogen atom having lost 0.75 eV, its electron affinity. [From Esaulov et al. (1978).]

ciple, the shape of the A-spectrum is a reflection of the energy spectrum of the detached electrons. There are, however, broadening effects, since the ejected electron carries momentum and can be ejected in any direction (roughly isotropic for s-wave), thus giving a range of H-atom velocities for a given electron energy. The problem of recovering the electron energy spectra from the neutral velocity spectra is discussed by Esaulov et al. (1978). We return to the subject of the energy spectra of the detached electrons later.

Risley et al. (1978) have measured the absolute cross sections for the production of excited H atoms in the detachment process by observing the intensities of the various Lyman lines (forbidden transitions are quenched by external fields) from the detachment products. A summary of their findings for H(np) production is shown in Fig. 11 for $E = 5$ keV. It is interesting to

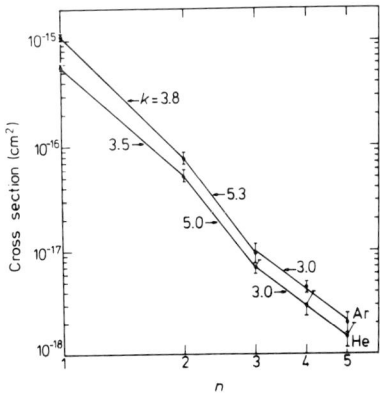

FIG. 11. Electron-detachment cross sections to the $n p_{1/2,3/2}$ states versus principal quantum number for 5-keV H⁻ on He and Ar [$\sigma(n) \propto n^{-k}$]. For $n = 4$ and 5 the cross sections include contributions from the $4d_{3/2}$ and $5d_{3/2}$ states. [From Risley et al. (1978).]

note that the H(np) production accounts for about 11% of the total detachment cross section at 5 keV. The probabilities of H(2s) production are considerably smaller than those for H(2p) formation.

The cross sections for $\sigma_{-11}(E)$ for the H$^-$–rare gas systems have been measured from about 200 to 500 keV and the results were summarized by Risley (1980), Tawara (1978), and Tawara and Russek (1973). For the He target, the fraction $\sigma_{-10}(E)/\sigma_{-11}(E)$ attains a broad maximum of about 10% for 5 keV $\lesssim E \lesssim$ 20 keV and decreases slowly for larger collision energies. The ratio drops rapidly for $E \lesssim$ 5 keV, falling to approximately 0.5% at 200 eV. Thus, for collision energies of 5 keV, processes other than direct detachment account for 20–25% of the total collisional detachment cross section.

The energy spectra of the detached electrons resulting from direct detachment in collisions of H$^-$ with rare gases have been studied experimentally by Geballe and Risley (1973), and such spectra have been determined indirectly from the H velocity spectra in time-of-flight experiments by Esaulov *et al.* (1978). In both cases there are broadening effects which cause some difficulties in the determination of the energy spectrum of the detached electrons with respect to the motion of the parent negative ion. Nevertheless, there are several salient features which have been deduced from the experiments: (*i*) the maximum in the detachment spectrum occurs for electron energies less than 1 eV and the distribution function is asymmetric, possessing a high-energy tail; (*ii*) the maximum and width of the distribution increase as the collision energy is increased; and (*iii*) the spectra depend only slightly on scattering angle. These features are all essentially reproduced by the previously discussed calculations of Herzenberg and Ojha (1979) and Gauyacq (1980a).

If one looks beyond the low-energy maximum in the spectra of detached electrons (i.e., at electrons with higher kinetic energies), discrete lines are seen to be superimposed on the continuum background of electrons from direct detachment. The electrons which comprise the rather sharp lines in the spectra come from autodetaching levels of the hydrogen negative ion [Eq. (2b)] and they have been the subject of a detailed study by Risley *et al.* (1974). The general subject of collisionally produced autodetachment is discussed by Edwards (1976). In addition to excitation of the projectile to its autodetaching levels, the electron energy spectra indicate that autodetachment from the negative ion resonances of the target which are populated by charge transfer [Eq. (2d)] may also be important in the autodetachment mechanism. An energy spectrum of the higher-energy detached electrons for the case of 3 keV H$^-$ incident on Ar is shown in Fig. 12.

The total cross section for H$^-$ + He collisions for excitation of H$^-$ to the 1s^2 autodetaching state is found to be rather flat for 0.2 $\lesssim E \lesssim$ 10 keV at about 0.03 Å^2, whereas excitation to the (2p^2) + (2s2p) levels decreases

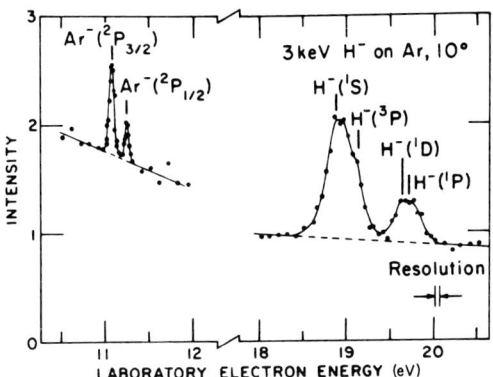

FIG. 12. Electron energy spectrum from 3-keV H⁻ on Ar observed at 10°. [From Risley et al. (1974).]

from 0.04 to about 0.01 Å² over the same range (Risley, 1973). Thus, for a collision energy of 5 keV, detachment via excitation to autodetaching states of H⁻ constitutes about 1% of the total detachment cross section.

B. $H^- - H$

Experimental results for this, the most elementary negative ion–atom system are somewhat scarce, especially at low collision energies. This is due in part to the experimental difficulty of producing hydrogen atoms from H_2 with a well-known and high dissociation fraction. The intermolecular potential curves for the ground electronic states of H_2 and H_2^- are shown in Fig. 13 and it is obvious that there are discrepancies among the several calculations for H_2^- (Ostrovskii, 1971; Bardsley and Cohen, 1978; Bardsley and Wadehra, 1979; Bardsley et al., 1966).

Hummer et al. (1960) measured $\sigma_{-10}(E)$ for $H^- + H$ for $0.5 < E < 40$ keV and found that the cross section decreased by about a factor of four over this energy range. Recent measurements by Geddes et al. (1980) have extended the $\sigma_{-10}(E)$ measurements to higher energies and found that this trend continues to the highest energies sampled ($E_{lab} \simeq 300$ keV). In addition to the detachment cross section, the charge transfer and ionization cross sections have been measured for the higher energies and a summary of the results is seen in Fig. 14. The results of calculations using several techniques for computing $\sigma_{-10}(E)$ are also shown in the figure. As the energy is increased, the perturbed stationary state method (Bardsley, 1967), the impulse approximation (Bates and Walker, 1967), and the Born approximation (Bell et al., 1978) are successively employed; the results are in fair agreement with the experimental observations.

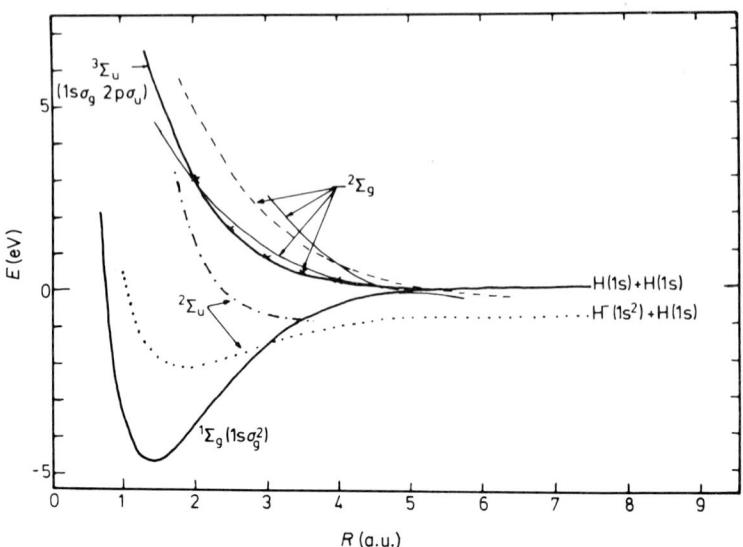

FIG. 13. Lowest molecular states of H_2 and H_2^-. H_2, $^1\Sigma_g$, and $^3\Sigma_u$ according to Kolos and Wolniewicz (1965); H_2^-, the two $^2\Sigma_g$ states in Ostrovskii's (1971) calculation correspond to the two pairs of poles of the S matrix of the H_2^- problem. $^2\Sigma_g$: (—) Ostrovskii (1971); (--) Bardsley and Cohen (1978); ($\times$) Bardsley and Wadehra (1979). $^2\Sigma_u$: ($\cdots$) Bardsley and Wadehra (1979); (-·-) Ostrovskii (1971). [From Esaulov (1980).]

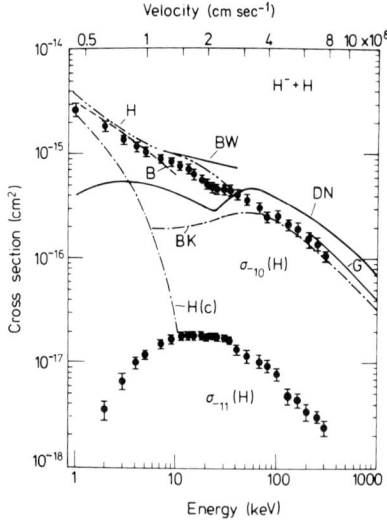

FIG. 14. Cross sections $\sigma_{-10}(H)$ and $\sigma_{-11}(H)$ for one- and two-electron loss by H^- in H. Experimental $\sigma_{-10}(H)$: (●) Geddes *et al.* (1980); Curve H(c), charge-transfer process. Hummer *et al.* (1960); curve H, charge-transfer plus electron production, Hummer *et al.* (1960). Theoretical $\sigma_{-10}(H)$: curve B, Bardsley (1967); curve BW, Bates and Walker (1967); curve DN, Dmitriev and Nikolaev (1963); curve BK, Bell *et al.* (1978); curve G, Gillespie (1977). (●) $\sigma_{-11}(H)$. [From Geddes *et al.* (1980).]

As E is increased, the detachment cross section is observed to decrease, which is in contrast to the previously discussed results for $H^- + He$. This observation gave credence to a collision model in which detachment could be described by a local complex potential. The physical differences between the $H^- + H$ and $H^- + He$ systems are the subject of a paper by Herzenberg and Ojha (1979). Basically, their arguments emphasize that even for internuclear separations less than the crossing radius, the extra electron in H_2^- cannot easily escape because of barriers associated with the $^2\Sigma_u$ shape resonance and the $^2\Sigma_g$ Feshbach resonance of H_2^-. Consequently, the electron wave function can remain somewhat localized at the molecule (in contrast to HeH^-) with decay rates $\Gamma_u(R)$ and $\Gamma_g(R)$ governed by a barrier penetration that depends only on the internuclear separation R.

Esaulov (1980) has measured the differential cross sections for the $D^- + H$ and $H^- + D$ systems for $100 \lesssim E \lesssim 1000$ eV, and examples for the charge transfer, detachment, and detachment with excitation differential cross sections are seen in Fig. 15 for a laboratory collision energy of 310 eV. The charge transfer differential cross section (which is dominant at this energy) is forward-collimated and detachment with excitation is seen to be relatively small; it is estimated to be 5% of σ_{-10} (310 eV). By integrating these differential cross-section measurements for several collision energies, it is observed that $\sigma_{-10}(E)$ may *not* increase as the projectile energy is lowered below 1 keV. This point, however, is not clear and certainly warrants further experimental investigation.

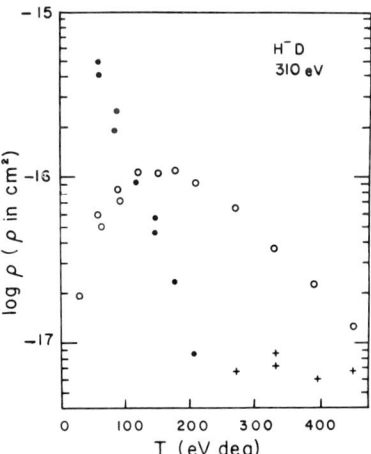

FIG. 15. Reduced differential cross section as a function of reduced scattering angle $T = E\theta$, for H^- on D at a lab energy of 310 eV: ●, charge transfer; ○, direct detachment; +, detachment with excitation. [From Esaulov (1980).]

In a series of papers, Mizuno and Chen (1969, 1971) have treated the $H^- + H$ system for low collision energies by using a semiclassical version of a complex potential theory. They use the stationary phase approximation to arrive at an expression for the deflection function $\theta(b)$ and the survival probability, $P_s(b, E)$. In contrast to the "classical" complex potential model [Eqs. (9) and (10)], the semiclassical functions each depend on both the real and imaginary portions of the complex potential. Their calculated differential cross sections are rich in structure at low collision energies due to rainbows and gerade–ungerade interference effects. There are no experiments with which to compare these low-energy results.

C. Halogen Negative Ions–Rare Gases

For collisional detachment of halogen negative ions by rare gas atoms, there are several investigations in which detailed measurements of $\sigma_{-10}(E)$ have been made for relative energies extending down to the threshold for detachment. These systems have received considerable experimental attention for low collision energies partly because of the relative ease with which well-defined and intense low-energy negative halogen-ion beams may be formed. Examples of $\sigma_{-10}(E)$ for several halogen ion–rare gas systems are given in Fig. 16, and a closer examination of $\sigma_{-10}(E)$ for the near-threshold region is given in Fig. 17. It is clear that the threshold for detachment is in the range 7–8 eV, or about twice the electron affinity of the halogen ion. Calculations of the intermolecular potentials for several of these systems have been made by Olson and Liu (1978, 1979), and the results for $ArCl^-$ and $ArCl$ are given in Fig. 18. The behavior of $\sigma_{-10}(E)$, as seen in Figs. 16 and 17, may be aptly described in terms of the earlier discussion presented for HeH^-: The detachment probability is very small for $b > b_x$ and the threshold for detachment is $E_{th} \simeq V(R_x)$, as defined earlier. For the detachment of H^- by He, it was shown in the zero-radius approximation that "undercrossing" transitions (i.e., detachment that occurs for $b > b_x$ which was attributed to dynamic coupling) contributed significantly to $\sigma_{-10}(E)$. For $E \simeq E_{th}$, the collision velocities of the halogen–rare gas systems are, however, smaller than the equivalent threshold velocities for $H^- + He$ collisions, and the undercrossing transitions predicted for $H^- + He$ may not be important in the detachment of the "p"-electron halogen negative ions by rare gases. This point has not yet been demonstrated, however.

If the effects of experimental broadening (which are not very large) are removed from the experimental results of Fig. 17, it is observed that the total detachment cross section increases in a manner that is approximately quadratic in the excess energy, i.e.,

$$\sigma_{-10}(E) \sim (E - E_{th})^2$$

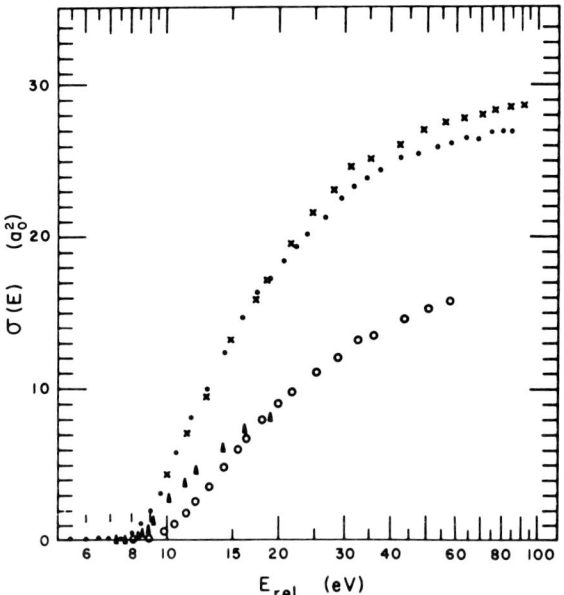

FIG. 16. Total detachment cross section for Cl^- + rare gas targets. The triangles are for He; open circles, Ne; solid dots, Ar; crosses, Kr. [From Smith et al. (1978).]

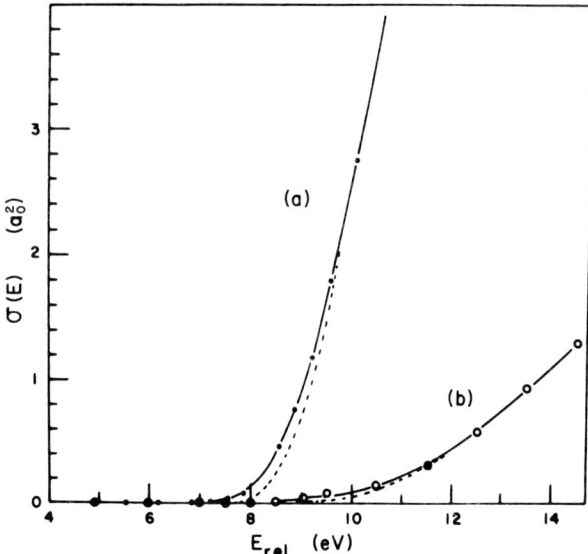

FIG. 17. Total detachment cross section in the near-threshold region for (a) Br^- + Ar and (b) Br^- + Ne. The dashed lines are a result of a numerical deconvolution, which should account for the apparatus broadening in the experiment. [From Smith et al. (1978).]

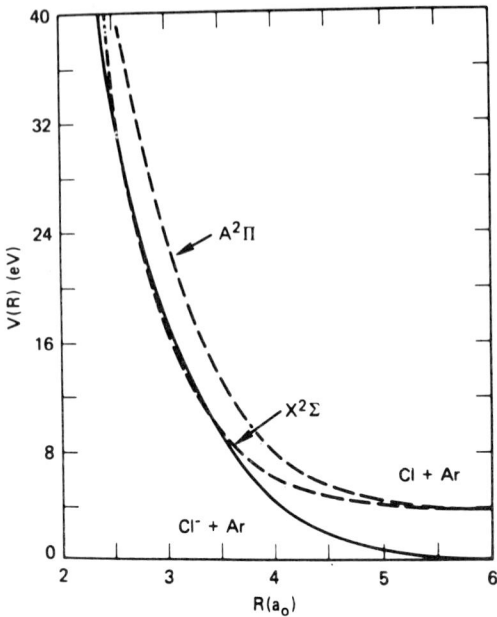

FIG. 18. Intermolecular potentials for the ground electronic states of ArCl$^-$ and ArCl. [From Olson and Liu (1978).]

This behavior has been found to provide a reasonable fit to the near-threshold behavior of $\sigma_{-10}(E)$ by several investigators (Bydin and Dukel'skii, 1957; Wynn et al., 1970; Smith et al., 1978; Haywood et al., 1981a) and serves as a useful definition of the threshold energy for collisional detachment. For the halogen ion–rare gas systems, the thresholds (as defined above) have been observed to be between 7.1 eV for Cl$^-$ + He and 8.6 eV for Br$^-$ + Ne.

Two halogen ion–rare gas systems have been found to behave in an anomalous fashion, however. The I$^-$ + Ne system exhibits a very small detachment cross section with $E_{\text{th}} \simeq 3.1$ eV (Bydin and Dukel'skii, 1957; de Vreugd et al., 1979b; Haywood et al., 1981a), suggesting that the NeI$^-$ state does not cross into the ground electronic state of NeI. This is in contrast to the examples of ArCl$^-$ given in Fig. 18 as well as to other halogen ion–rare gas systems studied thus far. For the Br$^-$ + He system it was found (Smith et al., 1978) that $\sigma_{-10}(E)$ became nonzero for $E \simeq 3.5$ eV and remained constant until the collision energy was increased beyond the predominant threshold of 7.4 eV. Calculations for the HeBr$^-$ and HeBr intermolecular potentials by Olson and Liu (1979) gave no indication as to why $\sigma_{-10}(E)$ should behave as observed for Br$^-$ + He.

Elastic differential cross sections for Cl$^-$–rare gas systems have been

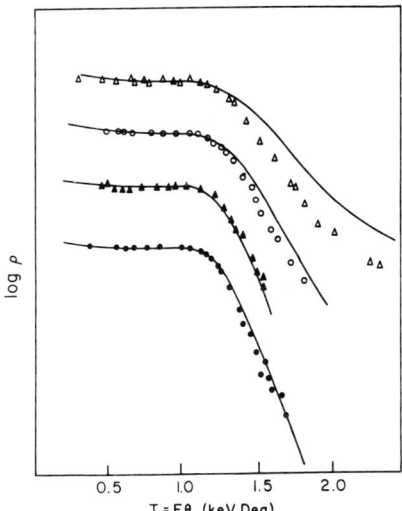

FIG. 19. Differential elastic scattering cross section for Cl⁻ + Ar at various relative collision energies (eV): △, 267; ○, 107; ▲, 53; ●, 40. The lowest energy result is from Champion and Doverspike (1976a) and the three highest energy results are from Fayeton et al. (1978). The solid line is a calculation based on a local complex potential model. It is clear that the calculation fails to reproduce the experimental observations for $E \gtrsim 100$ eV.

measured, and some typical results for Cl⁻ + Ar are illustrated in Fig. 19. The onset of detachment is clearly evident as the sudden decrease in the reduced differential cross section $\rho(\theta)$ for reduced scattering angles, $\tau \gtrsim 1200$ eV deg. Both the elastic differential and the total detachment cross sections can be well fitted by calculations based on a local complex potential model (Champion and Doverspike, 1976a; Smith et al., 1978) for $E \lesssim 100$ eV, but this model fails for collision energies $E > 100$ eV (Fayeton et al., 1978) as is clearly evident in Fig. 19.

The total detachment cross section has been measured for higher collision energies (up to several keV) by Hasted (1952, 1954), Bydin and Dukel'skii (1957), and Dukel'skii and Zandberg (1951). The cross sections are all observed to rise monotonically; this again indicates that a velocity-independent width alone cannot account fully for the electron detachment cross section.

De Vreugd et al. (1981a) have measured both the ionic and neutral differential cross sections for the halogen negative ion–rare gas systems over the laboratory energy range 500–3000 eV. These differential measurements also indicate that $P_d(b, E)$ may increase with increasing energy, and these authors suggest that rotational coupling of the discrete negative ion state to the continuum might account for this effect.

In addition to the elastic differential measurements depicted in Fig. 19, Fayeton et al. (1978) have measured the differential cross sections for direct detachment and detachment with excitation [Eqs. (2b)–(2d)] by integrating over energy-loss spectra such as in Fig. 20 for Cl^- + Ar. In this figure the peaks labeled A, B, and C correspond to elastic, direct detachment, and detachment with excitation, respectively. As seen from the positioning of the excited states of the reactants on the figure, the data cannot establish unambiguously whether the excitation energy resides in the halogen atom, its negative ion, or the target. For reactants with substantially different $Z(Cl^-$ + Ne; Cl^- + Xe), the assignment is less uncertain and it appears that the collision partner with the larger Z is excited. The three basic processes (elastic, direct detachment, and detachment–excitation) are, however, clearly resolved and the respective differential cross sections can easily be separated. The results for a collision energy of 2 keV are seen in Fig. 21. The curve labeled "b" in this figure corresponds to direct detachment, whereas curve "c" is the differential cross section for detachment with excitation. The rapid decrease in $\rho(\tau)$ for $\tau \gtrsim 2.5$ keV degree indicates that the direct detachment probability $P_d(b, E)$ may decrease for decreasing b, a feature common to rotational coupling [see Wijnaendts van Resandt et al. (1977), for example]. The suggestion of de Vreugd et al. (1981a) that rotational coupling is important in direct detachment is partially motivated by this observation.

It is clear from the differential cross section shown in Fig. 21 that the detachment with excitation cross section represents an appreciable fraction

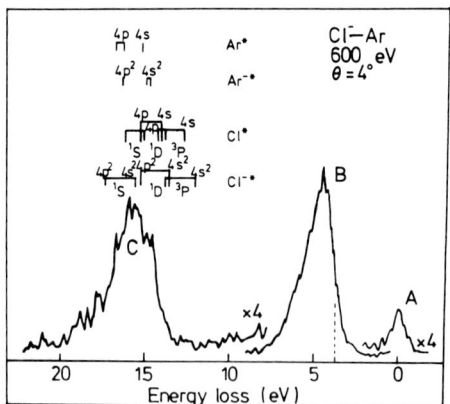

FIG. 20. Time-of-flight spectrum for collisions of Cl^- with Ar for $E = 600$ eV and $\theta = 4°$. Peak A is elastic scattering; B and C correspond to direct detachment and detachment excitation. The dotted line is the energy loss that a Cl atom would have (3.6 eV) if the detached electron were ejected with no kinetic energy. [From Fayeton et al. (1978).]

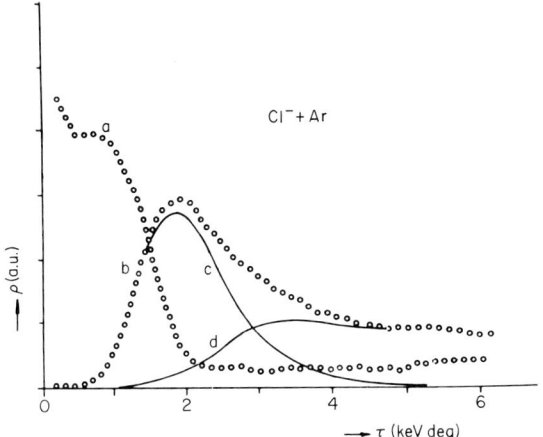

FIG. 21. Differential cross sections for Cl⁻ + Ar, 2000 eV: (a) elastic; (b) detachment; (c) direct detachment; (d) detachment with excitation. The delineation between (c) and (d) is from Fayeton et al. (1978), whereas the overall results (the open circles) are from de Vreugd et al. (1981a).

of the total detachment cross section. Neither the absolute cross section for "c" nor the fraction $\sigma^b_{-10}(E)/\sigma^c_{-10}(E)$ has been ascertained, however.

The energy spectra of the detached electrons which are ejected at an angle of 90° have been measured by de Vreugd et al. (1981b) in an experiment that uses a time-of-flight technique. The line spectra from autodetaching levels of Cl⁻ and Xe serve to calibrate the energy scale in the experiment. The results for Cl⁻ + Ar for a laboratory energy of 2 keV are seen in Fig. 22. The features of this distribution, which peaks for electron energies of about 1 eV are in basic agreement with the conclusions of Fayeton et al. (1978) in which the electron energy distribution is deduced from the velocity distribution of the neutral atoms that are a product of direct detachment (peak B of Fig. 20).

D. Halogen Negative Ions–Alkali Atoms

These systems differ from those previously discussed in that the alkali halides are highly polar molecules and form stable negative molecular ions [see Simons (1977), for example]. A (somewhat schematic) diagram for the intermolecular potential of NaCl⁻ and NaCl is given in Fig. 23. Recent differential cross-section measurements by de Vreugd et al. (1979a), in which fast neutral products of detachment were notable by their complete absence, indicate that the neutralization cross sections for halogen ion–alkali atom reactants are very small even for collision energies as high as several keV.

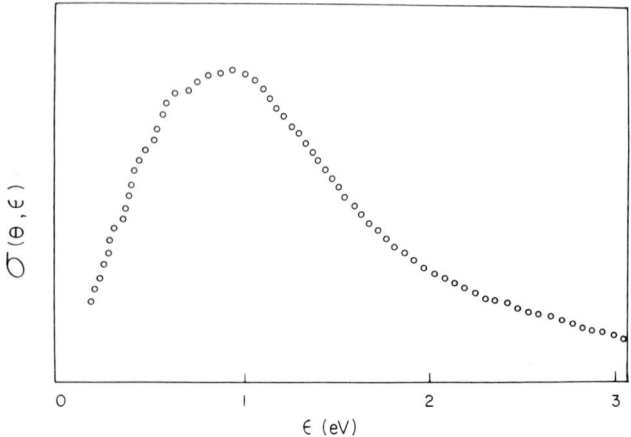

FIG. 22. Energy distribution for low-energy electrons detached in Cl⁻ + Ar collisions at 2000 eV. The electrons are observed at 90° with respect to the ion beam direction. [From de Vreugd et al. (1981b).]

The collisional dynamics for halogen ion–alkali reactants can best be illustrated by referring to the potential diagram of Fig. 23. For the example of Cl⁻ + Na, coupling between the ground electronic state of the negative molecular ion (labeled "a" in Fig. 23) and the continuum boundary (the NaCl ground state labeled "b") is not due to an obvious crossing of the ionic

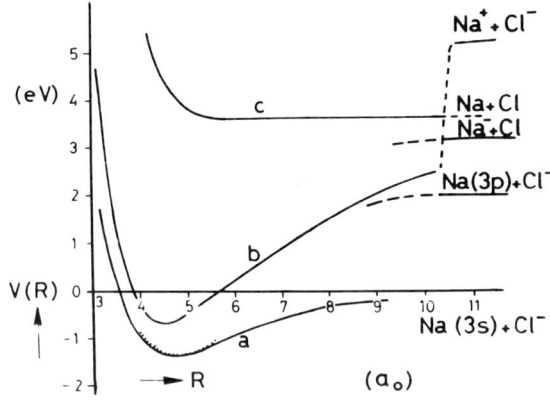

FIG. 23. Intermolecular potentials for the NaCl⁻ system. Curve (a) is the result of an inversion of differential elastic scattering data; (b) and (c) are the ionic and covalent potentials for NaCl. The dashed line around the equilibrium separation is from a calculation by Jordan and Wendoloski (1978). [From de Vreugd et al. (1979a).]

state into a continuum. In the Born–Oppenheimer approximation the ionic potential apparently remains below that for the neutral NaCl potential even for small internuclear separations. This result is compatible with the observation that the dipole moment of the alkali halide remains large as the internuclear separation is decreased. Crawford and Garrett (1977) point out that if the dipole moment of a molecule remains above 1.63 D (0.64 ea_0), then the molecule has a positive electron affinity (the dipole moment of NaCl for the equilibrium separation is 9 D). Calculations by Jordan and Wendoloski (1978) indicate that this condition is satisfied for fairly small values of the internuclear separation for alkali halides.

To the extent that some dynamic coupling of the ionic state to the neutral state exists, in e.g., collisions of Cl^- with Na, there are several product channels possible at low collision energies:

(i) $Cl^- + Na \rightarrow Cl^- + Na$

(ii) $Cl^- + Na \rightarrow Cl + Na + e$

(iii) $Cl^- + Na \rightarrow Cl^- + Na^+ + e$

The first channel is simply elastic scattering, which would produce rainbow and associated oscillations in the differential cross section. In the event that detachment may occur, the NaCl systems will, upon separating, exit by either channel (ii) or (iii) due to the ionic–covalent coupling at large separations. For collision energies greater than several electron volts, the probability of a transition from the outgoing ionic state to the covalent configuration is very small (Faist and Levine, 1976). Thus, even if electron detachment does occur, channel (iii) should dominate over (ii), thereby giving a "slower" Cl^- and ion–pair formation as a consequence of collisional detachment. In the work of de Vreugd et al. (1979a), no evidence for the occurrence of (ii) was found, i.e., no fast neutral halogen atoms could be detected as collision products of F^-, Cl^-, or $Br^- +$ alkali reactants. Moreover, doubly differential cross-section measurements indicate that the cross section for (iii) is also very small. Such evidence may be seen in Fig. 24, wherein the energy-loss spectra for $Cl^- +$ Na, K are displayed. For the Na target, no ions with an endothermicity required for (iii), the ionization potential of Na, are observed for the angle illustrated or for any other angle for $\tau \lesssim 3$ keV deg. For the $Cl^- + K$ system a small number of inelastically scattered ions are observed at the relatively high τ value of 1 keV deg. There is, therefore, indirect experimental evidence that (iii) and (ii) are improbable for low to moderate collision energies implying that $\sigma_{-10}(E)$ for these reactants and energies is quite small. However, no direct measurements of the total detachment cross sections have been made and no upper limit for the cross sections has been reported.

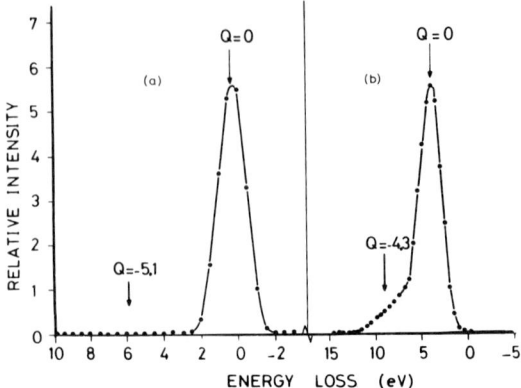

FIG. 24. Energy-loss spectra for (a) Cl$^-$ + Na; $E_{lab} = 51$ eV, $\theta_{lab} = 5°$ and for (b) Cl$^-$ + K; $E_{lab} = 71$ eV, $\theta_{lab} = 15°$. The arrows at (a) 5.95 eV and (b) 9.17 eV indicate the minimum laboratory energy loss for the ion pair formation X$^-$ + M → X$^-$ + M$^+$ + e; the energy loss Q for NaCl and KCl for this process is -5.14 and -4.34 eV, respectively. [From de Vreugd et al. (1979a).]

Since the detachment cross sections for the halogen ion–alkali systems are small, the elastic differential scattering cross section with its rainbow and associated oscillations can be inverted to find potential parameters (and hence "vertical" electron affinities) for the alkali halide anions. This has been done in the work of de Vreugd et al. (1979a), and the well depths are found to vary from 1.04 eV for KBr$^-$ to 1.92 eV for NaF$^-$.

E. H$^-$(D$^-$)–Alkali Atoms

Two mechanisms are responsible for the electron-loss cross section in low-energy collisions of H$^-$ with alkali atoms: electron detachment and charge transfer. Experiments for H$^-$(D$^-$) + Cs have been completed by Meyer (1980) for 150 eV < E < 2 keV and the results for the total electron-loss cross sections (detachment plus charge transfer) are given in Fig. 25. These experiments measure the total "fast" H or D flux from the reaction H$^-$(D$^-$) + Cs → H(D) + products, and consequently are unable to distinguish between the detachment and charge transfer channels.

Recent calculations of the intermolecular potentials for the NaH and NaH$^-$ systems have been reported by Olson and Liu (1980b) and by Karo et al. (1978), and the vertical electron affinity of the polar molecule (NaH) has been investigated by Griffing et al. (1975). The results of configuration interaction calculations by Olson and Liu (1980b) are shown in Fig. 26. As for the alkali halides, NaH$^-$ is stable and there is no apparent crossing of the ionic X$^2\Sigma$ state into the X$^1\Sigma$ continuum (representing NaH and a free elec-

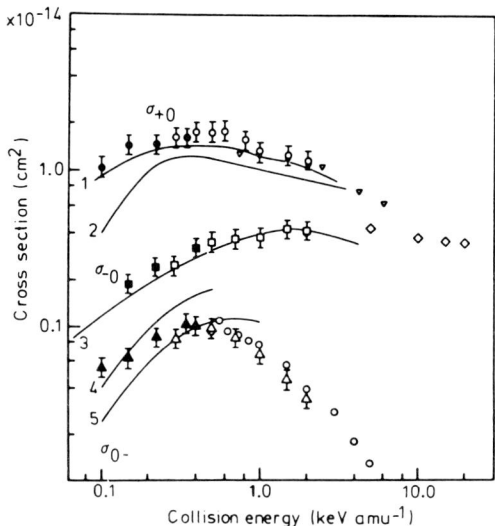

FIG. 25. Electron capture and loss cross section for $H^+(D^+)$, $H^0(D^0)$ and $H^-(D^-)$ projectiles incident on Cs versus projectile energy per atomic mass unit. Open symbols with error bars are results using H projectiles; solid symbols with error bars are for D projectiles (Meyer, 1980) ($\triangledown$) σ_{+0} experiment (Meyer, 1976); ($\diamondsuit$) σ_{-0} experiment (Leslie et al., 1971); ($\bigcirc$) σ_{0-} experiment (Nagata, 1979). Curve 1, σ_{+0} theory (Sidis and Kubach, 1978); curve 2, σ_{-0} theory (Olson et al., 1976); curve 3, σ_{-0} theory (C. Bottcher, 1980, private communication); curve 4, σ_{-0} theory (Olson and Liu, 1980b); curve 5, σ_{0-} theory (Olson, 1980). [From Meyer (1980).]

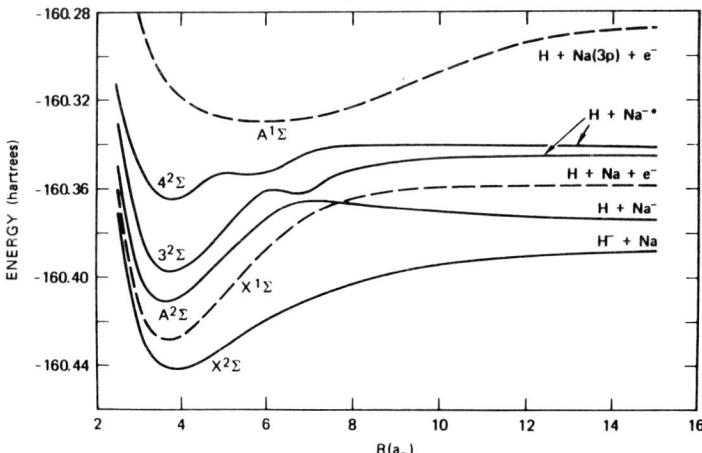

FIG. 26. Intermolecular potentials for NaH$^-$, NaH. [From Olson and Liu (1980b).]

tron) seen in Fig. 26. The calculations of Karo et al. (1978) do, however, indicate that such a crossing might occur in the region of $R \simeq 3a_0$.

The question arises as to the nature of the detachment mechanism, and one possibility is that detachment due to dynamic coupling occurs in a collision for an internuclear separation in the neighborhood of $3a_0$, where $\epsilon(R)$ is small. Alternately, Olson and Liu (1980b) suggest that the principal mechanism for neutralization and subsequent detachment may be due to charge transfer because of the long-range coupling between the $X^2\Sigma$ and $A^2\Sigma$ states of NaH^-. As the reactants approach, charge transfer can take place and electron detachment may follow due to a crossing of the $A^2\Sigma$ state into the $X^1\Sigma$ continuum at $R \simeq 8a_0$. The original charge transfer probability is significant for H^- + Na due to the rather small asymptotic separation (0.21 eV) of the two potential curves. The electron affinities of heavier alkali targets are slightly smaller and the separations (i.e., the endothermicity of charge transfer) are correspondingly larger, thus decreasing the charge transfer probability and presumably the electron-loss cross sections for heavier targets.

At collision energies above about 2 keV (where any adiabatic theory most certainly fails), there have been several measurements of either the neutralization cross section or, alternately, the "equilibrium fraction" defined for thick targets:

$$F_-^\infty = \sigma_{0-}/(\sigma_{0-} + \sigma_{-0})$$

where σ_{-0} and σ_{0-} are the neutralization and electron pick-up cross sections, respectively. A summary of these results, which are sometimes fairly discrepant, can be found in the review article by Schlachter (1977).

There is considerable activity, both experimental and theoretical, in the field of $H^-(D^-)$–alkali collisions. Definite measurements and calculations for electron detachment and charge transfer should be forthcoming in the near future.

F. O^-–Rare Gases

The interaction of O^- with the rare gases is complicated somewhat by the fact that the ground state of the negative ion is ^2P. Consequently there are two electronic states ($X^2\Sigma$, $A^2\Pi$) that govern the trajectories and hence the dynamics of collisional detachment. Potential energy diagrams for ArO^- and ArO, which should be similar to those for other rare gas targets, are given in Fig. 27. It is clear from this diagram that detachment may lead to neutral oxygen atoms in both the ^1D and ^3P levels. Esaulov et al. (1980) have measured the energy-loss spectra and elastic and inelastic differential cross sections for these systems for collision energies down to several hundred

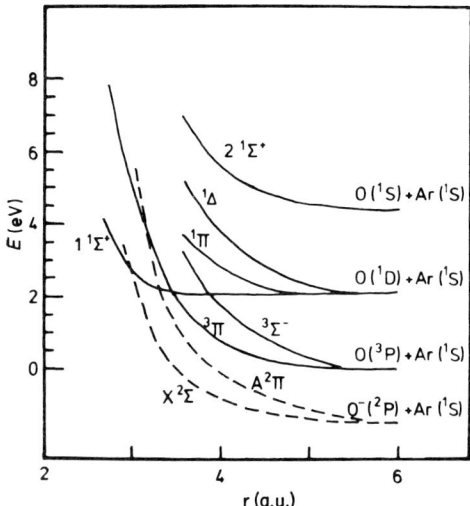

FIG. 27. Schematic diagram for ArO⁻ system. The ArO states are based on the calculations of Dunning and Hay (1977). [From Esaulov et al. (1980).]

electron volts. The lower energy results for the Ne and Ar targets are shown in Fig. 28. For detachment by Ar, the metastable ^{1}D state is seen to dominate the small-angle forward scattering, and Esaulov et al. suggest that this detachment mechanism could be used to provide a neutral oxygen beam consisting primarily of these metastable atoms. Conversely, it is important to note that a fast neutral oxygen beam derived from collisional detachment of negative ions does not necessarily produce a ground-state beam.

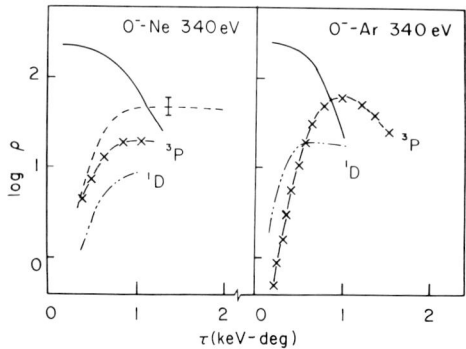

FIG. 28. Reduced differential cross sections for the O⁻ + Ne and Ar systems. The solid line is for the elastic scattering. The dashed curve is the total detachment, and the cross sections for each final state of oxygen are indicated. [From Esaulov et al. (1980).]

The difference in the threshold behavior for $O(^3P)$ and $O(^1D)$ production in the case of the Ne and Ar targets is attributed to the fact that the $^1\Sigma$ potential (Dunning and Hay, 1977) is rather flat for ArO (see Fig. 27), whereas it is more repulsive for NeO, thus causing larger angle deflections for $O(^1D)$ produced by collisional detachment from Ne.

Total detachment cross sections for some of the rare gas targets for these systems have been measured by several groups and a summary of this work is given in Fig. 29. If the representations of the ArO⁻ potentials in Fig. 27 are considered correct, then the crossing or merging of the ionic states with the $^1\Sigma$ ArO state at $R \simeq 3a_0$ suggests that the detachment cross section at several hundred electron volts should be about $28a_0^2$. This is essentially the result of measurements by Wynn et al. (1970) and Hasted (1952).

G. Detachment Rate Constants

There are several areas of physics and chemistry where the rate constants $k(T)$ for collisional detachment and the reverse process, three-body attachment, are important. Such areas include magnetohydrodynamics, ionospheric chemistry, flame chemistry, and gaseous dielectrics. The rate constants are determined from the cross section by the relation:

$$k(T) = \frac{8\pi\mu}{(2\pi\mu kT)^{1.5}} \int_0^\infty E\sigma_{-10}(E)e^{-E/kT} \, dE \qquad (17)$$

For $T \simeq 10^4$ K, $k(T)$ is extremely sensitive to the details of the detachment cross section in the region near threshold. The usual definition for $k(T)$ given above includes the implicit assumption that all degrees of freedom

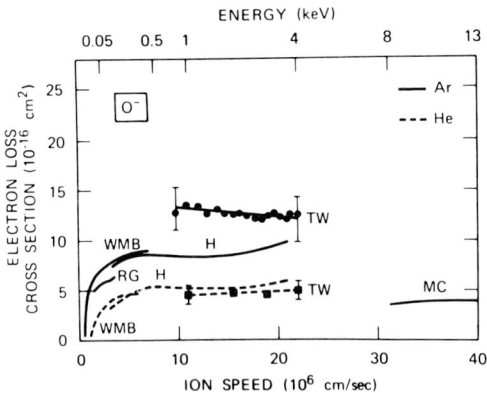

FIG. 29. $\sigma_{-10}(E)$ for O⁻ + He, Ar. TW, Bennett et al. (1975); H, Hasted (1952); WMB, Wynn et al. (1970); RG, Roche and Goodyear (1969); MC, Matić and Cobić (1971). [From Bennett et al. (1975).]

of the reactants are in thermodynamic equilibrium. For the collisional detachment of atomic negative ions by atoms, the determination of the rate constant from low-energy beam experiments in which the detachment cross sections are measured is straightforward, since there are no low-lying excited states of either reactant and the equipartition assumption is essentially correct.

The only measurements of $\sigma_{-10}(E)$ which have sufficient detail around the threshold energy so that meaningful calculations of $k(T)$ can be made are for the halogen–negative-ion rare gas systems (Smith et al., 1978; Haywood et al., 1981a). Even for these systems only an upper limit to $k(T)$ can be ascertained due to the fact that E_{th}/kT is so large for temperatures of interest (in most applications) and therefore it leads to very small detachment rate constants for these systems. This upper limit to $k(T)$ is primarily dictated by the minimum detachment cross section, which can be determined in the experimental measurement.

Experiments which use "flowing afterglow" or "flow-drift-tube" techniques (McFarland et al., 1972) can sample the energy range $0.04-\sim 3$ eV, but it appears that there are no measurements of $k(T)$ for direct detachment of atomic negative ions by atomic targets. There have been, however, numerous flow tube studies of associative detachment, ion–molecule reactions, and other processes for molecular reactants; the resulting rate constants have been compiled in an exhaustive study by Albritton (1978).

Shock tube techniques have been utilized to measure detachment rate constants for reactants involving halogen–negative ion and rare-gas targets (Mandl, 1976a,b; Mandl et al., 1970). The shock tube results for $k(T)$ are consistently several orders of magnitude higher than the rate constants determined from Eq. (17) and beam measurements of $\sigma_{-10}(E)$ (Smith et al., 1978; Champion and Doverspike, 1976b; Mandl, 1976b; Haywood, 1981a). There is not, as yet, any explanation for the rather serious disagreement between the two results.

Shui and Keck (1973) developed a phase space theory for collisional detachment and applied their technique to systems that include the halogen negative ions. In order for their calculations of $k(T)$ to be compatible with the results of shock tube measurements, it was necessary to assume that the threshold for detachment occurred at essentially the electron affinity of the halogen, which is contrary to the experimental observations.

IV. MOLECULAR REACTANTS

The collisional detachment of negative ions for systems that include a molecular reactant (or reactants) has not been as extensively investigated as has detachment for atomic reactants. Nevertheless, there are several cases where detailed experiments have been reported for molecular targets. There are essentially no theoretical studies of collisional detachment for molecular systems due to the complexity of such systems and the complete absence of adequate potential surface calculations. (Not even the $H^- + H_2$ potential surface is well known.) Hence, the discussion which follows is primarily a presentation of a few selected experimental observations without detailed discussions of the dynamics of collisional detachment.

A. Atomic Negative Ions–Molecular Targets

The most extensively studied ion–molecule systems involve the negative ion H^-. Several measurements of $\sigma_{-10}(E)$ have been reported for various targets and the results published prior to 1974 have been reviewed in some detail by Risley and Geballe (1974) and will not be repeated here. Examples of the total cross sections and the ratios $\sigma_{-11}(E)/\sigma_{-10}(E)$ are given in Figs. 30 and 31 for the energy range 200 eV–10 keV. It is interesting to note that the two-electron process $[\sigma_{-11}(E)]$ remains very small for $H^- + H_2$ even at 10 keV. Recent measurements of $\sigma_{-10}(E)$ for higher energies have been reported for the H_2 and N_2 targets by Heinemeier et al. (1976). Recent cross-section measurements for lower energies (2–100 eV) have been reported for the N_2 target by Champion et al. (1976).

The effect of substituting D^- for H^- in the $\sigma_{-10}(E)$ measurements is also seen in Fig. 30. When the cross sections (for H^- and $D^- + X$) are plotted as a function of relative velocity, they appear to be indistinguishable. On the other hand, there is an "isotope effect" observed at lower ($E \lesssim 100$ eV) energies (Champion et al., 1976) for the N_2 target. For $H^-(D^-) + N_2$ the detachment cross sections are not the same functions of the relative collision energy, nor are they the same functions of the relative collision velocity. [It is useful to recall that the relation between E_{rel} and V_{rel} depends on the masses of the isotopes. There are then three choices of variables with which one can display data for different isotopes: E_{lab}, E_{rel}, and V_{rel}.] The current theoretical understanding of collisional detachment by molecular targets is inadequate to predict the effects of isotopic substitution on the detachment cross section. We return later to a further example of isotopic substitution for the reactants $Cl^- + H_2, D_2$.

The results for $H^- + O_2$ seen in Fig. 30 are derived from beam attenuation experiments. The apparent increase in the detachment cross section as the collision energy is decreased may be due to charge transfer, which would also decrease the intensity of the transmitted H^- ion beam. Bailey and Mahadevan (1970) have measured the detachment cross section for $H^- + O_2$ (by detecting the detached electrons) over the energy range 5–35 eV.

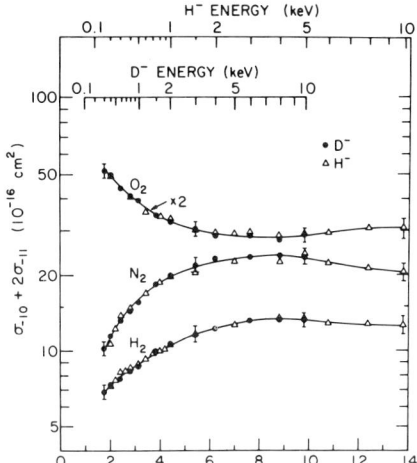

FIG. 30. Detachment cross sections ($\sigma_{-10} + 2\sigma_{-11}$) versus ion velocity in units of 10^7 cm sec^{-1} for molecular targets. The cross section for O_2 has been multiplied by two. [From Risley (1974).]

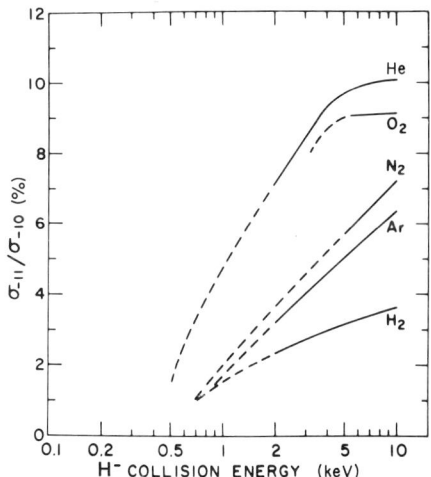

FIG. 31. $\sigma_{-11}/\sigma_{-10}$ versus H$^-$ energy. He, Ar, and H$_2$ data taken from Williams (1967) for σ_{-11} and σ_{-10}. For O_2 and N_2 data are from Fogel', Ankudinov, and Slabospitskii (1957) for σ_{-11}, and from Stier and Barnett (1956) for σ_{-10}. The dashed line is an extrapolation to lower energies based on the differential measurements of McCaughey and Bednar (1972). The uncertainty in the measured values of the ratio is less than 20%. [From Risley and Geballe (1974).]

They find that $\sigma_{-10}(E)$ does *not* decrease with increasing energy for $E \lesssim 350$ eV. This is contrary to the results of Fig. 30. The detachment cross section reported by Bailey and Mahadevan (1970) at 200 eV is σ_{-10} (200 eV) $\simeq 10$ Å^2, which is about 40% of the value reported by Risley (1974) at the same energy. Bailey and Mahadevan also measured the charge transfer cross section for H$^-$ + O$_2$ and found it to be about 10 Å^2 at 200 eV. Thus, the apparent discrepancy in the detachment cross section is resolved if one includes charge transfer in the ordinate of Fig. 30.

One of the more interesting features of collisional detachment by molecular targets has been reported by Risley (1977) for the system H$^-$ + N$_2$. N$_2$ does not form a stable negative molecular ion like O$_2$ does, but there are unstable states (or resonances) of N$_2^-$ that have been identified in electron scattering experiments (Schultz, 1973). In an experiment where the kinetic energy spectrum of electrons detached in H$^-$ + N$_2$ collisions was measured, it was demonstrated that such a state of N$_2^-$ (the $^2\Pi_g$ "shape" resonance) is involved in the dynamics of detachment via charge transfer to the temporarily bound N$_2^-$, followed by decay of the negative molecular ion:

$$H^- + N_2 \rightarrow H + N_2^{-*} \rightarrow H + N_2 + e$$

The kinetic energy distribution of the detached electrons shows regular oscillations associated with N$_2^-[^2\Pi_g(v')] \rightarrow$ N$_2[^1\Sigma_g(v)]$ transitions as may be seen in Fig. 32. The oscillatory structure appears to be superimposed upon a rather broad distribution typical of direct detachment by atomic targets, suggesting that detachment via charge transfer may be minor when compared to direct detachment. It is, however, difficult to delineate the two in the spectrum due to the resolution inherent in the experiment. Aside from the relative importance of unstable molecular negative ion states to the detachment process, such charge transfer mechanisms are necessarily included in any experimental measurements of $\sigma_{-10}(E)$ since the lifetimes of these resonances are quite small, typically $\simeq 10^{-14}$ sec.

For the Cl$^-$ + N$_2$ system, Annis et al. (1980) have reported the results of experiments in which the time-of-flight spectra for product chlorine atoms have been measured. These experiments give further evidence that resonances of N$_2^-$ [the $^2\Pi_g$ and a'—see Schultz (1973) for a discussion of molecular negative ion resonances] are very important in the detachment mechanism. Experimental results for a Cl$^-$ energy of 200 eV are exhibited in Fig. 33 for several scattering angles. The two groups of chlorine atoms observed in the time-of-flight spectra have (most probable) energy losses Q that are compatible with:

$$Cl^- + N_2 \rightarrow Cl + N_2^-(^2\Pi_g) \quad \text{or} \quad Cl + N_2^-(a')$$

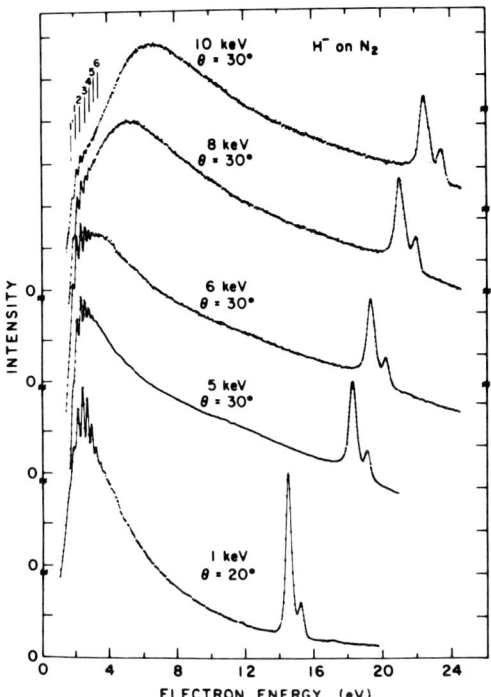

FIG. 32. Energy distribution of secondary electrons produced in collisions of H^- with N_2 observed at 150°. H^- lines are shifted in the laboratory reference frame because of reaction kinematics. [From Risley (1977).]

The bottom panel of Fig. 33 also contains a spectrum for the argon target for which there is only direct detachment, i.e., no appreciable target or projectile excitation is observed. For the negative ion–molecule experiments, one cannot ascertain whether (1) the detached electron carries away the excess energy or (2) the target, N_2, is left rather highly excited as a result of the detachment. The structure of the spectra and the angular dependence of the amplitudes of the two peaks as seen in Fig. 33 strongly suggest that direct detachment (at least as understood in ion–atom collisions) is certainly a minor contributor to detachment for these reactants. This observation differs from the suggestion of Risley (1977) for the $H^- + N_2$ system that charge transfer to the N_2^- resonance plays a minor role in the detachment for that system. A complete answer to this question of the importance of the resonances in negative ion–molecule detachment can be obtained only with detailed measurements of the energy spectra of the detached electrons (preferably in coincidence with neutral products of a given energy loss).

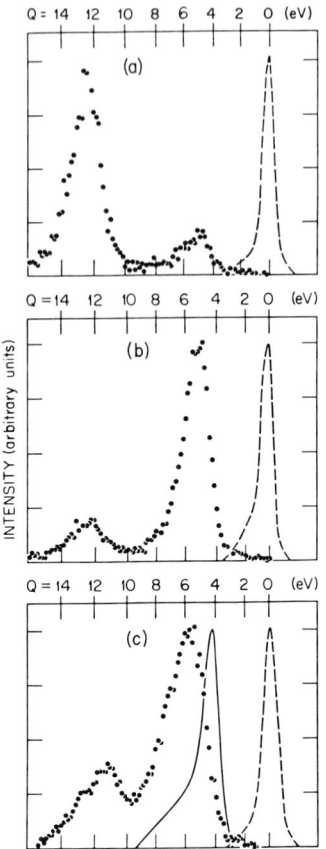

FIG. 33. Time-of-flight energy-loss spectra of Cl^0 formed by detachment in Cl^- (E_{lab} = 198 eV) + N_2 collisions. Also shown are spectra for elastic scattering from Ar (dashed lines) and collisional detachment from Ar (solid lines): (a) θ_{LAB} = 1.6°, E_{LAB} = 198 eV; (b) θ_{LAB} = 3°, E_{LAB} = 198 eV; (c) θ_{LAB} = 4.5°, E_{LAB} = 198 eV. [From Annis et al. (1980).]

For collision energies near the threshold for detachment, the negative ion–molecule systems most extensively studied involve halogen negative ions in collisions with various molecular targets (Doverspike et al., 1980). The results for $\sigma_{-10}(E)$, for $E_{rel} \gtrsim 100$ eV, for Br^- are shown in Fig. 34. It is interesting to note from the figure that the apparent threshold for detachment is 7–8 eV (except for O_2), which is essentially identical with the observation for the rare gas targets. It is rather surprising that these detachment thresholds for halogen negative ions and molecular targets are consistently found to be several electron volts larger than the electron affinity of the

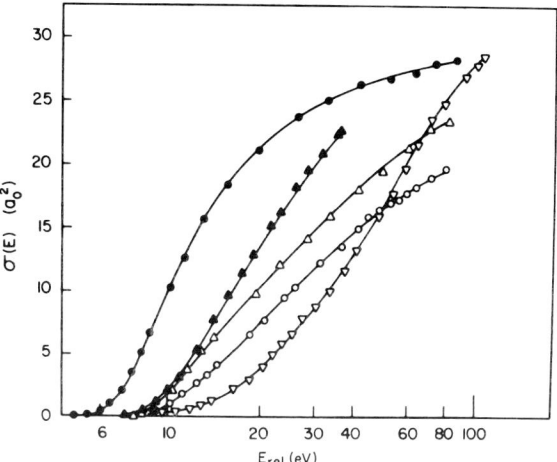

FIG. 34. Absolute detachment cross sections for the reactants indicated: ●, $Br^- + O_2$; ▲, $Br^- + CH_4$; △, $Br^- + N_2$; ○, $Br^- + CO$; ▽, $Br^- + CO_2$. [From Doverspike et al (1980).]

negative ion. The total detachment cross sections vary smoothly with energy and offer no evidence (supportive or otherwise) concerning the role of molecular negative ion resonances in the detachment process.

Near-threshold measurements for $\sigma_{-10}(E)$ for the $O^- + O_2$ system have been reported by Roche and Goodyear (1969) and Bailey and Mahadevan (1970). These measurements indicate that the threshold for detachment is quite close to the electron affinity of oxygen. The charge transfer cross section ($O^- + O_2 \rightarrow O_2^- + O$) is also found to be large with a pronounced peak for $E_{rel} \simeq 6$ eV. The collisional detachment cross sections for O^- and several molecular targets have been measured and a summary for O_2 and N_2 targets is given in the work of Bennett et al. (1975).

The low-energy total detachment cross sections for $Cl^- + H_2$ and D_2 are shown in Fig. 35 as a function of the relative collision energy. There is a substantial isotope effect observed which is typical of theoretical predictions involving dynamic coupling of the discrete state to the continuum and contrary to a complex potential description. On the other hand, it may be that the isotope effect arises because of subtle differences between the potential surfaces for the two systems, but this is certainly not apparent at this time. Cheung and Datz (1980) have measured the time-of-flight spectra for neutral chlorine atoms resulting from detachment in $Cl^- + H_2(D_2)$ collisions and, as in the case for $Cl^- + N_2$, charge transfer to the temporarily bound negative ion resonances of $H_2^-(D_2^-)$ appears to be an important

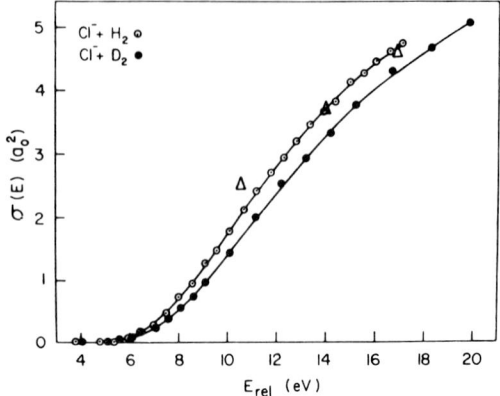

FIG. 35. Absolute total detachment cross sections for $Cl^- + H_2, D_2$. The three open triangles are for $Cl^- + H_2$ and are taken from Bydin and Dukelskii (1957). [From Doverspike et al. (1980).]

mechanism for electron detachment. Moreover, it appears that ion–molecule reactions

$$Cl^- + H_2 \to (HCl^-)^* + H \to H + Cl + e + H \quad \text{or} \quad HCl + e + H$$

may also be a source of free electrons, at least for low relative collision energies (Cheung and Datz, 1980; Herbst et al., 1982). Hence, it is obvious that a complete theoretical description of detachment in such negative ion–molecule systems will be considerably more involved than in the case of ion–atom collisions.

B. Molecular Negative Ions

For molecular negative ions, the collisional detachment of O_2^- appears to have been investigated most extensively. Low-energy experiments (collision energies less than several hundred electron volts) have been reported for the detachment of O_2^- by several target gases by Hasted and Smith (1956), Roche and Goodyear (1969), Bailey and Mahadevan (1970), and Wynn et al. (1970). It is interesting to compare the threshold behavior of $\sigma_{-10}(E)$ for $O_2^- + O_2$ with that for $O_2^- + He$. In the former case, $\sigma_{-10}(E)$ is found to exhibit a detachment threshold in the vicinity of $E_{rel} \simeq 3.5$ eV, whereas the latter reactants show an apparent threshold at $E_{rel} \simeq 0$, as may be clearly seen in Figs. 36 and 37. The electron affinity of O_2 is about 0.44 eV, and it is suggested by Wynn et al. (1970) that excited vibrational states of O_2^- in the primary ion beam are the source of the low threshold observed for the He target. As in the case for detachment of negative halogen ions by molecular targets, the high threshold observed for $O_2^- + O_2$ may be indicative of a

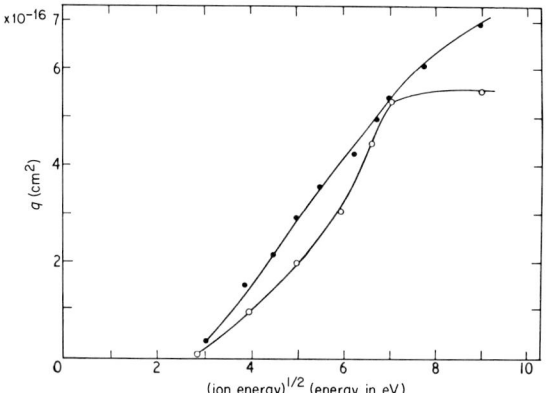

FIG. 36. Detachment cross sections for O_2^- in O_2: $O_2^- + O_2 \rightarrow O_2 + O_2 + e$; $\bigcirc$, Bailey; ●, present results. [From Roche and Goodyear (1969).]

curve (surface) crossing of the $O_2 + O_2^-$ surface with that of $O_2 + O_2$. Alternatively, the inelastic or reactive cross section for low collision energies may be completely dominated by charge transfer collisions for the case of $O_2^- + O_2$. Bailey and Mahadevan (1970) also measured this charge transfer cross section down to $E_{rel} = 2$ eV and found it to increase with decreasing energy. The charge transfer cross section was observed to be larger than the detachment cross section for all energies investigated ($E < 300$ eV) and rose to a value of 20 Å^2 at $E_{rel} \simeq 2$ eV.

For higher collision energies, detachment cross sections for O_2^- and several target gases have been measured by Bennett et al. (1975) and Hasted and Smith (1956).

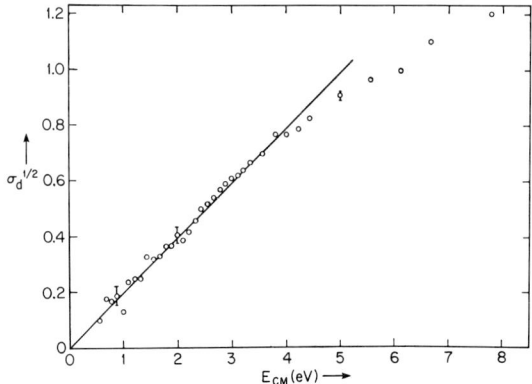

FIG. 37. $\sigma_d (10^{-16}$ cm$^2)^{1/2}$ versus relative collision energy, for O_2^-–He. [From Wynn et al. (1970).]

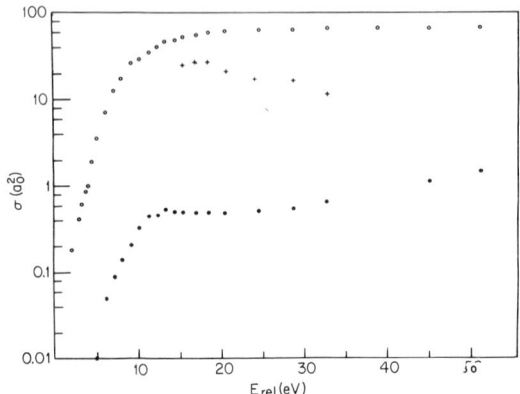

FIG. 38. Cross sections for the various decomposition channels for collisions of UF_6^- with argon as a function of the relative collision energy. The three channels are detachment (solid dots), $UF_5 + F^-$ (open circles) and $UF_5^- + F$ (crosses). [From Haywood et al. (1981b).]

The detachment of OH^- by He has also been investigated by Wynn et al. (1970) and an apparent threshold energy that was several electron volts in excess of the electron affinity of OH (~ 1.8 eV) was observed.

Other than the experiments of Bennett et al. (1975), which extend the energy range up to 3 keV, there appear to be no experimental results for the collisional detachment of molecular negative ions at high collision energies.

One last system which clearly illustrates that detachment is not necessarily the dominant inelastic process for negative ion collisions deserves mention. The collisional detachment and dissociation cross sections for collisions of UF_6^- with the rare gases have been measured by Haywood et al. (1981b) for laboratory energies less than 500 eV, and the detachment cross section was found to be quite small. For this energy range $\sigma_{-10}(E)$ was never more than several percent of the total "decomposition" cross section. The dominant decomposition channels involved collision-induced dissociation, yielding the products $UF_5 + F^-$ and $UF_5^- + F$, and the results for the argon targets are shown in Fig. 38. A statistical theory was used with some success to predict the observed branching ratios for the collisional products, including the low yield of detached electrons.

V. Summary

The various mechanisms which can lead to the collisional detachment of negative ions have been studied in some detail for energies ranging from below the threshold for detachment up to several hundred kilo-electron volts. Examples of these studies have been presented in this review. A qualitative discussion of these experimental observations for low collision energies has been based primarily on features of the intermolecular potentials for the negative ion–atom reactants. In addition, for the simpler reactants there are several theoretical approaches that have been used with some success to describe the dynamics of the discrete–continuum interaction associated with direct detachment. It has been shown that direct detachment almost always dominates the inelastic scattering in collisions of negative ions with atoms or molecules.

For cases where molecular reactants are involved in collisional detachment (either molecular negative ions or molecular targets), experimental studies are far from extensive and theoretical descriptions of detachment for such systems are essentially nonexistent. One interesting feature observed in collisional detachment by molecular targets is that the negative ion resonances of the molecular target may serve as stepping-stones to detachment.

ACKNOWLEDGMENTS

It is a pleasure for the author to express his gratitude to his colleagues L. D. Doverspike and J. B. Delos for their many interactions and contributions. The support of the Division of Chemical Sciences, Office of Basic Energy Sciences of the Department of Energy is also gratefully acknowledged.

REFERENCES

Albritton, D. L. (1978). *At. Data Nucl. Data Tables* **22**, 1.
Annis, B. K., Datz, S., Champion, R. L., and Doverspike, L. D. (1980). *Phys. Rev. Lett.* **45**, 1554.
Bailey, T. L., and Mahadevan, P. (1970). *J. Chem. Phys.* **52**, 179.
Bardsley, J. N. (1967). *Proc. Phys. Soc.* **91**, 300.
Bardsley, J. N., and Cohen, J. S. (1978). *J. Phys. B* **11**, 3645.
Bardsley, J. N., and Wadehra, J. M. (1979). *Phys. Rev. A* **20**, 1398.
Bardsley, J. N., Herzenberg, A., and Mandel, F. (1966). *Proc. Phys. Soc.* **89**, 305.
Bates, D. R., and Walker, J. C. G. (1967). *Proc. Phys. Soc.* **90**, 333.
Bell, K. L., Kingston, A. E., and Madden, P. J. (1978). *J. Phys. B* **11**, 3357.
Bennett, R. A., Moseley, J. T., and Peterson, J. R. (1975). *J. Chem. Phys.* **62**, 2223.
Bernstein, R. B. (1966). *Adv. Chem. Phys.* **10**, 75.
Bydin, Yu. F., and Dukel'skii, V. M. (1957). *Sov. Phys.—JETP* (Engl. Transl.) **4**, 474.
Champion, R. L., and Doverspike, L. D. (1976a). *Phys. Rev. A* **13**, 609.
Champion, R. L., and Doverspike, L. D. (1976b). *J. Chem. Phys.* **65**, 2482.
Champion, R. L., Doverspike, L. D., and Lam, S. K. (1976). *Phys. Rev. A* **13**, 617.
Chen, J. C. Y., and Peacher, J. L. (1968). *Phys. Rev.* **167**, 30.
Cheung, J. T., and Datz, S. (1980). *J. Chem. Phys.* **73**, 3159.
Corderman, R. R., and Lineberger, W. C. (1979). *Annu. Rev. Phys. Chem.* **30**, 347.

Crawford, O. H., and Garrett, W. R. (1977). *J. Chem. Phys.* **66**, 4968.
Demkov, Yu. N. (1964). *Sov. Phys.—JETP (Engl. Transl.)* **19**, 762.
Demkov, Yu. N. (1980). *Electron. At. Collisions, Proc. Int. Conf., 11th*, p. 645.
Devdarianni, A. Z. (1973). *Sov. Phys.–Tech. Phys. (Engl. Transl.)* **18**, 255.
de Vreugd, C., Wijnaendts van Resandt, R. W., Los, J., Smith, B. T., and Champion, R. L. (1979a). *Chem. Phys.* **42**, 305.
de Vreugd, C., Wijnaendts van Resandt, R. W., and Los, J. (1979b). *Chem. Phys. Lett.* **64**, 175.
de Vreugd, C., Wijnaendts van Resandt, R. W., Delos, J. B., and Los, J. (1981a, 1981b). To be published, *Chem. Phys.*
Dewangan, J. P., and Walters, H. R. J. (1978). *J. Phys. B* **11**, 3983.
Dmitriev, I. S., and Nikolaev, V. S. (1963). *Sov. Phys.—Tech. Phys.* **11**, 919.
Doverspike, L. D., Smith, B. T., and Champion, R. L. (1980). *Phys. Rev. A* **22**, 393.
Dukel'skii, V. M., and Zandberg, E. I. (1951). *Zh. Eksp. Teor. Fiz.* **21**, 1270.
Dunning, T. H., and Hay, P. J. (1977). *J. Chem. Phys.* **66**, 3767.
Edwards, A. K. (1976). *Phys. Electron. At. Collisions, Int. Conf., 9th*, p. 790.
Esaulov, V. (1980). *J. Phys. B* **13**, 4039.
Esaulov, V. A., Gauyacq, J. P., and Doverspike, L. D. (1980). *J. Phys. B* **13**, 193.
Esaulov, V., Dhuicq, D., and Gauyacq, J. P. (1978). *J. Phys. B* **11**, 1049.
Faist, M. B., and Levine, R. D. (1976). *J. Chem. Phys.* **64**, 2953.
Fogel', Ia. M., Ankudinov, V. A., and Slabospitskii (1957). *Sov. Phys.–JETP (Engl. Transl.)* **5**, 382.
Franklin, J. L., and Harland, P. W. (1974). *Annu. Rev. Chem.* **25**, 485.
Gauyacq, J. P. (1979). *J. Phys. B* **13**, L387.
Gauyacq, J. P. (1980a). *J. Phys. B* **13**, 4417.
Gauyacq, J. P. (1980b). *J. Phys. B* **13**, L501.
Geballe, R., and Risley, J. S. (1973). *Electron. At. Collisions, Abstr. Pap. Int. Conf., 8th*.
Geddes, J., Hill, J., Shah, M. B., Goffe, T. V., and Gilbody, H. B. (1980). *J. Phys. B* **13**, 319.
Gillespie, G. H. (1977). *Phys. Rev. A* **15**, 563.
Griffing, K. M., Kenney, J., Simons, J., and Jordan, K. D. (1975). *J. Chem. Phys.* **63**, 4073.
Hasted, J. B. (1952). *Proc. R. Soc. London*, **A212**, 235.
Hasted, J. B. (1954). *Proc. R. Soc. London, Ser. A* **222**, 74.
Hasted, J. B., and Smith, R. A. (1956). *Proc. R. Soc. London, Ser. A* **235**, 349.
Haywood, S. E., Bowen, D. J., Champion, R. L., and Doverspike, L. D. (1981a). *J. Phys. B* **14**, 261.
Haywood, S. E., Doverspike, L. D., Champion, R. L., Herbst, E., Annis, B. K., and Datz, S. (1981b). *J. Chem. Phys.* **74**, 2845.
Heinemeier, J., Hvelplund, P., and Simpson, F. R. (1976). *J. Phys. B* **9**, 2669.
Herbst, E., Brown, D. R., Champion, R. L., and Doverspike, L. D. (1982). To be published.
Herzenberg, A., and Ojha, P. (1979). *Phys. Rev. A* **20**, 1905.
Hotop, H., and Lineberger, W. C. (1975). *J. Phys. Chem. Ref. Data* **4**, 539.
Huq, M. S., Esaulov, V., Doverspike, L. D., and Champion, R. L. (1982). To be published.
Hummer, D. G., Stebbings, R. F., and Fite, W. L. (1960). *Phys. Rev.* **119**, 668.
Janousek, B. K., and Brauman, J. I. (1979). *Gas Phase Ion Chem.* **2**.
Jordan, K. D., and Wendoloski, J. J. (1978). *Mol. Phys.* **35**, 223.
Karo, A. M., Gardner, M. A., and Hiskes, J. R. (1978). *J. Chem. Phys.* **68**, 1942.
Kolos, W., and Wolniewicz, L. (1965). *J. Chem. Phys.* **43**, 2429.
Lam, S. K., Delos, J. B., Champion, R. L., and Doverspike, L. D. (1974). *Phys. Rev. A* **9**, 1828.
Leslie, T. E., Sarver, K. P., and Anderson, L. W. (1971). *Phys. Rev. A* **4**, 408.
Lopantseva, G. B., and Firsov, O. B. (1966). *Sov. Phys.—JETP (Engl. Transl.)* **23**, 648.
McCaughey, M. P., and Bednar, J. A. (1972). *Phys. Rev. Lett.* **28**, 1011.

McFarland, M., Dunkin, D. B., Fehsenfeld, F. C., Schmeltekopf, A. L., and Ferguson, E. E. (1972). *J. Chem. Phys.* **56**, 2358.
Mandl, A. (1976a). *J. Chem. Phys.* **64**, 903.
Mandl, A. (1976b). *J. Chem. Phys.* **65**, 2484.
Mandl, A., Kirel, B., and Evans, E. W. (1970). *J. Chem. Phys.* **53**, 2363.
Mason, E. A., and Vanderslice, I. E. (1958). *J. Chem. Phys.* **28**, 253.
Massey, H. S. W. (1976). "Negative Ions," 3rd ed. Cambridge Univ. Press, London and New York.
Matić, M. Cobié, B. (1971). *J. Phys.* **B 4**, 111.
Melius, C. F., and Goddard, W. A. (1974). *Phys. Rev. A* **10**, 1541.
Meyer, F. W. (1976). Ph.D. Thesis, University of Wisconsin, Madison.
Meyer, F. W. (1980). *J. Phys.* **B 13**, 3823.
Mizuno, J., and Chen, J. C. Y. (1969). *Phys. Rev.* **187**, 167.
Mizuno, J., and Chen, J. C. Y. (1971). *Phys. Rev. A* **4**, 1500.
Nagata, T. (1979). *Proc. Int. Conf. Phys. Electron. At. Collisions, 11th*, p. 512.
Olson, R. E. (1980). *Phys. Lett. A* **77**, 143.
Olson, R. E., and Liu, B. (1978). *Phys. Rev. A* **17**, 1568.
Olson, R. E., and Liu, B. (1979). *Phys. Rev. A* **20**, 1344.
Olson, R. E., and Liu, B. (1980a). *Phys. Rev. A* **22**, 1389.
Olson, R. E., and Liu, B. (1980b). *J. Chem. Phys.* **73**, 2817.
Olson, R. E., Shipsey, E. J., and Browne, J. C. (1976). *Phys. Rev. A* **13**, 180.
Ostrovskii, V. N. (1971). *Vestr. Leningr. Gos. Univ.* **10**, 16.
Risley, J. S. (1973). Ph.D. Thesis, University of Washington, Seattle (Univ. Microfilm, Ann Arbor, Mich.).
Risley, J. S. (1974). *Phys. Rev. A* **10**, 731.
Risley, J. S. (1977). *Phys. Rev. A* **16**, 2346.
Risley, J. S. (1980). *Electron. At. Collisions, Proc. Int. Conf., 11th*, p. 619.
Risley, J. S., and Geballe, R. (1974). *Phys. Rev. A* **9**, 2485.
Risley, J. S., Edwards, A. K., and Geballe, R. (1974). *Phys. Rev. A* **9**, 1115.
Risley, J. S., de Heers, F. J., and Kerkdijk, C. D. (1978). *J. Phys.* **B 11**, 1783.
Roche, A. E., and Goodyear, C. C. (1969). *J. Phys. B* **2**, 191.
Schlachter, A. S. (1977). *Proc. Symp. Prod. Neutral. Negative Ions Beams, BNL 50727*, p. 11.
Schultz, G. J. (1973). *Rev. Mod. Phys.* **45**, 423.
Shui, V. H., and Keck, J. C. (1973). *J. Chem. Phys.* **59**, 5242.
Sidis, V., and Kubach, C. (1978). *J. Phys.* **B 4**, 2687.
Simons, J. (1977). *Ann. Rev. Phys. Chem.* **28**, 15.
Smith, B. T., Edwards, W. R., Doverspike, L. D., and Champion, R. L. (1978). *Phys. Rev. A* **18**, 945.
Smith, F. T., Marchi, R. P., Aberth, W., Lorentz, D. C., and Heinz, O. (1967). *Phys. Rev.* **161**, 31.
Stier, P. M., and Barnett, C. F. (1956). *Phys. Rev.* **103**, 896.
Tawara, H. (1978). *At. Nucl. Data Tables* **22**, 491.
Tawara, H., and Russek, A. (1973). *Rev. Mod. Phys.* **45**, 178.
Taylor, R., and Delos, J. B. (1982). *Proc. Roy. Soc. A***379**, 179, 209.
Wijnaendts, van Resandt, R. W., de Vreugd, C., Champion, R. L., and Los, J. (1977). *Chem. Phys.* **26**, 223.
Wijnaendts van Resandt, R. W., de Vreugd, C., Champion, R. L., and Los, J. (1978). *Chem. Phys.* **29**, 151.
Williams, J. F. (1967). *Phys. Rev.* **154**, 9.
Wynn, M. J., Martin, J. D., and Bailey, T. L. (1970). *J. Chem. Phys.* **52**, 191.

Ion Implantation for Very Large Scale Integration

HEINER RYSSEL

Fraunhofer-Institut für Festkörpertechnologie
Munich, West Germany

1. Introduction	191
2. Range Distributions of Implanted Ions	193
2.1. Range Profiles of Implanted Ions	194
2.2. Lateral Spread of Implanted Ions	200
2.3. Sputtering during Implantation	200
2.4. Knock-On Implantation and Atomic Mixing	204
2.5. Implantation of Molecular Ions	209
3. Annealing of Implanted Layers	210
3.1. Temperature Dependence of Annealing	210
3.2. Orientation Dependence of Annealing	211
3.3. Dopant Dependence of Annealing	214
3.4. Dependence of Annealing on Atmosphere	214
3.5. Residual Defects after Annealing	216
3.6. Process Modeling	219
3.7. Laser Annealing	225
4. Nondoping and Other New Applications of Implantation	231
4.1. Ion-Beam Gettering	232
4.2. Damage-Enhanced Etching	236
4.3. LOCOS Process Using Nitrogen Implantation	241
4.4. Ion-Beam Lithography	244
4.5. Various New Applications	248
5. Application of Implantation to Devices	251
5.1. MOS Devices	251
5.2. Schottky and Junction FET Devices	257
5.3. Charge-Coupled Devices	258
5.4. Bipolar Devices	260
6. Conclusions	265
References	266

1. Introduction

Ion implantation was invented very early in semiconductor history, and the basic patent of Shockley (*1*) in 1954 describes virtually all aspects of ion implantation. Nevertheless, it took a long time for this technique to become well recognized in the semiconductor world. An increasing number of experiments were performed in the 1950s and 1960s, and many of these old investigations are described in the literature [e.g., Mayer *et al.* (*2*), Dearnaley *et al.* (*3*), Ryssel and Ruge (*4*), Namba and Masuda (*5*)], but it was not until about 1968 that the first applications to real devices took place. At Hughes Aircraft the metal-oxide semiconductor (MOS) transistor with self-aligned gate was invented [Bower *et al.* (*6*)] and shortly afterward the threshold adjust [Aubuchon (*7*), McDougal *et al.* (*8*)]. This was the beginning of the breakthrough for the ion implantation technique. Within just a few years all semiconductor manufacturers were applying implantation as a routine tool for MOS integrated circuit production. In the early days none of the trials to fabricate bipolar transistors was successful, until in 1971 at Bell Telephone Laboratories arsenic was used for the first time as the emitter dopant [Payne *et al.* (*9*), Reddi and Yu (*10*)].

Since that time ion implantation has evolved into the major doping technology for high-performance integrated circuits. The reasons for this success are its extremely tight process control with respect to dopant concentration and profile shape (one merely measures the current and voltage of the ions); the possibility of implanting virtually all ions at all interesting concentrations (arsenic doping, e.g., is otherwise possible only using ampul, doped-oxide, or polysilicon doping, and low concentrations are impossible to control by thermal diffusion); and the fact that one can produce very shallow distributions with steep gradients toward the bulk silicon.

The drawbacks, however, have always been the relatively expensive equipment and the damage introduced by the process. But since the development of medium- and high-current implanters, the throughput of an implanter has increased considerably, thus reducing the cost problem. The annealing of the implantation damage has also turned out to be a controllable problem; moreover, a very new technique, that of laser annealing, promises to solve completely all remaining annealing problems (*11–13*).

In the beginning of large-scale integration (LSI) in the mid-1970s, it was thought that ion implantation would overcome all critical doping steps because of its tighter control of profile parameters, due to exact dose and energy control and the assumed absence of lateral spread of the implanted ions.

Now, with very large scale integration (VLSI), structures are becoming smaller and smaller. Semiconductor technology is approaching its physical limits, set by Debye length, and its voltage limits, set by noise and reliability margins. Therefore many new problems have arisen in semiconductor technology. Many technological steps have to be improved or developed to obtain the high spatial resolution and yield that are necessary. Among these are mainly low-temperature and dry processes such as ion-beam etching; electron-beam, X-ray, or ion-beam lithography; high- and low-pressure chemical vapor deposition (CVD); molecular epitaxy; and last but not least, ion implantation. In this article, however, aspects related to ion implantation only are treated.

The physical limits of MOS and bipolar transistors are well known (*14–16*). If one assumes operating voltages of 1 V, the minimum size for a MOS single-transistor cell is approximately 0.4 μm^2, and about 2 μm^2 is the minimum for a bipolar transistor cell. The reduction in area is a factor of 625 for MOS cells compared to a standard 16 K random-access memory (RAM), and approximately 100 compared to a standard oxide-isolated transistor with 5-μm dimensions (*16*). Of course, it takes considerable effort to reach these small dimensions. The junction depth required to realize such structures is about 50 nm for the source and drain of MOS devices. For bipolar transistors an emitter width of 50 nm, a base width of 50 nm, and a buried collector depth of 150 nm are desirable. The minimum lateral dimension will be 200 nm.

Ion implantation for doping applications always requires a high-temperature treatment to anneal the radiation damage produced by the implantation itself. The temperatures applied during this treatment are generally between 800 and 1100°C, and are lower only in special cases. One major problem is the profile modifications occuring during high-temperature processes. For extremely shallow structures, with implantation energies down into the keV range, these effects are hardly measurable, and it will be necessary to calculate the resulting profiles using some sort of process modeling. To do this, an exact knowledge of many parameters, such as segregation at the SiO$_2$–Si interface, oxidation-rate enhancement, and diffusion coefficients at low temperatures, is required. Other problems connected with scaling down the dimensions involve a knowledge of the exact profile, which is not Gaussian even before the heat treatment. Also, lateral effects that may not yet be significant will play an important role. Therefore, three-dimensional process modeling has to be done, and this is not only difficult from the viewpoint of the physical models involved, but also from that of the mathematics required.

In the second part of this article theoretical considerations concerning range profiles, lateral spread, sputtering, knock-on and molecular implantation are given and are compared to experimental results. The third part is devoted to the annealing of implanted layers including damage, with additional attention to diffusion, segregation, and some problems concerning process modeling. In the fourth part nondoping applications of ion implantation are discussed. In the last part some device applications are presented, demonstrating the superior possibilities of ion implantation.

2. Range Distributions of Implanted Ions

In this section, no treatment of the theory of ion ranges in solids will be given [for this see the literature, e.g., Mayer *et al.* (*2*), Dearnaley *et al.* (*3*)], but limits are discussed that were thought to be of no importance just a few years ago. These are mainly real-range distributions in crystals, including channeling, the lateral spread of implanted ions, profile modification due to sputtering, and knock-on implantation.

2.1. Range Profiles of Implanted Ions

The oldest description of implantation profiles according to the LSS theory (17) is given by a Gaussian curve. This is described by two moments, the projected range R_p and the projected standard deviation or straggling ΔR_p. Together with the implanted dose $N_\Box$, they describe the implanted profile by

$$C(x) = [N_\Box/(2\pi)^{1/2} \Delta R_p] \exp[-(x - R_p)^2/2 \Delta R_p^2] \qquad (2.1)$$

Many experimental investigations have shown, however, that this simple description is not adequate for most ions in silicon and other semiconductors. It has been argued that this might be due to channeling because of the crystalline structure of the usual semiconductors. It has been found, however, that the profiles of many ions are asymmetrical in amorphous targets as well, and thus higher moments have to be used to construct range distributions.

The best description of implantation profiles in the range up to 300 keV is given by Pearson-IV distributions with four moments (18, 19). The advantages of Pearson distributions in comparison to other distributions, e.g., joint half-Gaussian distributions (20) or Edgeworth distributions (21) are that the Pearson distributions have no negative values and have a single maximum. Moreover, they can also model residual channeling tails that are still present even if a proper misalignment has been used (22).

The Pearson distribution of type IV centered around the projected range R_p is given by

$$C(x) = K[b_2(x - R_p)^2 + b_1(x - R_p) + b_0]^{1/(2b_2)}$$

$$\times \exp\left[-\frac{(b_1/b_2) + 2a}{(4b_2b_0 - b_1^2)^{1/2}} \arctan \frac{2b_2(x - R_p) + b_1}{(4b_2b_0 - b_1^2)^{1/2}}\right] \qquad (2.2)$$

where K is a constant necessary for normalizing the distribution. The four other constants a, b_0, b_1, and b_2 are given by

$$\begin{aligned} a &= -[\Delta R_p \gamma(\beta + 3)]/A, & b_0 &= -[\Delta R_p^2(4\beta - 3\gamma^2)]/A \\ b_1 &= a, & b_2 &= -(2\beta - 3\gamma^2 - 6)/A \end{aligned} \qquad (2.3)$$

where

$$A = 10\beta - 12\gamma^2 - 18$$

The four constants a, b_0, b_1, and b_2 can be expressed by four moments μ_1, μ_2, μ_3, and μ_4 of the distribution $C(x)$.

The first moment μ_1 is well known as the average projected range

$$\mu_1 = R_p = \int_{-\infty}^{\infty} xC(x)\,dx \qquad (2.4)$$

The three higher moments μ_i are given by

$$\mu_i = \int_{-\infty}^{\infty} (x - R_p)^i C(x)\,dx, \qquad i = 2, 3, 4 \qquad (2.5)$$

It is customary to use the standard deviation ΔR_p, which is defined by the square root of the second moment μ_2, and dimensionless expressions for the higher moments:

$$\begin{aligned} \text{standard deviation,} \quad & \Delta R_p = (\mu_2)^{1/2} \\ \text{skewness,} \quad & \gamma = \mu_3/\Delta R_p^{\;3} \\ \text{kurtosis,} \quad & \beta = \mu_4/\Delta R_p^{\;4} \end{aligned} \qquad (2.6)$$

The skewness γ indicates the tilting of the profile and the kurtosis β indicates the flatness at the top of the profile.

The relation between the third and fourth moments has to be chosen to satisfy

$$\beta \geq \beta_{\min} = [48 + 39\gamma^2 + 6(\gamma^2 + 4)^{3/2}]/(32 - \gamma^2) \qquad (2.7)$$

in order to give a Pearson distribution of type IV.

Since no theoretical calculations of the fourth moment are known, one has to use empirical values. Only for the third moment have estimated values been published (23).

In the following figures examples of measured range distributions are given and compared with theoretical calculations. In Figs. 2.1 and 2.2, boron profiles in silicon and SiO_2 are compared with Pearson-IV fits. These profiles have been measured using the $^{10}B(n, \alpha)$ ^7Li reaction (24). It is clearly seen that the Pearson distributions perfectly match the experimental profiles. The moments required to describe these distributions are given in Figs. 2.3 and 2.4 for range and range straggling, and in Figs. 2.5 and 2.6 for skewness and kurtosis. Theoretical calculations according to Gibbons et al. (23) and Biersack (25) are included for comparison. For boron in SiO_2 the measured data for the first two moments agree well with the theory, assuming a density of 2.27 g cm^{-3}. For the higher moments slight deviations from theory for energies above 60 keV are found. The behavior at 30 keV can be explained by the profile broadening due to limited detector resolution (4). Similar results were found for boron in Si_3N_4 and arsenic in SiO_2 and Si_3N_4 (26).

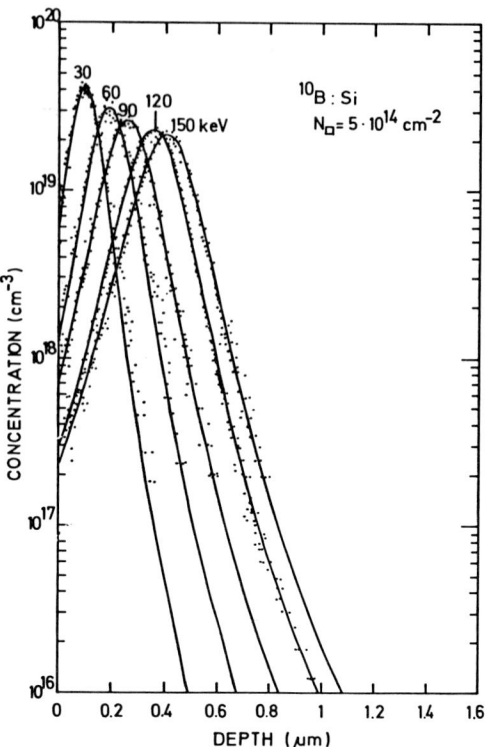

FIG. 2.1. Comparison between measured boron profiles in silicon and Pearson IV distributions. No annealing was performed after implantation (26).

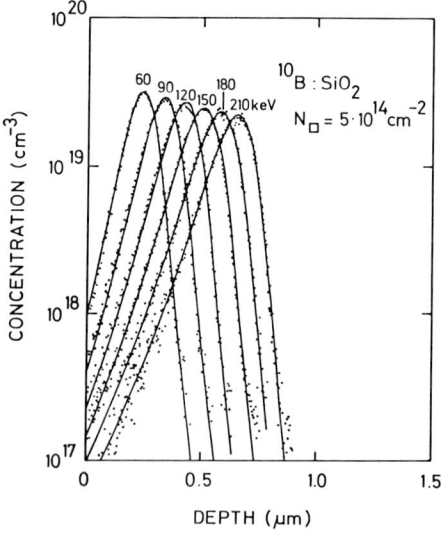

FIG. 2.2. Comparison between measured boron profiles in SiO$_2$ and Pearson IV distributions. No annealing was performed after implantation (26).

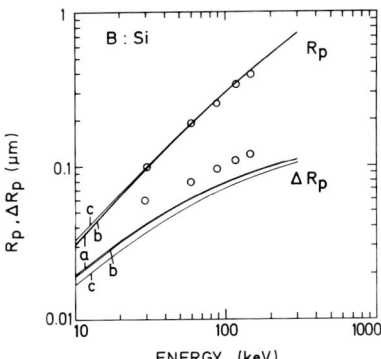

FIG. 2.3. Range and range straggling of boron in silicon. (○) Experimental data; solid lines theory: (a) ^{10}B, Biersack (25); (b) ^{11}B, Biersack (25); (c) ^{11}B, Gibbons et al. (23).

For boron in silicon the deviations are small for the range but quite large for the other moments. This might be due to the crystalline nature of silicon, e.g., to a residual amount of channeling. This, however, can be included in the profile description by the Pearson distribution. Gibbons and Mylroie (20) have shown that a joint half-Gaussian distribution using three moments, or an Edgeworth expansion using four moments, improves the profile shape significantly. In the calculation of Gibbons et al. (23) an estimated value of the third moment is given, together with an expression similar to Eq. (2.7) for the fourth moment resulting from an Edgeworth expansion. This distribution, however, oscillates and shows negative concentrations.

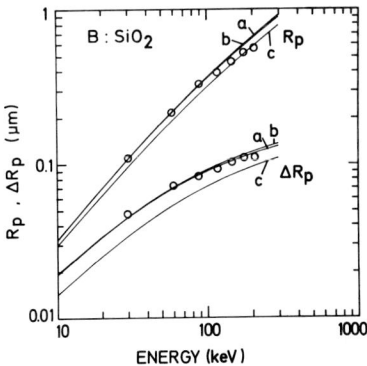

FIG. 2.4. Range and range straggling for boron implanted in SiO_2. (○) Experimental data; (a) ^{10}B, Biersack (25); (b) ^{11}B, Biersack (25); (c) ^{11}B, Gibbons et al. (23).

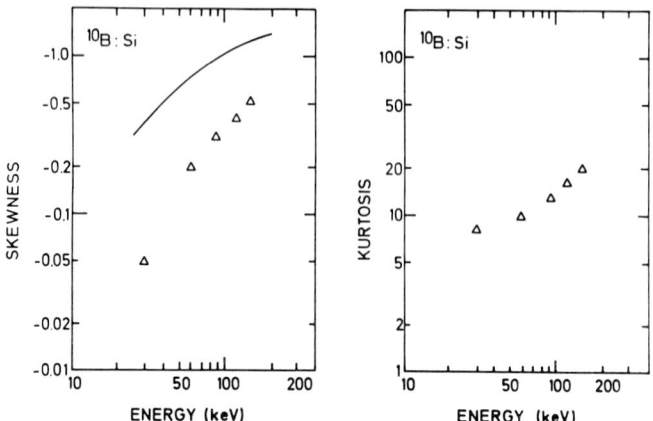

FIG. 2.5. Higher moments of boron implants in silicon. Solid line: theoretical data according to Gibbons *et al.* (*23*).

In order to suppress channeling and obtain reproducible profiles, not only a proper tilting (~ 7–$10°$) but also a proper rotation has to be performed. This is shown in Fig. 2.7 (*22*) for $\langle 111 \rangle$-oriented silicon. For $\langle 100 \rangle$- and $\langle 111 \rangle$-oriented silicon, tilting using the $\langle 110 \rangle$-flat as tilting axis is recommended for optimal suppression to avoid a planar channeling. The conditions for proper tilting and rotation are not so stringent if a scattering oxide or nitride layer is used.

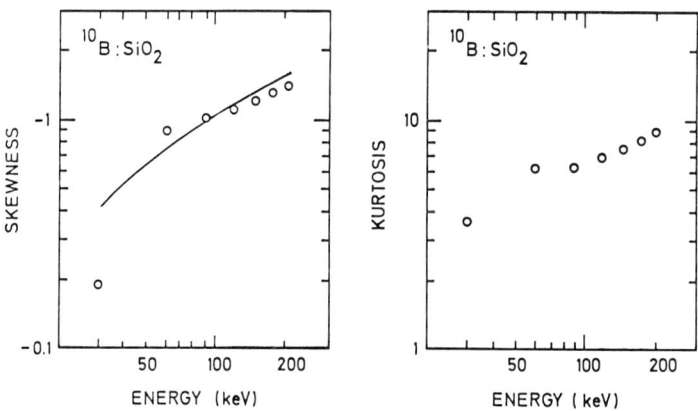

FIG. 2.6. Skewness and kurtosis of implanted boron in SiO_2. The moments were obtained by fitting measured profiles with Pearson IV distributions. Solid line according to Gibbons *et al.* (*23*).

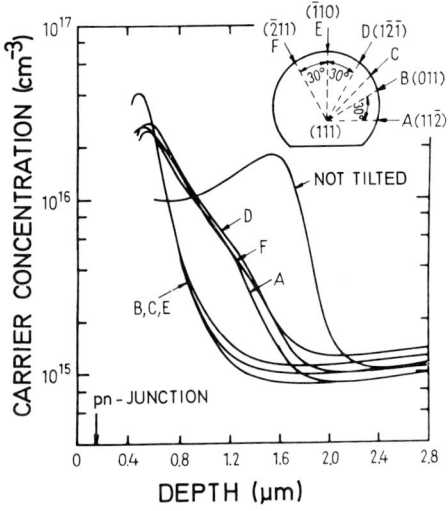

FIG. 2.7. Dependence of phosphorus profiles on rotation (22).

Another question is whether the channeling effect can be used to obtain deep distributions, e.g., for buried layers. For this, a parallel ion beam (obtainable either by mechanical scanning of the wafers or by an electrostatic double-deflection scanner) is required. The crystal orientation, however, has to be exact within $\pm 0.1°$, which is very difficult to obtain. The influence that even very small deviations have on the profile shape is depicted in Fig. 2.8 for a phosphorus implant with 450 keV into $\langle 111 \rangle$ silicon. Therefore, an application of channeling in a VLSI production environment seems very unlikely.

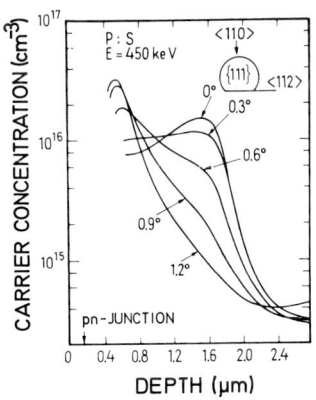

FIG. 2.8. Dependence of phosphorus profiles on tilting (22).

2.2. Lateral Spread of Implanted Ions

In the beginning days of ion implantation, the lateral spread was thought to be negligible in comparison to both the lateral dimensions of the devices and the lateral diffusion. With VLSI circuits, however, dimensions are shrinking into the micron and submicron range and the lateral spread has to be taken into consideration. According to the basic calculations of Matsumura and Furukawa (27), the lateral spread has been found to be in the range of the standard deviation. Assuming a Gaussian distribution for the spatial shape of the ion distribution, they could derive an expression for two-dimensional implantation profiles. For implantation through a window of a masking layer with infinite thickness they found the profile to be

$$C(x, y) = \frac{N_\square}{(2\pi)^{1/2} \Delta R_p} \left[\exp\left(-\frac{(x - R_p)^2}{2 \Delta R_p^2}\right) \right]$$
$$\times \frac{1}{2} \left[\text{erfc} \frac{y - a}{\sqrt{2} \Delta R_{p,L}} - \text{erfc} \frac{y + a}{\sqrt{2} \Delta R_{p,L}} \right] \quad (2.8)$$

where $N_\square$ is the implanted dose, $2a$ is the width of the window in the masking layer, R_p is the projected range, ΔR_p is the spread, and $\Delta R_{p,L}$ is the lateral spread of the ions.

For real masking layers with an arbitrary shape, Runge (28) found

$$C(x, y) = \frac{N_\square}{2\pi \Delta R_{p,L} \Delta R_p}$$
$$\times \int_{-\infty}^{+\infty} \left[\exp\left(-\frac{(x - \xi)^2}{2 \Delta R_{p,L}^2} - \frac{(x - d_{0x}(\xi) - R_p)^2}{2 \Delta R_p^2}\right) \right] d\xi \quad (2.9)$$

where $d_{0x}(\zeta)$ is the local thickness of the masking layer.

In Figs. 2.9–2.11 examples of theoretical calculations of the lateral spread are given. In contrast to Eq. (2.9), a Pearson distribution has been assumed in the vertical direction and a Gaussian distribution in the lateral direction, due to the lack of theoretical data. The calculations have been performed using a numerical program for process modeling [ICECREM (29)] that will be discussed in more detail later in this article.

2.3. Sputtering during Implantation

For shallow devices, low energies and high doses have to be used to obtain the desired junction depth and sheet resistivity. In that case sputtering has to be considered.

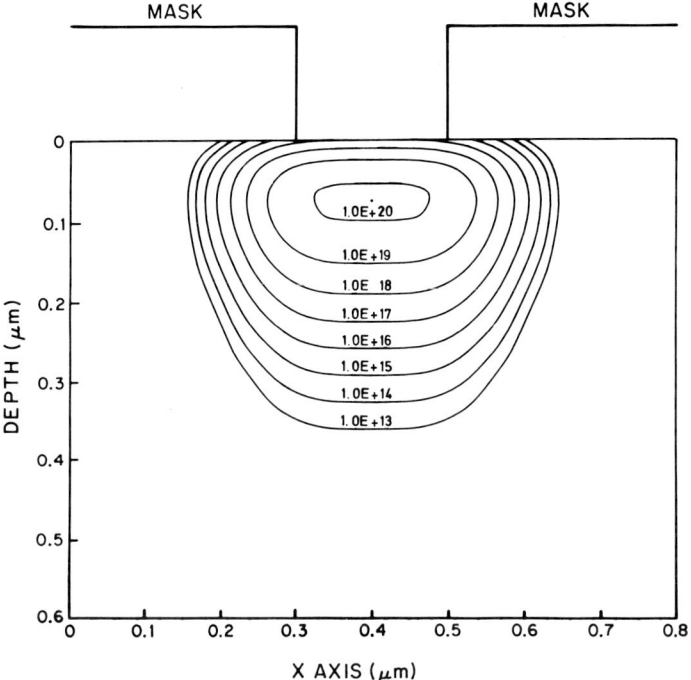

FIG. 2.9. Isoconcentration lines of an arsenic implantation ($N_\Box = 10^{15}$ cm^{-2}, $E = 150$ keV) through a 0.2-μm-wide vertical window in a 2-μm-thick masking layer.

The modification of implantation profiles by sputtering can be calculated using some simplifying assumptions:

(1) the sputtering yield is constant,
(2) no knock-on takes place,
(3) the volume change due to the damage may be neglected, and
(4) the profile is Gaussian.

Under these assumptions, the following formula is valid (4):

$$C(x) = \frac{N}{2S}\left[\text{erf}\frac{x - R_p + N_\Box(S/N)}{\sqrt{2}\,\Delta R_p} - \text{erf}\frac{x - R_p}{\sqrt{2}\,\Delta R_p}\right] \quad (2.10)$$

where N is the atomic density of silicon ($N = 5 \times 10^{22}$ cm^{-3}), S is the sputtering yield, R_p is the range, ΔR_p is the range straggling, and $N_\Box$ is the implanted dose.

The saturation profile is given by

$$C(x) = (N/2S)\,\text{erfc}[(x - R_p)/\sqrt{2}\,\Delta R_p] \quad (2.11)$$

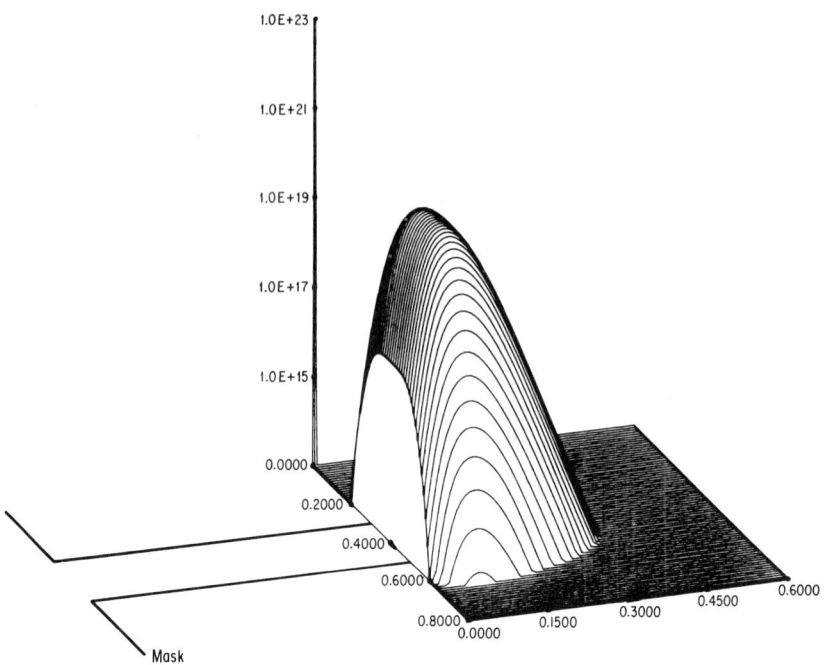

FIG. 2.10. The same profile as in Fig. 2.9 in three-dimensional view.

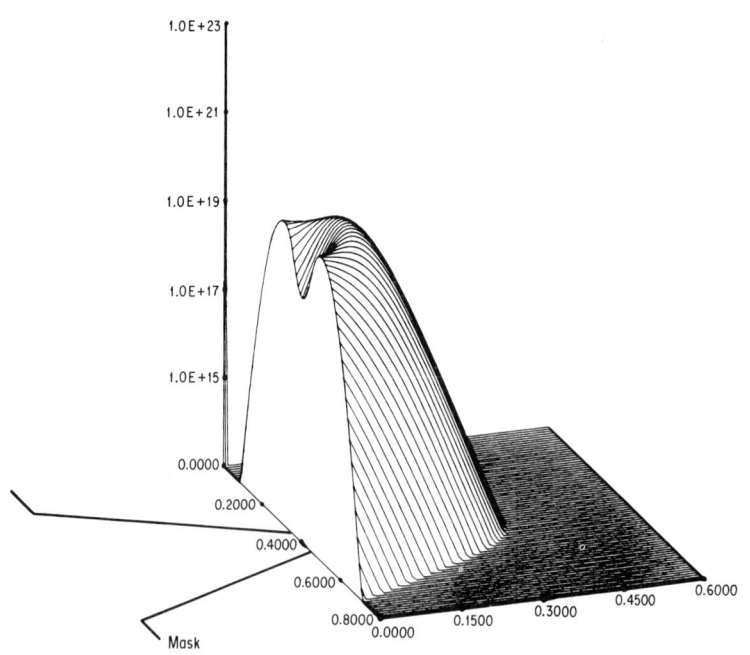

FIG. 2.11. Arsenic implantation ($N_\square = 10^{15}$ cm^{-2}, $E = 150$ keV) through a tapered window (width 0.1 μm, mask thickness 0.4 μm).

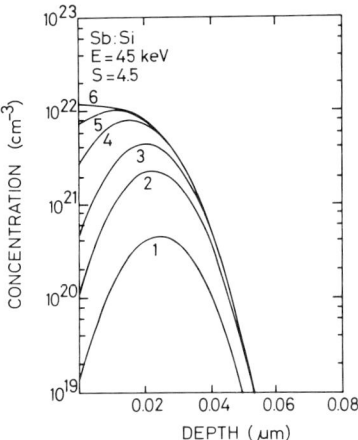

FIG. 2.12. Sputtering-modified doping profile, depending on the implanted dose:

1 = $N_\square = 1 \times 10^{15}$ cm^{-2} 4 = $N_\square = 2 - 10^{16}$ cm^{-2}
2 = $N_\square = 5 \times 10^{15}$ cm^{-2} 5 = $N_\square = 3 \times 10^{16}$ cm^{-2}
3 = $N_\square = 1 \times 10^{16}$ cm^{-2} 6 = $N_\square = 5 \times 10^{16}$ cm^{-2}

with the maximum concentration at the surface:

$$C_{max} = (N/2S)\,\text{erfc}(-R_p/\sqrt{2}\,\Delta R_p) \approx N/S \quad \text{for} \quad R_p > 3\,\Delta R_p \quad (2.12)$$

If one assumes non-Gaussian profiles or nonconstant sputtering parameters, numerical calculations are required.

In Fig. 2.12, a theoretical example for antimony in silicon is given. It is seen that doses in excess of 10^{16} cm^{-2} are required in the case of a sputtering yield of 4.5 to obtain profile modifications. The saturation profile is obtained for doses in excess of 5×10^{16} cm^{-2}. In Fig. 2.13, the sputtering yields of different ions in silicon according to Anderson et al. (30) are given in comparison to theoretical calculations. The sputtering yield is high for heavy ions and depends also on the energy. This is shown in Fig. 2.14, where the energy dependence of the yield according to a formula of Sigmund (31) has been calculated:

$$S = \frac{3}{4}\frac{S_n(E)\alpha(M_2/M_1)}{\pi^2 C_0 U_0} \quad (2.13)$$

with $C_0 = 0.826$ nm^2; $\alpha(M_2/M_1)$ is a numerically calculable function of the ratio of ion mass M_1 to target mass M_2, U_0 is the surface binding energy (7.81 eV for silicon), and $S_n(E)$ is the energy deposited at the surface in a cascade.

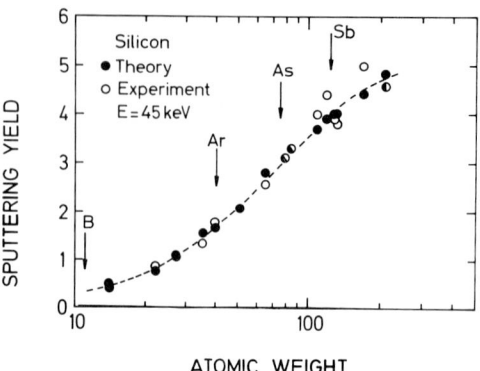

FIG. 2.13. Sputtering yield of ions with different mass at 45 keV in silicon (*30*). The theoretical curve is after (*31*).

Table 2.1 shows the sputtered thickness after implanting different ions at the energy for maximum sputtering. One sees that the effect is slight for all doses below 10^{16} cm^{-2}; above this dose it may be more significant.

2.4. Knock-On Implantation and Atomic Mixing

The implantation of ions, whether through a masking layer or through a window in a thick masking layer, results in knock-on or recoil implantation of ions. If the mass of the implanted ions is not too different from the mass of the atoms of the masking layer, a large fraction of the energy of the primary particles can be transferred to the atoms of the masking layer. These particles are then implanted into the substrate. The knock-on effect has been known

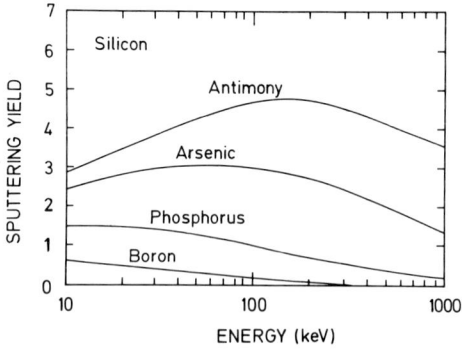

FIG. 2.14. Energy dependence of sputtering yield of boron, arsenic, phosphorus, and antimony in silicon according to Eq. (2.13).

TABLE 2.1

SPUTTERED THICKNESS FOR MAXIMUM SPUTTERING
YIELD, FOR DIFFERENT IONS IMPLANTED
INTO SILICON[a]

Ion	Energy (keV)	Dose (cm^{-2})		
		10^{15}	10^{16}	10^{17}
B	10	0.1 nm	1 nm	10 nm
P	10	0.3 nm	3 nm	30 nm
As	50	0.6 nm	6 nm	60 nm
Sb	150	1 nm	10 nm	100 nm

[a] The lowest energy used for the calculations is 10 keV.

for several years (32), but was thought to be detrimental, at least in silicon technology. In Fig. 2.15 an example is given in which arsenic ions were implanted through a tapered SiO_2 mask. In the areas where the SiO_2 was thick enough to stop all ions, or where no SiO_2 was present, no defects were visible after damage etching. But in the area where silicon and oxygen atoms were knock-on implanted, defects were revealed by etching. Up to now, however, no negative influence of knock-on ions has been found on device parameters, and very often implants are performed through thin scattering layers. This is due to the fact that the distribution of the recoil-implanted atoms is more shallow than the distribution of the primary implanted atoms.

Theoretical calculations of the distribution of recoil implants have been performed by Moline et al. (33), Fischer et al. (34), and recently by Sigmund (35) and Hirao et al. (36). The formalism is too complicated to present here in detail, no simple analytical profile description exists, nor do any tabulated data. The latter would of course be impossible, since there is a great variety of different primary ions and target and layer atoms that are recoil-implanted. Many measurements have been reported on knock-on implants (33, 36–38). In Fig. 2.16 a typical example is given for arsenic implanted through a 650-Å-thick layer of Si_3N_4 with 355 keV to a dose of 10^{16} cm^{-2} (36). The extremely shallow recoil profile with the maximum at the surface is clearly seen. The primary ions come to rest deeper in the crystal than the recoil-implanted atoms.

All experiments and theories show the same phenomena: A very shallow distribution of the recoil-implanted ions and a deeper distribution of the primary atoms. This is caused by the facts that only high-energy primaries can knock a secondary in the semiconductor and that all angles occur during this process.

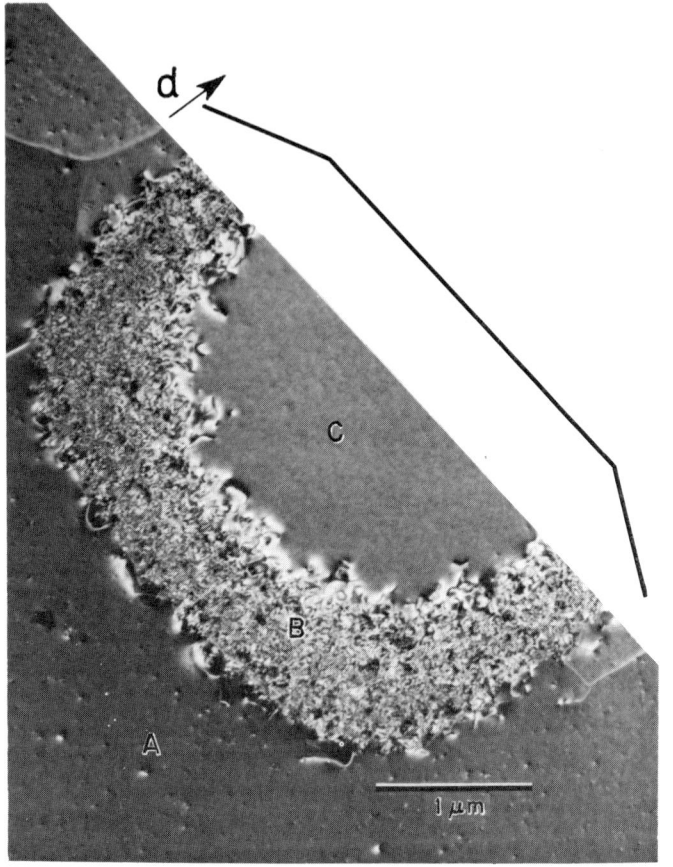

FIG. 2.15. Knock-on oxygen by arsenic implantation through a tapered SiO_2 layer (32).

Some applications of the knock-on effect were found recently for the control of the barrier height of Schottky diodes (39, 40). Also, doping experiments with arsenic in silicon have been performed, implanting 100-keV Si ions at a dose of 3×10^{14} cm^{-2} through 100 Å of evaporated arsenic, resulting in a resistivity of the doped silicon layer of 300 $\Omega/\square$ (41). Similar results have been found for Sb (42).

Knock-on implantations into insulators have been used by Ito et al. (43) to produce nonvolatile memories. They produced traps at the Si_3N_4–SiO_2 interface by recoil implantation of tungsten with 200-keV Si ions. The technique is depicted schematically in Fig. 2.17, whereas the complete fabrication scheme of these memories is given in Fig. 2.18. After deposition

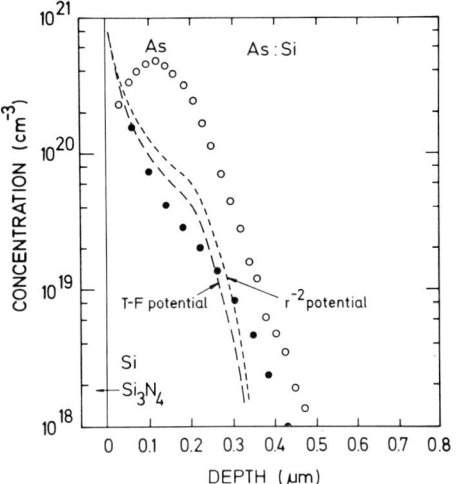

FIG. 2.16. Comparison of experimental and theoretical profiles for arsenic implantation at $E = 355$ keV to a dose of $N_\square = 1 \times 10^{16}$ cm^{-2} through Si$_3$N$_4$ of 650-Å thickness. Recoil nitrogen: ●, experimental; --, theoretical (36). Arsenic: ○, experimental (36).

of a tungsten layer on top of a SiO$_2$ layer, silicon was implanted to recoil-implant the tungsten into the SiO$_2$ layer. Subsequently the tungsten was etched off, Si$_3$N$_4$ and polysilicon were deposited, and transistors were fabricated using phosphorus silica glass (PSG). Doses above 10^{14} cm^{-2} were effective in producing electron and hole traps. Memory retention was tested

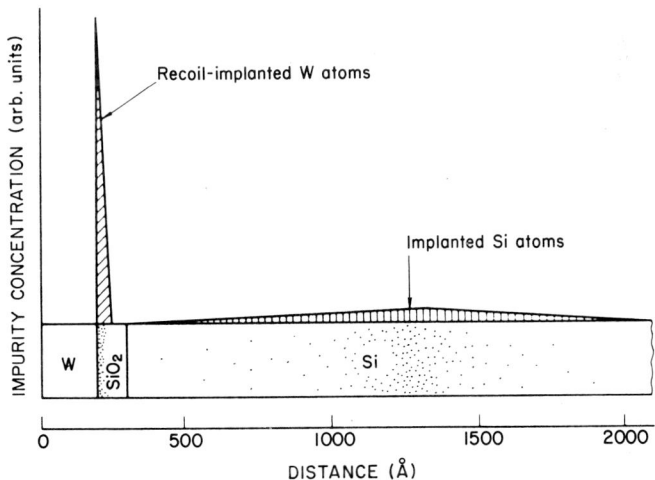

FIG. 2.17. Schematic profile of recoil-implanted tungsten atoms in a SiO$_2$ film (43).

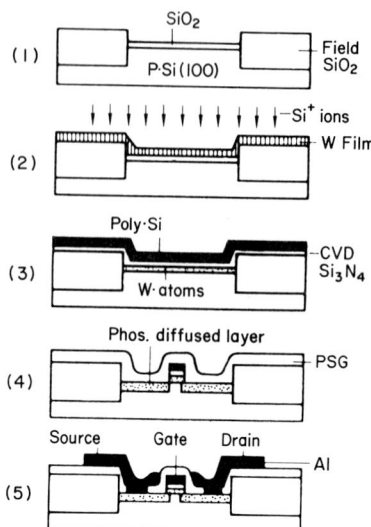

FIG. 2.18. Process (in five steps) for an MNOS memory field-effect transistor incorporating recoil implantation (43).

for 5×10^5 sec (≈ 6 days) without significant change in the levels. The essential memory characteristics, speed, and designed voltages for write and erase operation as well as repetition and memory retention, have been improved in comparison to conventional MNOS memories.

The damage produced by the knocking ions, however, must be considered. It will usually be in the area of the p–n junction and may severely degrade its properties. But there is undoubtedly no other implantation process capable of producing profiles as shallow as those obtainable using this process with the maximum of the distribution at the surface.

Very recently a new type of effect has been found to occur at interfaces: atomic mixing (85–88). During the implantation of ions through an interface between a thin film and a substrate, not only are atoms from the thin film recoil-implanted into the substrate, but substrate atoms also are transported into the thin film. This effect can be explained by a simple model (87). Each incident ion initiates a collision cascade in a volume around the ion track. Within that volume there is some atomic mobility for a very short time interval following the impact, resulting in an intermixing of the atoms near the interface. At the same time reactions between the mixing atoms can also take place. Most of these studies have been made using metal–silicon reactions, e.g., platinum (85, 88), molybdenum (86), nickel and hafnium (88). The formation of silicide induced by this effect at room temperature might offer a future method for production of low-temperature ohmic contacts or Schottky diodes.

2.5. Implantation of Molecular Ions

Another method of producing extremely shallow profiles is the use of molecular ions (44). Such molecules have a higher mass, thus reducing the dose required to obtain an amorphous layer. Amorphous layers show a better regrowth behavior than nonamorphous layers. This is especially important for boron, where the amorphous dose is 8×10^{16} cm^{-2} (45). The use of BF$_2$ molecules therefore offers the possibility of reducing the amorphous dose at room temperature, without cooling or predamaging as an additional process step. This is illustrated in Fig. 2.19, where the annealing behaviors of BF$_2$ and boron-implanted silicon are compared (44).

Depending on the mass ratio of the dopant and the molecule, this method offers a means of producing low-energy range distribution from a given accelerator. The effective energy of the dopant is given by

$$E_{\text{dopant}} = (m_{\text{dopant}}/m_{\text{molecule}})E_{\text{molecule}} \qquad (2.14)$$

For this calculation it was assumed that the molecule splits into its components when it strikes the surface of the sample.

This technique is currently used very often to provide for extremely shallow base contacts. It is possible, e.g., with a 30-keV BF$_2$ implantation, to obtain an effective boron energy of 7 keV, which results in $R_p = 26$ nm and

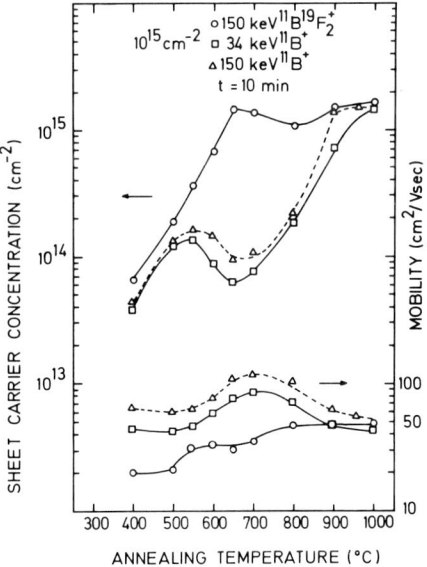

FIG. 2.19. Carrier concentration and mobility versus anneal temperature for room-temperature implantations of 1×10^{15} cm^{-2} of ^{11}B and ^{11}B^{19}F$_2$.

$\Delta R_p = 13$ nm. With 5 keV, the lowest energy at present available with implanters, one obtains an effective energy of about 1 keV, approaching the energy range of sputter profiling and ion-beam deposition.

3. ANNEALING OF IMPLANTED LAYERS

During implantation many crystal defects are produced. At high doses amorphous layers can also form. A temperature treatment is required to anneal out these defects, and to render the implanted ions electrically active. The annealing process depends mainly on the temperature, but also on the implanted ions, the orientation of the crystal, and the atmosphere. The annealing process always leads to a modification of the implanted profile, because of diffusion and segregation. For extremely shallow doping distributions with small lateral dimensions an experimental determination of such three-dimensional profiles is not possible. Thus, process modeling has to be used to obtain information on the doping profile after annealing.

In this section, the dependence of the annealing process on temperature, orientation, doping concentration, and atmosphere is briefly discussed. Finally, the techniques of laser annealing and process modeling are treated.

3.1. Temperature Dependence of Annealing

Simple defects anneal out at low temperatures, e.g., vacancies at 70–150 K, vacancy-group-V-element defects between 400 and 500 K, and vacancy-group-III-element defects at about 500 K. Heavily damaged or amorphous regions require temperatures between 800 and 850 K (550–600°C), and in special cases temperatures up to 1300 K. After annealing at 850 K all standard doping elements of silicon obtain their complete electrical activation except boron. Boron shows a reverse annealing effect at doses or implantation temperatures that result in no amorphous layer. This abnormal behavior was already shown in Fig. 2.18. If BF_2 is implanted, or the implantation is done at liquid-nitrogen temperatures, an amorphous layer forms at a dose of 10^{15} cm^{-2}, and an activation behavior similar to that of the other dopant ions is found.

For a complete recovery of the carrier mobility, slightly higher temperatures are required; and to restore a good carrier lifetime, temperatures of about 1200–1300 K are necessary. This can be seen from measurements of the emitter efficiency of bipolar transistors or from measurements of the leakage current following silicon self-implants in the p–n area of diodes.

3.2. Orientation Dependence of Annealing

The annealing behavior is strongly orientation-dependent. This effect was studied in detail by Csepregi (46, 47). The $\langle 100 \rangle$-oriented wafers recrystallize much faster than $\langle 111 \rangle$-oriented wafers. This is shown in Fig. 3.1 for self-implanted silicon samples. Moreover, in the case of $\langle 111 \rangle$ orientation, a second maximum develops at the border between the undamaged substrate and the amorphous layer. The recrystallization rate is about 8 nm min^{-1} in $\langle 100 \rangle$ material and 2.5 nm min^{-1} in $\langle 110 \rangle$ material at 550°C; the activation energy is 2.3 eV in both cases. For $\langle 111 \rangle$ material the process is more complicated, and no activation energy can be given.

This orientation-dependent annealing depends strongly, in turn, on the thermal history of the wafer. This is shown in Fig. 3.2. If the annealing is done immediately at high temperatures, a very poor annealing effect is the

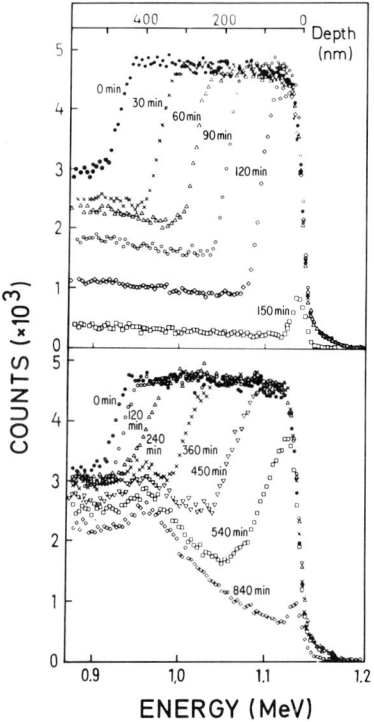

FIG. 3.1. Channeling spectra of self-implanted silicon (50 and 250 keV, 8×10^{15} cm^{-2}), preannealing at 400°C, annealing at 550°C. Upper curve $\langle 100 \rangle$-, lower curve $\langle 111 \rangle$-oriented sample (46).

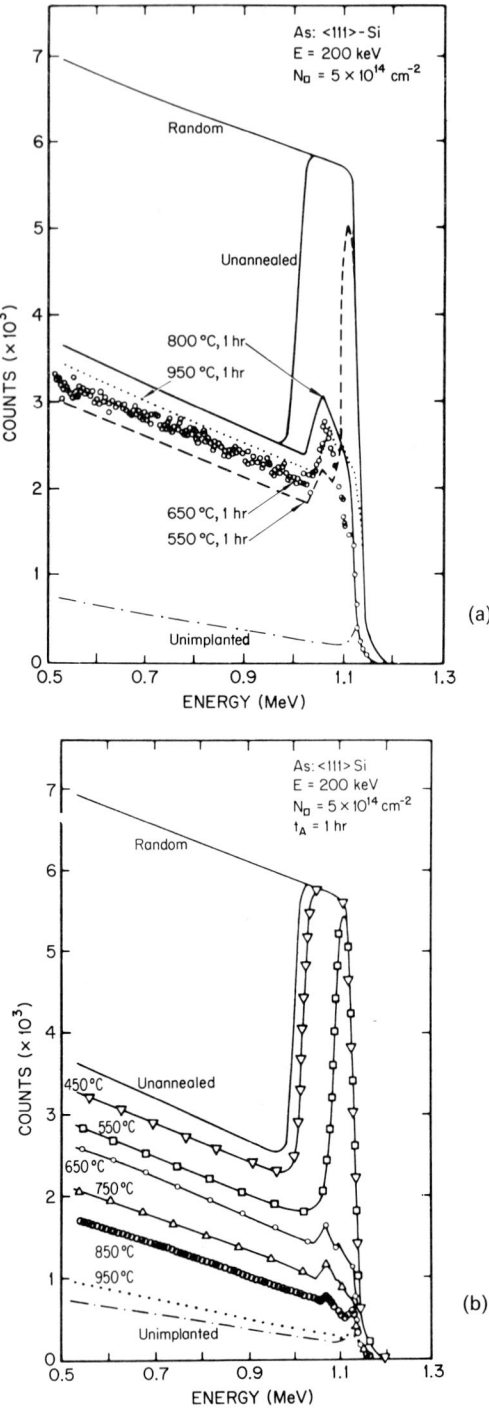

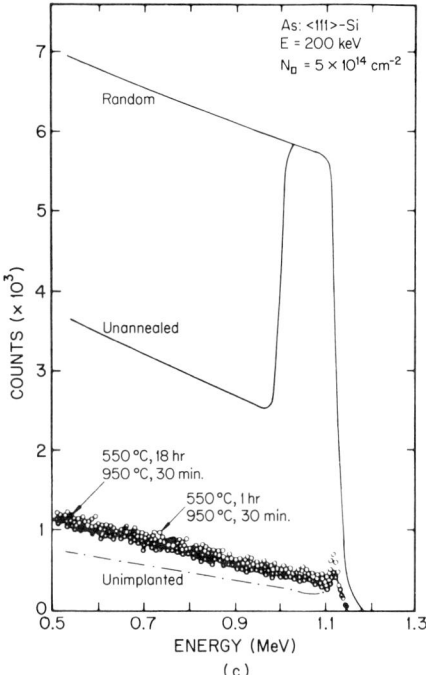

FIG. 3.2. Random and channeling spectra of $\langle 111 \rangle$ silicon implanted at 77 K with arsenic (200 keV, 5×10^{14} cm^{-2}) (47). (a) One-step annealing, (b) annealing in 100°C steps, (c) 550°C preannealing.

result (Fig. 3.2a). If the samples are annealed with steadily increasing temperatures, perfect annealing is obtained (Fig. 3.2b). A preannealing at 550°C with a subsequent high-temperature annealing step also results in perfect annealing (Fig. 3.2c).

The orientation dependence of the regrowth rates can be explained by geometrical arguments, assuming that atoms can only be transferred from the amorphous to the crystalline phase at positions where at least two nearest-neighbor atoms at the interface are already in crystalline positions. This is a purely geometrical argument, requiring a minimum of two atoms on lattice sites for epitaxial growth. On this basis growth will not proceed along the $\langle 111 \rangle$ direction because alternate planes have atoms with only one bond along this direction. Growth along the $\langle 111 \rangle$ direction involves nucleation, and hence leads to nonuniform interface and twin formation (48).

A comparison between theory and experiment is given in Fig. 3.3. A zero regrowth rate is predicted at an exact $\langle 111 \rangle$ orientation. However, this simple model does not account for the nonzero $\langle 111 \rangle$ regrowth rate.

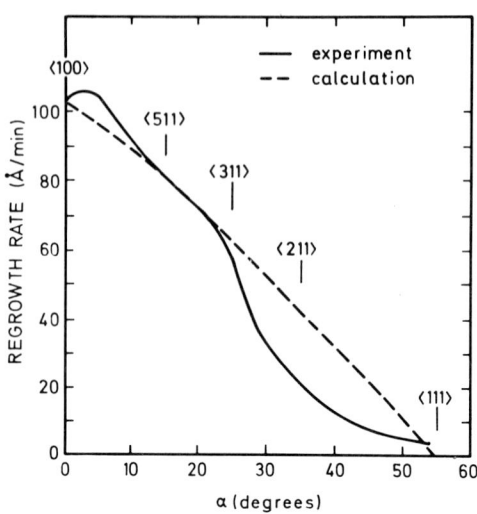

FIG. 3.3. Plot of regrowth rate at 550°C versus substrate orientation angle from the $\langle 100 \rangle$ direction (48).

3.3. Dopant Dependence of Annealing

The annealing dependence on the dopant species was also first investigated by Csepregi (49). For implanted concentrations of B, Ar, or P exceeding about 10^{20} cm^{-3}, there is an increase in the growth rate over that of amorphous layers that are only Si-implanted. The increase is associated with the local impurity concentration in the proximity of the amorphous–crystalline interface. The growth rate increases by a factor of 6–25. For electrically inactive impurities, the growth rate decreases by a comparable amount. This was measured for O, C, N, Ne, Ar, and Kr (50). Since the enhanced or retarded recrystallization rate depends on the impurity concentration, it is not constant during the regrowth process. The reasons for this behavior are not yet well understood.

3.4. Dependence of Annealing on Atmosphere

In silicon technology, oxidizing treatments are very often performed, either for the driving-in of a dopant or to obtain a masking layer for the next technological step. In the case of ion implantation it was found that oxidizing annealing leads to the formation of many dislocations that grow into the bulk of the wafer. This is also the case if no amorphous layer was formed during the implant. An example of such dislocations after a boron implant (30 keV, 10^{14} cm^{-2}, $\langle 100 \rangle$ orientation) is given in Fig. 3.4. The wafer was annealed at 1100°C in a wet atmosphere (51). Detailed studies of this effect

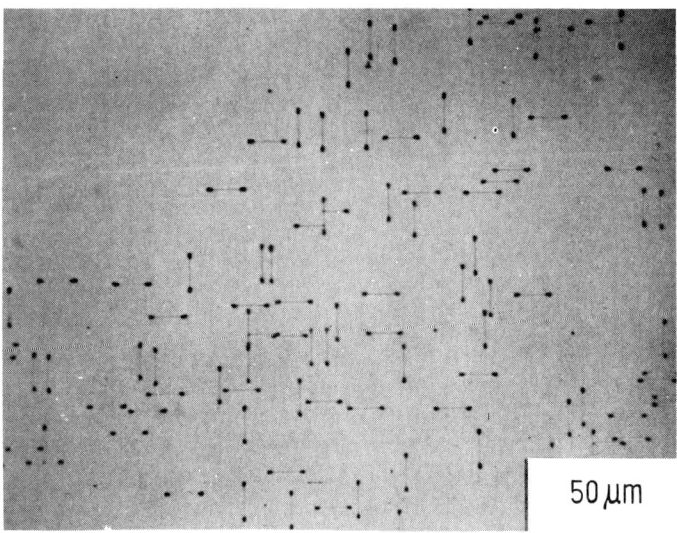

FIG. 3.4. Boron-implanted silicon (30 keV, 10^{14} cm^{-2}, $\langle 100 \rangle$ orientation) after wet oxidation at 1100°C (51). (This figure was originally presented at the Spring, 1980 Meeting of The Electrochemical Society, Inc., held in St. Louis, Missouri. Reprinted by permission, The Electrochemical Society, Inc.)

have been performed by many other groups (52–55), in connection with devices (56) as well.

If an inert preannealing is performed at 900–1000°C before the oxidation, virtually no dislocations remain. This can be done either in two separate steps, or merely by changing the gas composition. The importance of such an inert annealing or preannealing for device characteristics was impressively shown by Seidel et al. (56). They studied the emitter–base currents and current gains as a function of the oxygen content in the annealing atmosphere of bipolar transistors. They found that the best results are obtained for an oxygen content below 0.1%, even for through-oxide implants as well. This is especially true for $\langle 100 \rangle$ silicon, whereas $\langle 111 \rangle$ silicon is more tolerant in regard to oxidizing treatments.

The question of whether or not such defects are detrimental to device characteristics depends on the individual case. If an epitaxial layer has to be grown, defect-free material is required. This can be obtained by removing the layer containing defects by thermal oxidation or by etching before deposition. In both cases one must diffuse the implanted ions; otherwise they are also removed. Usually, however, the defects have no effect on devices. This is the case if they do not grow during prolonged annealing and drive-in diffusion and do not penetrate junctions.

3.5. Residual Defects after Annealing

After annealing many defects remain. Some are formed during annealing, although one speaks of epitaxial recrystallization. These defects are mainly dislocations and stacking faults. The best method for revealing such defects is electron transmission microscopy. In the following figures some examples are given. At low arsenic doses (5×10^{15} cm^{-2} at 80 keV) prismatic dislocation loops grow during annealing at 1000°C for 60 min (Fig. 3.5). At higher doses (10^{16} cm^{-2}), only large half-loops form. These dislocations are of the misfit type (Fig. 3.6).

Implanting through thin SiO$_2$ layers also results in many defects; these are partly caused by knock-on particles. An example showing dislocation loops and oxide particles is given in Fig. 3.7. A half-masked sample showing only large loops in the bare case, and a thick dislocation network in the through-oxide case, appears in Fig. 3.8 for another arsenic implant.

No amorphous layer is usually formed in the case of boron implantation. Nonetheless, defects grow during annealing; however, the density after inert annealing at 900–1100°C is very low.

Very detailed studies of the defects resulting from the annealing of phosphorus implants at a dose of 5×10^{14} cm^{-2} were made by Tamura (59). He found a great variety of defects, depending on the implantation and annealing temperatures. A schematic description of his results is given in Fig. 3.9. After implantation at room temperature point defects develop that

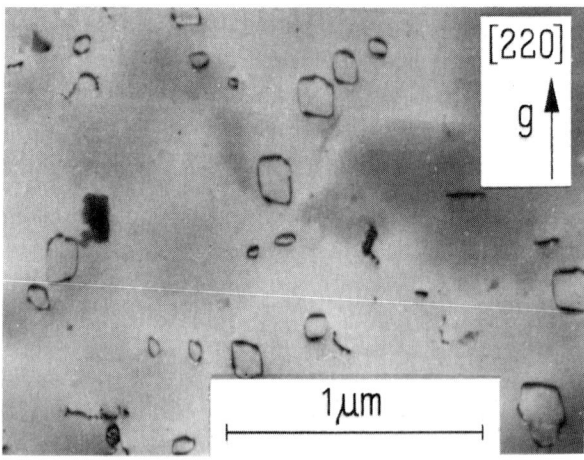

FIG. 3.5. Prismatic dislocation loops in arsenic-implanted silicon (80 keV, 5×10^{15} cm^{-2}) annealed at 1000°C for 60 min (57).

ION IMPLANTATION FOR VLSI 217

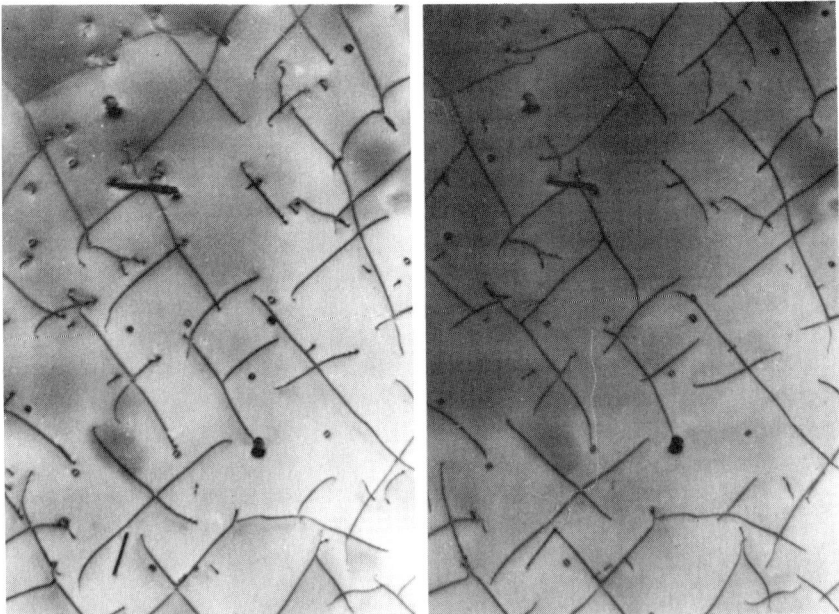

FIG. 3.6. Stereo TEM photograph of dislocation lines after arsenic implantation (80 keV, 10^{16} cm^{-2}) and annealing at 1000°C for 60 min (57).

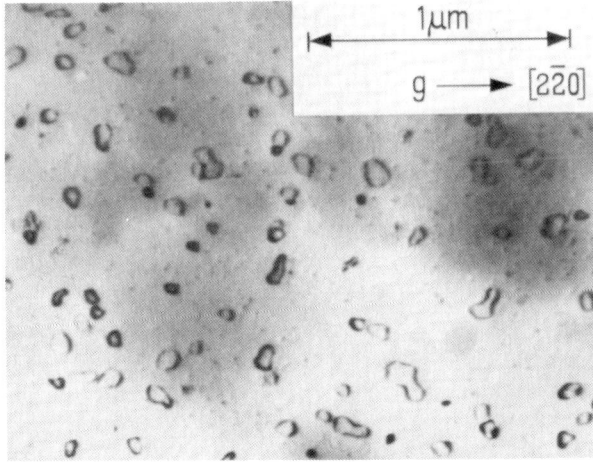

FIG. 3.7. Interstitial dislocation loops in silicon implanted with arsenic (40 keV, 2×10^{16} cm^{-2}) through 10 nm of SiO$_2$ and annealed at 976°C for 30 min (57).

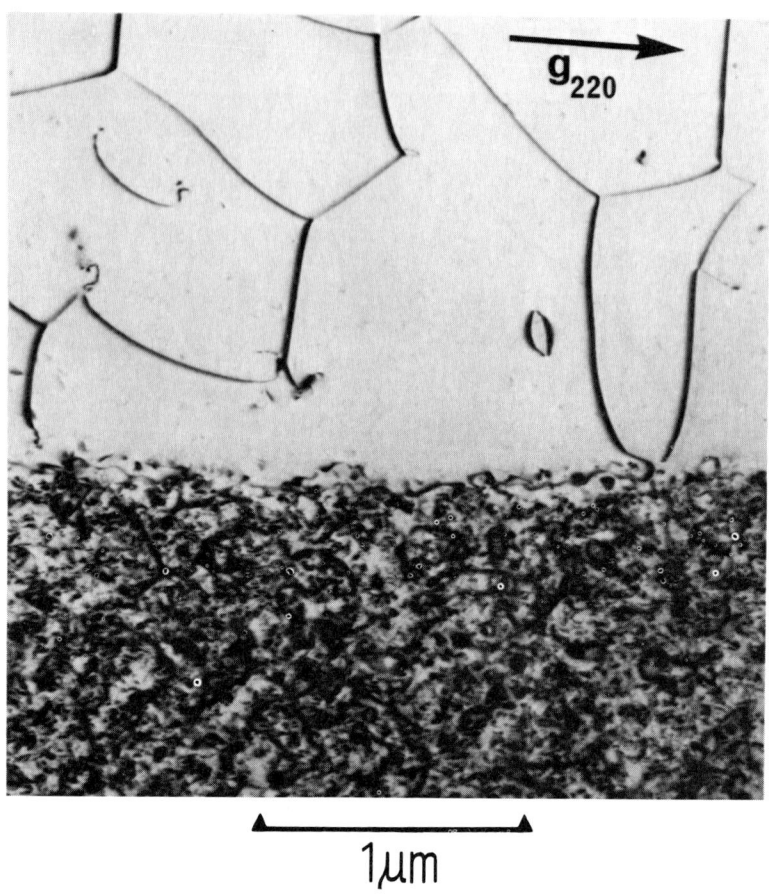

FIG. 3.8. Arsenic implantation through 43 nm SiO_2 (lower part) and into bare silicon (upper part) (58).

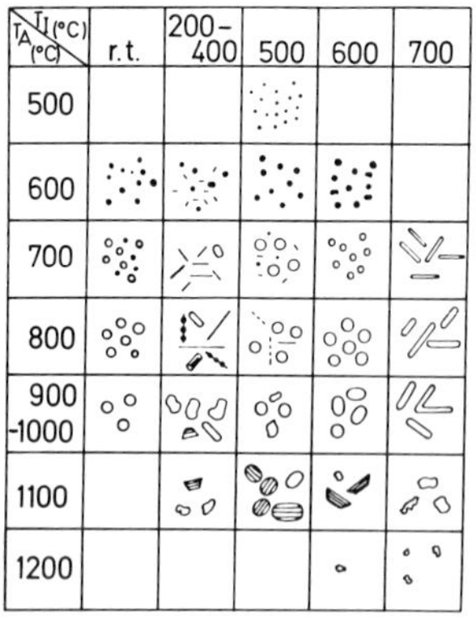

FIG. 3.9. Defects in phosphorus-implanted silicon, depending on implantation and annealing temperature (59).

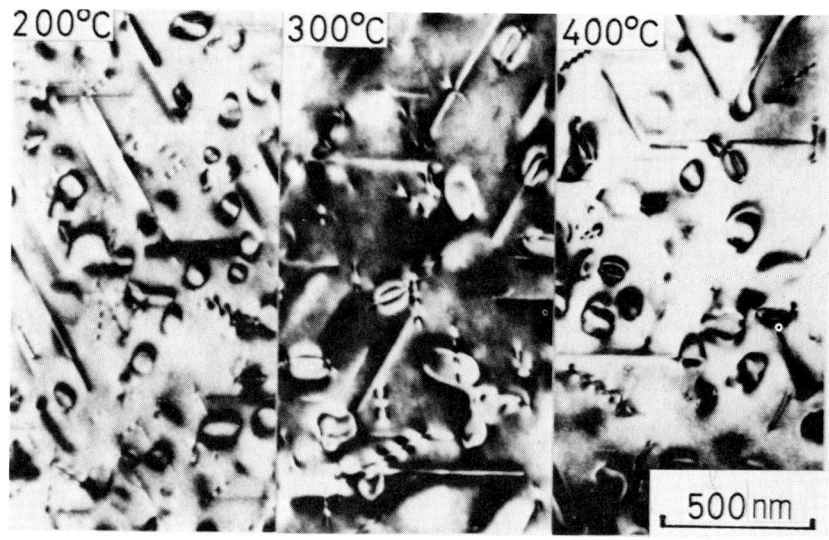

FIG. 3.10. Defects after phosphorus implantation at different substrate temperatures; annealing temperature 800°C (59).

form dislocation loops at higher temperatures and dissolve at annealing temperatures above 1000°C. Implantations at elevated temperatures result in many different defects which hardly anneal out. The appearance of rodlike defects can be correlated with a reverse annealing similar to the case of boron. A photograph of defects after implanting at 200–400°C, with subsequent annealing at 800°C, is shown in Fig. 3.10.

3.6. Process Modeling

During annealing not only the recrystallization of the lattice and the incorporation of the dopant ions on electrically active sites are important, but also the profile modifications due to diffusion, segregation, and oxidation.

Diffusion coefficients depend on orientation, atmosphere, and dopant concentration (60, 61). To obtain shallow structures low-temperature anneals are required, but in this case damage-enhanced diffusion still has to be considered. Also, it was found recently that at low temperatures the diffusion coefficient might be larger than that expected from an Arrhenius plot. This shows that even these basic parameters are not yet well known. Ion implantation is a very good method to study these phenomena.

For high concentrations the diffusion coefficient becomes concentration-dependent and additionally complicates the profile control of implanted layers.

For shallow structures the influence of the surface is much more important than for the relatively deep ones used now. Segregation at the SiO_2–Si interface is very important, but no precise data are known. At the low temperatures involved in annealing implanted layers the segregation coefficient can no longer be considered time-independent.

In Fig. 3.11, the most recent data on the segregation coefficient of boron are given (*61–63, 89, 90*). It can be seen that the spread is still much too large to draw any conclusions from these data for the calculation of diffusion processes. No comparable data at all are known for arsenic or phosphorus, not to mention any other dopant ions.

The oxidation rate depends on atmosphere and doping concentration. Whereas the first dependence is well understood, the second has been recently described by a model that explains the oxidation process as being Fermi-level-dependent (*91*). Using this model the large enhanced oxidation rate in highly n-doped material as well as the low effect with p-doped material can be explained.

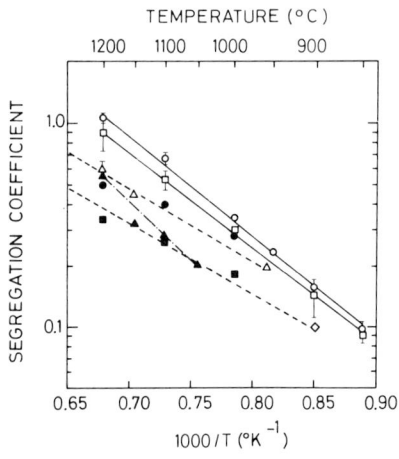

FIG. 3.11. Measured boron segregation coefficients in $\langle 100 \rangle$ and $\langle 111 \rangle$ silicon:

Plane	Curve	Ref.	Plane	Curve	Ref.
$\langle 100 \rangle$	○	*61*	$\langle 111 \rangle$	□	*61*
	●	*63*		■	*63*
	△	*62*		▲	*89*
				◇	*90*

With dimensions shrinking in vertical and lateral directions, it is becoming more and more difficult to measure doping distributions resulting from implantation and annealing. Usually, only model experiments using special structures can be performed to obtain an insight into the doping process. For device modeling, however, an exact knowledge of the profile in the lateral and vertical directions is required. This can be supplied only by a theoretical calculation using a process modeling program, e.g., SUPREM (*92*) or ICECREM (*29*). Such programs can model the diffusion behavior of implanted distributions by inert and oxidizing annealing in one dimension using suitable models for the concentration dependence of diffusion (field, vacancy concentration, clustering) and oxidation; they take into account the segregation coefficient and the coupled diffusion of different species, and also allow modeling of other processes such as epitaxy. One problem, however, is the still-insufficient knowledge of the relevant parameters.

For one-component diffusion, assuming a *n*-type semiconductor, the concentration-dependent diffusion coefficient may be written (*29*)

$$D = D_i \frac{d_F + \beta(n/n_i)}{1 + \beta} d_{Cl} \tag{3.1}$$

where D_i is the intrinsic diffusion coefficient, d_F is the field enhancement factor caused by the internal electrical field, β is the ratio of the diffusion coefficient for diffusion via negatively charged vacancies (for *n*-type semiconductors; for *p*-type semiconductors it applies to positively charged vacancies) in the doped case to that for the intrinsic case, and d_{Cl} is the cluster factor.

Assuming charge neutrality, the field enhancement factor is given by (*93*)

$$d_F = 1 + [C/(C^2 + 4n_i^2)^{1/2}] \tag{3.2}$$

Under the same assumption, n/n_i in Eq. (3.2) can be calculated to be

$$n/n_i = [C + (C^2 + 4n_i^2)^{1/2}]/2n_i \tag{3.3}$$

For the cluster factor, different models exist (*29, 60*). Particularly simple is the model of Ryssel *et al.* (*29*), where d_{Cl} is given by

$$d_{Cl} = \partial C_s/\partial C = [1 + m^2(C_s/C^*)^{m-1}]^{-1} \tag{3.4}$$

Here C_s is the concentration of the dopant on lattice sites (which diffuses and is electrically active), *m* is the order of the cluster, and C^* is related to the solubility. In Fig. 3.12, as an example, the concentration and temperature dependence of the arsenic and boron diffusion coefficients are given. For arsenic $C^* = 4 \times 10^{20}$ cm^{-3} and $m = 2$; for boron a temperature-dependent C^* of 1.5×10^{19} cm^{-3} at 800°C, of 6×10^{19} cm^{-3} at 900°C, and of 1.1×10^{20} cm^{-3} at 1000°C has been assumed, together with $m = 12$.

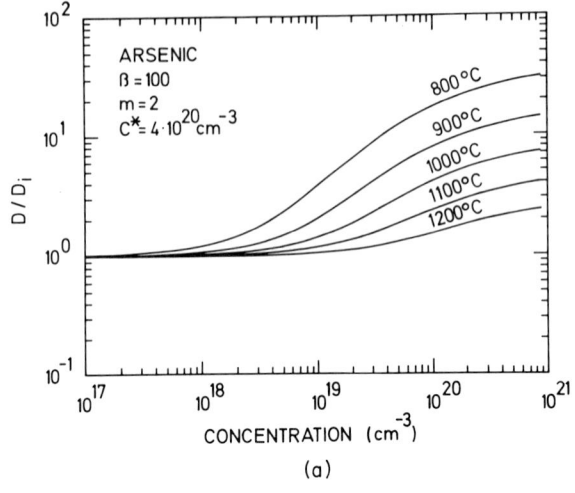

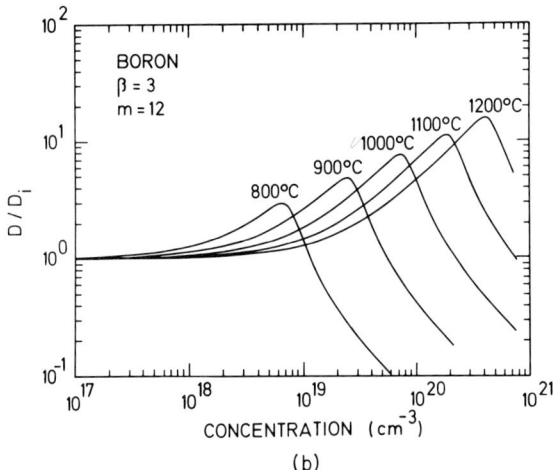

FIG. 3.12. Concentration and temperature dependence of arsenic (a) and boron (b) diffusion coefficients.

In the following figures some examples of simulations are given in part together with experimental profiles. Figure 3.13 shows measured arsenic profiles annealed at 1000°C for 60 and 240 min after 150-keV implantation through 520-Å SiO_2 with 9×10^{15} cm^{-2}. Included are also calculations according to the model of Eq. (3.1). Arsenic shows an extremely strong con-

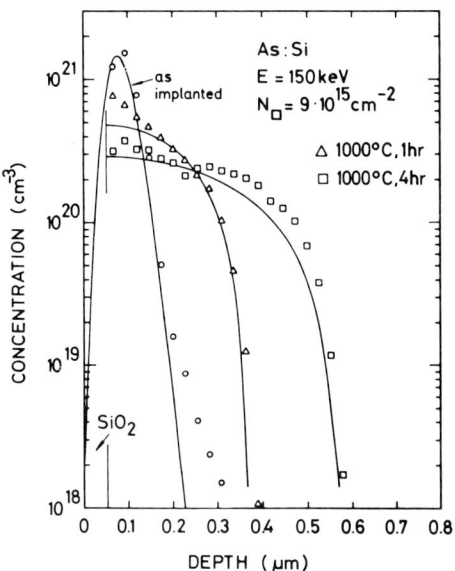

FIG. 3.13. Arsenic profiles after annealing at 1000°C for 60 and 240 min. Solid lines: theoretical Eq. (3.1).

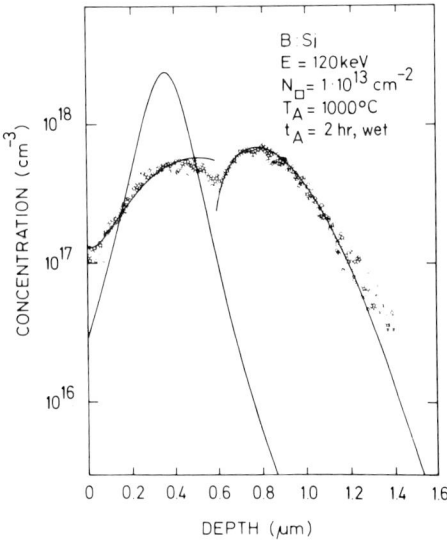

FIG. 3.14. Boron profile after wet oxidation at 1000°C for 2 hr.

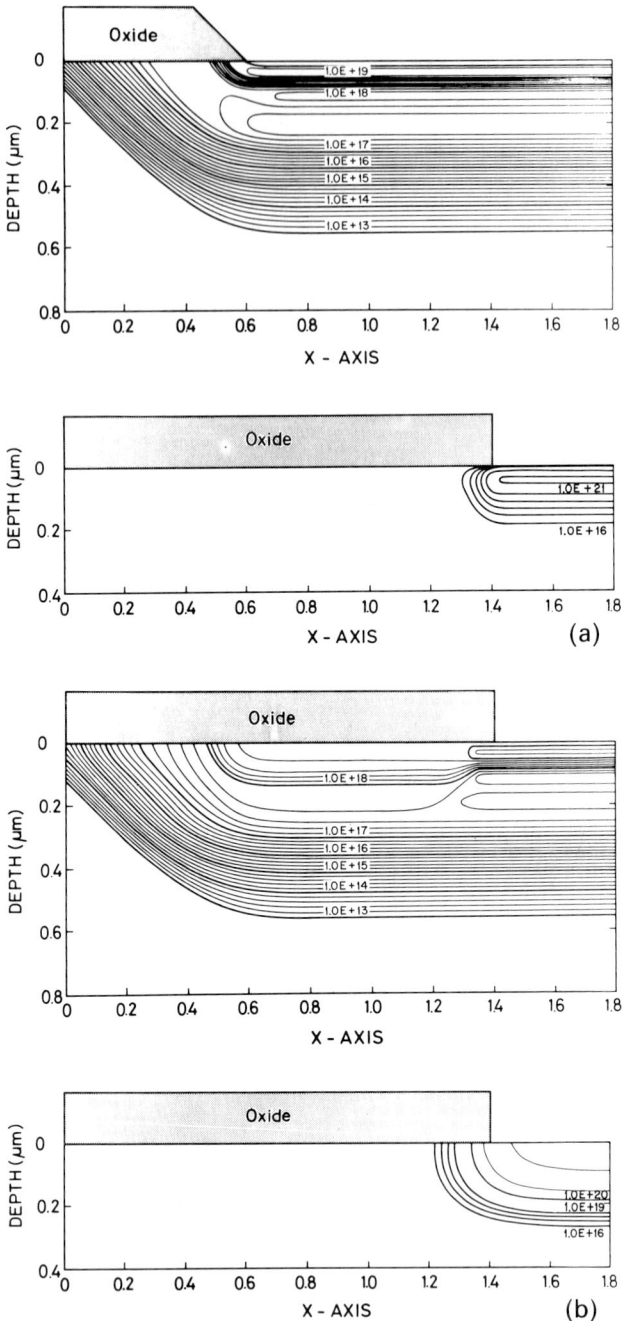

FIG. 3.15. Two-dimensional diffusion profiles of a simulated bipolar transistor. (a) Implanted boron (upper curve) and arsenic (lower curve) profiles; (b) the same profiles after diffusion at 950°C for 30 min.

centration-dependent diffusion coefficient that is very well modeled by the simulation. In Fig. 3.14 a boron profile in a SiO_2–Si structure obtained after implanting at 120 keV into bare silicon with 10^{13} cm^{-2}, and a subsequent wet oxidation at 1000°C for 120 min, is given. The solid line shows a fit to the experiment using an automatic optimization procedure included in the program ICECREM (29). The program determined the diffusion coefficients in SiO_2 and Si to be 3.5×10^{-15} and 3.5×10^{-14} cm^2 sec^{-1}, respectively; the segregation coefficient was found to be 0.42. Figure 3.15 shows the two-dimensional diffusion of arsenic and boron in silicon. Boron was implanted at 60 and 10 keV with doses of 10^{13} and 8×10^{13} cm^{-2} to form the active base and the base contacts, respectively, of a bipolar transistor. The implantation was performed through a tapered SiO_2 mask with a slope of 45°. For the emitter, arsenic was implanted through a vertical window with 80 keV and a dose of 5×10^{15} cm^{-2}; the annealing was performed at 950°C for 30 min. One clearly sees the effect of coupled diffusion at the intersection of the concentrations. For future work on process modeling of VLSI circuits, it will also be necessary to include oxidizing treatments in such a two-dimensional program.

3.7. Laser Annealing

A very new method for annealing implantation damage is laser annealing. Usually ruby or Nd:YAG lasers are used with wavelengths of 0.69 or 1.06 μm, respectively. Pulses of 20–100 nsec with an energy density of a few J cm^{-2} result in a melting of some thousand angstroms of the surface layer and a subsequent epitaxial regrowth that may be absolutely free of crystal defects. This was shown to work with very good results for silicon, germanium, GaAs, and other compound semiconductors (94–98). The melting also results in a redistribution of dopant ions since the diffusion coefficients of dopants in liquid silicon are of the order of 10^{-4} cm^2 sec^{-1}. The transition of an amorphous or heavily damaged layer occurs above a threshold energy density that depends on the thickness of the damaged layer. The threshold characteristic is caused by the necessity of melting the damaged layer down to the single-crystalline substrate. This is illustrated in Fig. 3.16 for the recrystallization of amorphous silicon layers with a pulsed ruby laser (97). Below the threshold a polycrystalline layer is formed. Depending on the energy density of the pulse or on the number of pulses, a different redistribution takes place. In Fig. 3.17 redistributed profiles of an arsenic implantation ($E = 150$ keV, $N_\Box = 10^{16}$ cm^{-2}), after annealing with a Nd:YAG laser having different energy densities, are given (98). The pulse duration was

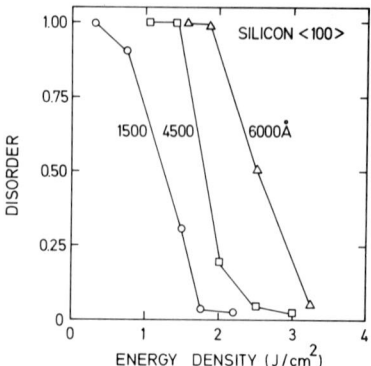

FIG. 3.16. Residual disorder measured by the channeling technique for silicon samples with different amorphous surface-layer thicknesses after single-pulse ruby laser irradiation (97). (This figure was originally presented at the Spring, 1980 Meeting of The Electrochemical Society, Inc., held in St. Louis, Missouri. Reprinted by permission, The Electrochemical Society, Inc.)

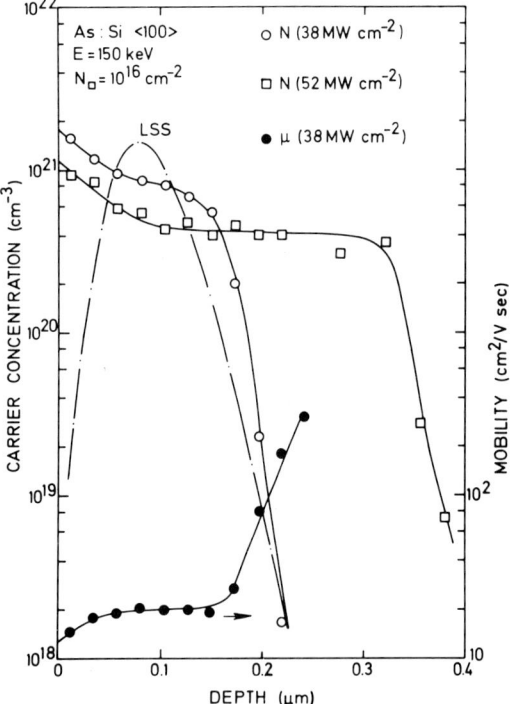

FIG. 3.17. Comparison of the carrier concentration profiles for arsenic-implanted silicon after laser annealing with 38 and 52 MW cm^{-2} and a scanning speed of 2.3 cm sec^{-1}. For the sample annealed with 38 MW cm^{-2} the mobility is also given.

120 nsec, the repetition rate 10 kHz, the scanning speed 2.3 cm sec^{-1}, the spot size 150 μm, and the power densities 38 and 52 MW cm^{-2}, respectively. The profiles show a pronounced redistribution of the arsenic. This can be explained by a melting of the top layer. In molten silicon, the diffusion coefficient of impurities is very high (of the order of 10^{-4} cm^2 sec^{-1}). The diffusion during the melt phase results in a very flat profile. The melting depth is about 0.15 μm for 38 MW cm^{-2}, and 0.33 μm for 52 MW cm^{-2}, respectively. The profiles show an accumulation of arsenic toward the surface. This can be explained by the liquid–solid segregation of the arsenic. At the surface the carrier concentration is about 1.1×10^{21} cm^{-3} for an irradiation with 52 MW cm^{-2}, and 1.8×10^{21} cm^{-3} for an irradiation with 38 MW cm^{-2}, respectively. For the sample annealed with 38 MW cm^{-2}, the mobility profile is also given in Fig. 3.17. At the surface it is as low as 15 cm^2 V^{-1} sec^{-1} due to the extremely high carrier concentration at the surface caused by the segregation of the arsenic.

The concentrations obtainable by this fast melt–regrowth process are metastable, since the solubility is exceeded. Lietoila et al. (99) found that a thermal posttreatment as low as 350°C for 40 min is sufficient to relax a supersaturated arsenic doping concentration by 6%.

An example in which the relaxation of the supersaturated arsenic was done using a cw CO_2 laser is shown in Fig. 3.18. Directly after the laser annealing with a Nd:YAG laser (power density 52 MW cm^{-2}, 7 cm sec^{-1}), a segregation peak at the surface can be seen, and a flat concentration toward the bulk of the crystal. After the CO_2 laser irradiation, the segregation peak disappears, and the concentration in the flat part of the profile is reduced, depending on the irradiation time, until the equilibrium solubility is reached.

Such supersaturated concentrations could provide a very efficient means to reduce sheet resistivity, e.g., in polysilicon lines; however, the long-term stability of supersaturated dopant concentrations at low temperatures has not yet been investigated.

The annealing of implanted distributions is limited for VLSI applications, since the redistribution of the dopants during the melting phase leads to a profile broadening. This is especially disadvantageous for bipolar transistors, in which a narrow base width is required. It could be used, however, as a substitution for certain epitaxial processes, and for annealing of implanted source and drain areas, where a lateral diffusion of the dopants can be reduced using a selective irradiation, e.g., as shown in Fig. 3.19. One very important but still unsolved problem is the different coupling of the laser light into the semiconductor in device structures where different dopant levels, crystalline structures, and oxide thicknesses influence the absorbed power, which sometimes results in cracking of oxide layers due to thermal stress and the different melting points of the oxide and silicon.

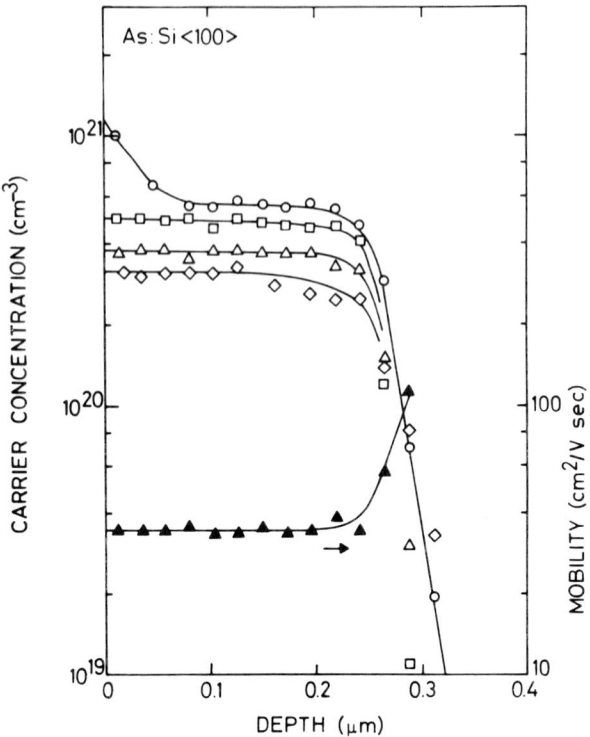

FIG. 3.18. Nd:YAG laser-annealing arsenic profiles after postirradiation with a CO_2 laser for different times. $E = 150$ keV; $N_\square = 10^{16}$ cm^{-2}; ○, Nd:YAG laser annealed (52 MW cm^{-2}, 7 cm sec^{-1}), post-treatment with CO_2 laser (100 W cm^{-2}); □, 22 sec; ▲ and △, 122 sec; ◇, 12 min.

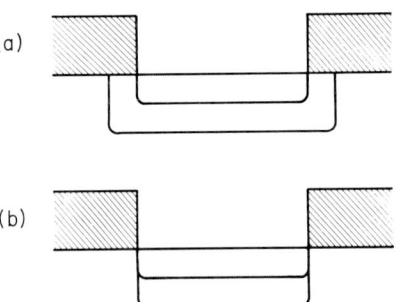

FIG. 3.19. Reduction of the lateral diffusion by laser annealing: (a) furnace annealing, (b) laser annealing. Dotted: directly after implantation; solid line: after annealing.

Well suited for the precise doping of shallow structures is the method of Gibbons et al., using a scanned cw argon laser ($\lambda = 0.49$ μm) (*100*). Without melting the material, electrical activation similar to that for thermal annealing by solid-phase epitaxy is obtained without any redistribution of the implanted ions. Therefore, a very precise tailoring of the profile is possible. Similar results can be obtained using a cw CO_2 laser with a wavelength of 10.6 μm. Although the energy of the emitted light is smaller than the band gap, an absorption of the light and heating of the sample is possible due to free-carrier absorption.

With this alternative way of laser annealing (similar results are possible using electron beams), diffusion is avoided using a very short (from a few

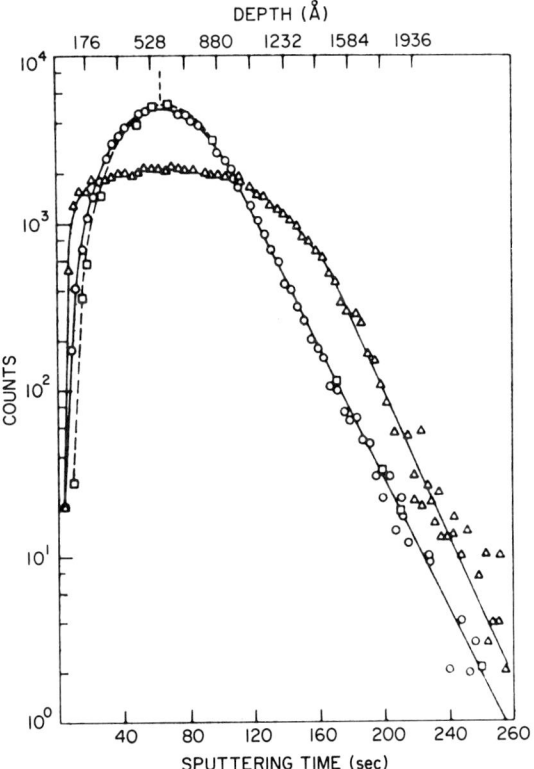

FIG. 3.20. Arsenic concentration profiles in arsenic-implanted silicon: as implanted (□), after laser anneal (○) and thermal anneal (△) (1000°C, 30 min): ---, Pearson IV distribution with LSS range statistics (*101*). (This figure was originally presented at the Spring, 1980 Meeting of the Electrochemical Society, Inc., held in St. Louis, Missouri. By permission, The Electrochemical Society, Inc.)

milliseconds to several seconds) thermal heating of the crystal up to temperatures close to the melting point. In contrast to annealing with pulsed lasers, no perfect defect-free crystals result, but rather a certain number of crystalline defects remains, as is the case with standard thermal annealing. The short-term process, however, reduces problems with the diffusion of unwanted impurities (lifetime killers such as gold and iron) and avoids a thermal diffusion of the dopant.

In Fig. 3.20, a comparison between a cw-argon-laser-annealed and a thermally annealed arsenic implantation is given. Whereas in the case of thermal annealing a remarkable diffusion has taken place, the laser-annealed profile is identical to the theoretical profile according to the LSS (*17*, *23*) theory. The next example shows the possibilities of cw CO_2 laser annealing. In Fig. 3.21 carrier concentration profiles of a boron-implanted sample are

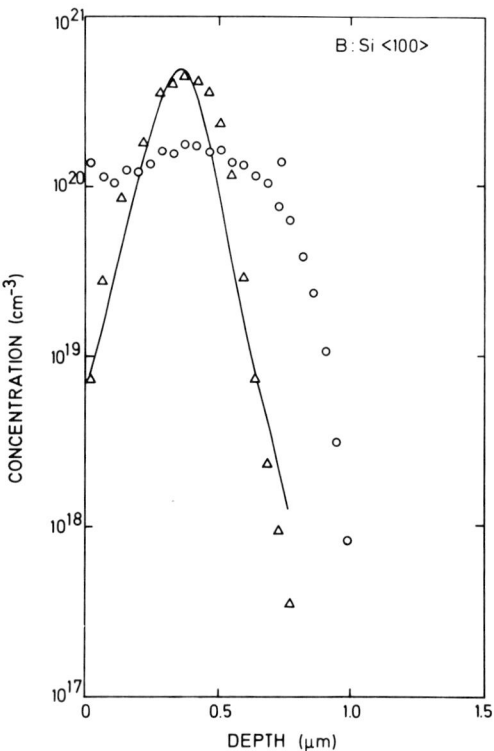

FIG. 3.21. Comparison of measured hole concentration with atomic concentration profiles. $E = 120 \text{ keV}$; $N_\square = 10^{16} \text{ cm}^{-2}$: —, not annealed, measured by the $^{10}B(n, \alpha)^7Li$ nuclear reaction; △, Hall CO_2 laser annealed; ○, Hall 1000°C, 30 min furnace annealed.

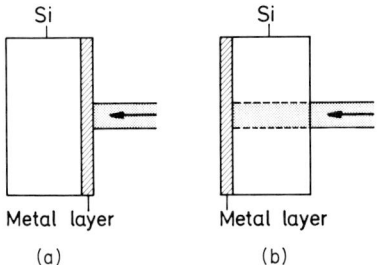

FIG. 3.22. Formation of contacts by laser irradiation: (a) irradiation of the metal layer and heating of the interface by heat conduction; (b) irradiation through the semiconductor with a CO_2 laser with direct heating.

given, again in comparison to that of a thermally annealed sample and an unannealed profile.

From Figs. 3.20 and 3.21 one can see that also in the case of a cw laser, without melting the crystal, the solubility is higher than that obtained using standard annealing temperatures between 900° and 1000°C. This makes this technique also worthwhile for the treatment of polysilicon lines.

As a new technique, laser-induced alloying of metallic layers could also evolve (*102, 103*), reducing the temperature stress of the noninvolved part of the crystal. This is shown schematically in Fig. 3.22. Two concepts are possible: irradiation of the metallic layer with subsequent melting and alloying from the front side, or irradiation and direct heating of the interface from the back surface of the semiconductor. The semiconductor has to be transparent to the laser light; therefore, a CO_2 laser has to be used. In this latter case, not only alloying contacts but also bonding to a header is possible (*102*).

4. Nondoping and Other New Applications of Implantation

For many years only the doping action of implanted ions was of any interest. The radiation damage produced as a by-product was thought to be detrimental to the desired effects of ion implantation. In recent years, however, new nondoping applications of ion implantation have been found that also utilize the radiation damage. Especially for VLSI circuits, several of these new applications have some distinct advantages. Following discussion of these, some other new doping applications will also be dealt with later in this section.

4.1. Ion-Beam Gettering

The gettering action of ion-beam-induced damage was discovered several years ago (64, 65). Most measurements, however, were done using the backscattering technique, which detects only the amount of gettered impurities, and not the electrical action of the gettering. The optimum temperature was found to be 1000°C; above this temperature the standard phosphorus gettering was shown to be better (66).

Only a few experiments were performed in which electrical parameters were measured. Hsieh et al. (67) reduced the leakage currents of silicon photodiode array cameras by argon-damaging the wafers, and Nassibian et al. (68) made several studies measuring the generation lifetime, using MOS capacitors after gettering with argon. Another study was made by Ryssel et al. (69), using the minority carrier recombination lifetime. Both investigations show that ion-implantation damage gettering is a relatively low-temperature process, well suited for VLSI technology.

Usually gettering is accomplished using the back side of the wafers. Ryssel et al. (69) compared back-side gettering to front-side gettering. The front-side gettering was done selectively in the area of the junction. The depth of the gettering implant, however, has to be much shallower than the $p-n$ depth of the active devices. The advantage of front-side gettering is that one avoids scratches on the front side of the wafer during handling in the implanter. Lifetime measurements between 800 and 1000°C have indicated that both methods give about the same result, but with a better reproducibility for front-side gettering. Therefore, all results presented have been obtained using front-side gettering.

The best gettering effect for all ions is obtained between 850 and 900°C. In Fig. 4.1 the results for all ions investigated are shown for these temperatures. It is seen that argon and BF_2 give the best results. Although C, N, and O form stable compounds with silicon they are much less effective in gettering. These experiments were also performed in a wet oxidizing ambient between 800 and 950°C. The results were very similar, but with a slight tendency toward higher lifetimes in the case of inert gettering.

In Fig. 4.2, the lifetime depending on gettering temperature for argon and BF_2 is shown in more detail. Between 800 and 850°C there is a very steep increase in lifetime for $\langle 100 \rangle$ silicon, with a similar decrease in lifetime in the case of higher temperatures. For $\langle 111 \rangle$ silicon the behavior is similar, but with an optimum lifetime at 900°C and a smaller absolute lifetime. For comparison, measurements after phosphorus gettering are also included, showing the same general behavior. The maximum lifetime is always obtained after a temperature treatment at 850–900°C for these experiments

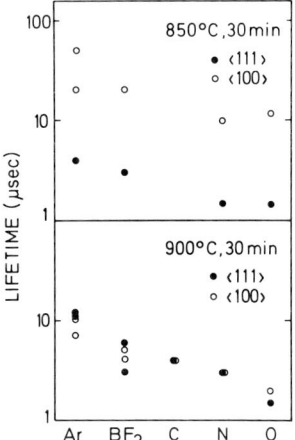

FIG. 4.1. Comparison of lifetimes for different ions at 850 and 900°C for $\langle 100 \rangle$ and $\langle 111 \rangle$ silicon.

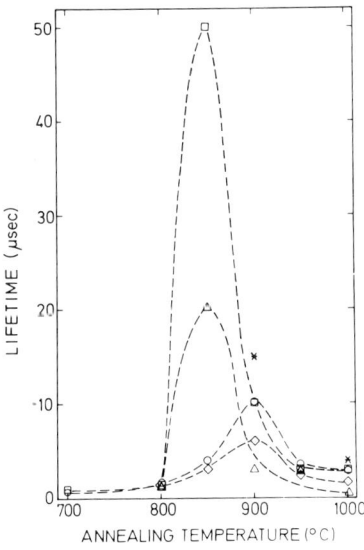

FIG. 4.2. Lifetime versus gettering temperature for argon and BF_2 in $\langle 100 \rangle$ and $\langle 111 \rangle$ silicon: $E = 150$ keV; $N_\square = 10^{16}$ cm^{-2}; $t_A = 30$ min; $\diamondsuit$, $BF_2 \langle 111 \rangle$; $\triangle$, $BF_2 \langle 100 \rangle$; $\bigcirc$, $Ar \langle 111 \rangle$; $\square$, $Ar \langle 100 \rangle$; $*$, phosphorous diffusion (50 min at 1000°C, 30 min at 300°C).

and it shows no clear orientation dependence. There is a trend, however, to a higher lifetime with $\langle 100 \rangle$ silicon. This is in contrast to results obtained by backscattering measurements. All experiments described up to now were done using float-zone silicon.

A comparison of float-zone and Czochralski material is given in Fig. 4.3. Different current densities were also used for these measurements. There is no distinct difference between both materials. The reduction of the current density seems to shift the maximum lifetime in the direction of higher temperatures. A very important problem in using the minority carrier lifetime to measure the gettering effect of the damage is the great influence of the temperature treatment on the wafer before the gettering. Therefore all results given here have to be seen qualitatively, and similar results can only be obtained using the same heat schedule.

In a different set of experiments the effect of multiple gettering was investigated. After a first gettering step at 900 or 950°C using argon or BF_2, 1500 Å of silicon were etched off, and a second gettering implant with argon was performed. The material used was $\langle 100 \rangle$ float-zone silicon. The resulting

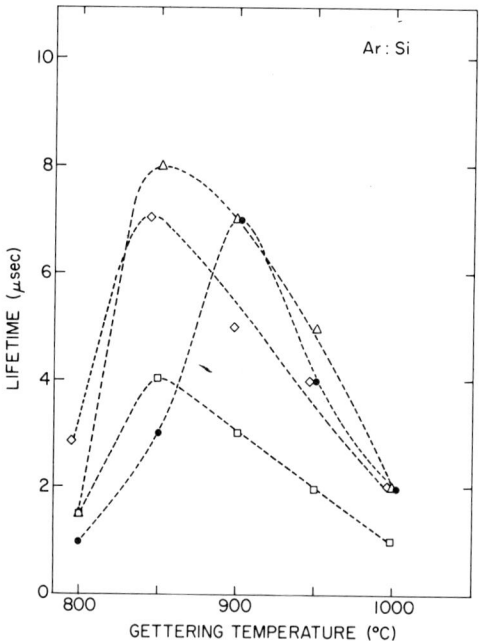

FIG. 4.3. Comparison of lifetime for float-zone (FZ) and Czochralski (CZ) silicon after gettering with argon: $E = 150$ keV; $N_\square = 1 \times 10^{16}$ cm^{-2}; current density of 60 μA cm^{-2}: $\square$, $\langle 111 \rangle$ FZ; $\triangle$, $\langle 100 \rangle$ CZ; $\diamond$, $\langle 100 \rangle$ FZ; current density 24 μA cm^{-2}: $\bullet$, $\langle 100 \rangle$ CZ.

lifetimes are given in Fig. 4.4. The best results are again obtained using argon. After an optimum argon gettering at 900°C, no further improvement is obtained. In the case of nonoptimum gettering, as is the case for all BF_2 implants or the first gettering with argon at 950°C, an improvement is obtained after the second gettering with argon at 900°C. The reference part of the wafer received the same heat treatment, but only the second gettering implant. It is seen that for the samples first gettered using BF_2 a better lifetime is obtained in the reference part of the wafer, indicating a permanent degradation of the lifetime due to the BF_2 implant. If the first gettering was done with argon, the reference did not achieve an equally good lifetime. The reason is that every heat treatment without a gettering step degrades the lifetime irreparably.

An important question for the application of ion-damage gettering is, at what point this process should be employed in the production sequence of devices. To clarify this problem, $\langle 111 \rangle$ float-zone wafers were gettered with argon damage at 900°C for 30 min with half of the wafer masked against the implantation. After this, 1500 Å were etched off from half the wafers after a 90° rotation. Diodes were then produced using the standard procedure, resulting in four groups of diodes. The first and second groups were gettered with and without etching of the gettered layer, respectively; the third was etched without a gettering implant; and the last one was left untreated as a reference quarter. The results were as follows: Gettering before diode fabrication without etching off the gettered layer resulted in a lifetime of 1–2 μsec. The reference diodes showed a lifetime of 4–5 μsec, and the diodes etched after the first gettering gave results identical to those of the reference diodes. This result clearly indicates that gettering without etching before the diode fabrication degrades the lifetime, and that gettering before a high-temper-

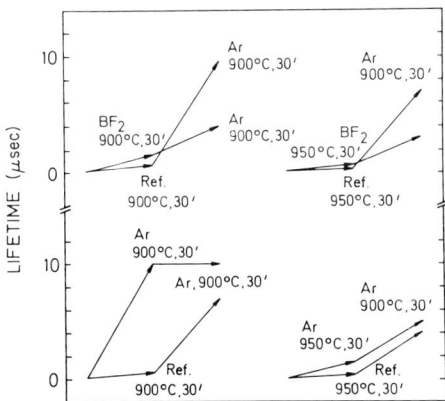

FIG. 4.4. Multiple gettering.

ature step with etching is useless because the same lifetime is obtained without this step.

Golja and Nassibian (70) used measurements of the generation lifetime to investigate the gettering of argon implants. The wafers were n-type, 8-Ω cm, $\langle 100 \rangle$-oriented silicon wafers. A 0.15-μm MOS oxide was grown at 1080°C in dry oxygen, followed by a 15-min nitrogen anneal. The wafers were implanted into half of the back surface with a dose of 10^{16} cm^{-2} at 180 keV. Annealing in dry nitrogen for 60 min at temperatures between 950 and 1100°C resulted in lifetimes of 3–7 μsec, whereas the control wafer had a lifetime of less than 2 μsec (Fig. 4.5). The optimum gettering temperature for this process is 1050°C. These results show the same general trend but another optimal gettering temperature. The solubility of generation impurities in damaged areas results in a reduction of these centers and an increase in the lifetime. The length of the gettering treatment was also varied (Fig. 4.6), showing that short temperature treatments result in the highest lifetime. This indicates that in the case of prolonged annealing the gettering damage produced by the argon anneals out and the impurities redissolve in the bulk, thus reducing the lifetime.

4.2. Damage-Enhanced Etching

In semiconductor technology, the definition of structures in layers of SiO_2, Si_3N_4, or polycrystalline silicon is usually made by photolithography, which is followed by chemical or plasma etching of the film. If the etched structure is coated by another film, the step coverage of this second film at the windows is determined by the wall contour of the window. The step

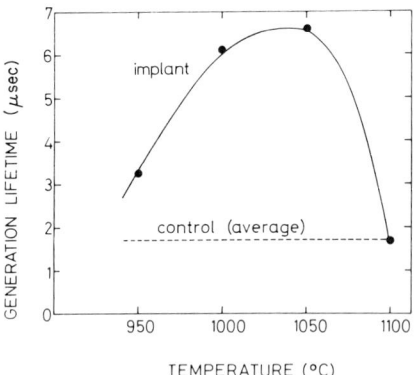

FIG. 4.5. Generation lifetime against the gettering anneal temperature (70); anneal time 60 min.

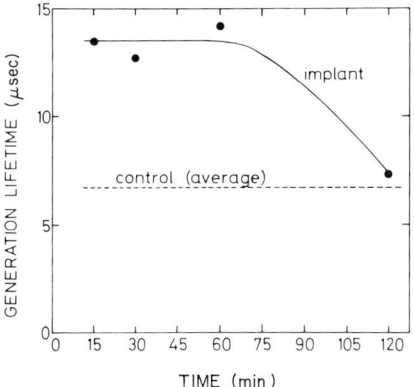

FIG. 4.6. Generation lifetime against the gettering time; anneal temperature 1050°C (70).

coverage is especially important with metal layers, since at a steep edge the metal is thinner, which can cause severe problems with reliability.

An often-used method to taper windows is the reflow of phosphorus–silicon–glass (PSG). PSG is deposited at low temperatures from silane, oxygen, and phosphine in a nitrogen atmosphere, and reflowed at about 900°C, resulting in very smooth window edges (71). New evaporation techniques using magnetron sputtering have also brought about a significant improvement. The most reliable method is, however, the tapering of the slope of the window walls.

Moline et al. (72) reported on the tapering of SiO_2 using Ar ions; North et al. (73) also used Ar ions to taper PSG glass, and Bell and Hoepfner (74) used Ar ions from an ion-etching machine for the tapering of SiO_2 layers. Akasaka et al. (75) and Parry and Bristol (77) studied the enhanced etching rate of implanted Si_3N_4.

In Fig. 4.7 results obtained with phosphorus-doped SiO_2 (PSG) are shown (73). Argon ions with 30 or 50 keV were used, with doses up to 3×10^{14} cm^{-2}. The lowest angle obtained was about 30° for a 1% phosphorus content. For higher concentrations this angle is larger. The increase for doses about 10^{14} cm^{-2} is believed to be due to an inhomogeneous phosphorus content of the SiO_2.

A detailed study of the tapering of SiO_2, Si_3N_4, and polysilicon by damage implantation with argon, arsenic, and boron ions was made by Götzlich et al. (76).

The implantation of Si_3N_4 layers with arsenic ions (80 keV) reduced the taper angle to 11° for a dose of 10^{16} cm^{-2} (Fig. 4.8). Even for the higher doses, no saturation of angle decrease could be seen; that is, the etching rate seemed to be continuously enhanced while the ion dose increased. The same de-

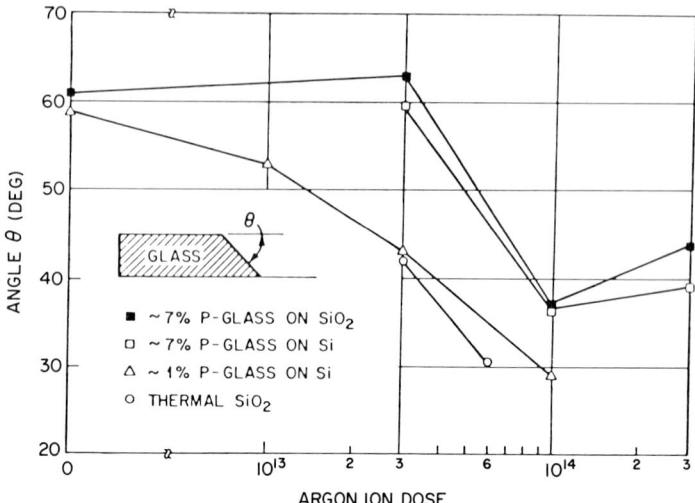

FIG. 4.7. Angle of window walls as a function of the argon-ion dose. In those cases where the window wall is not a uniform slope, the angle of the steepest part of the wall (excluding the damaged layer) was plotted (73). (By permission, The Electrochemical Society, Inc.)

pendence was observed for the implantation of 50-keV argon ions; the slope angle decreased continuously to 14° for the 10^{16}-cm^{-2} dose.

In Fig. 4.9 the slope angles of window walls in SiO_2 after implantation with argon and arsenic ions for different ion doses are shown. By implanting argon ions with 50 keV or arsenic ions with 80 keV the slope angle can be reduced to about 12°; whereas by implanting argon ions with 200 keV the

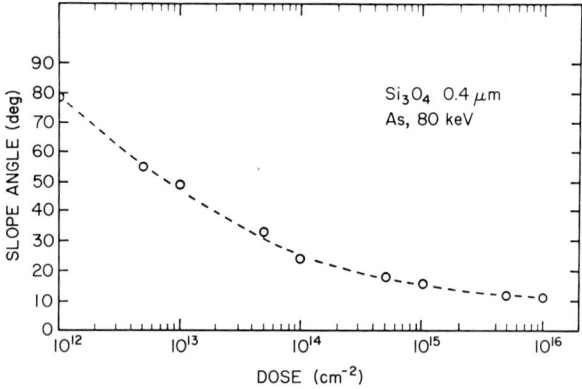

FIG. 4.8. Slope angles of window walls in Si_3N_4 versus ion dose after implantation with 80-keV arsenic.

ION IMPLANTATION FOR VLSI 239

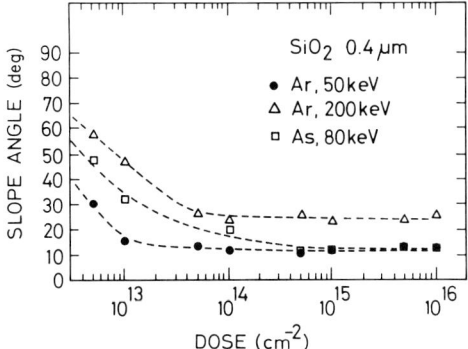

FIG. 4.9. Slope angles of window walls in SiO_2 versus ion dose after implantation with 50- and 200-keV argon and 80-keV arsenic.

minimum angle is about 24°. In all cases there is a saturation effect that depends on ion dose, species, and energy. This saturation effect should be correlated with the saturation of defects produced by ion implantation. High-energy implants always show a step on top of the etched slope for the higher doses correlated to the width of the damaged layer.

The results of the experiments with polysilicon layers are shown in Fig. 4.10. These measurements showed no effect of the ion implantation on the etching rate for doses up to 10^{13} cm^{-2}, e.g., a steep slope of 80°–90° was observed. For ion doses greater than 10^{13} cm^{-2}, there was a reduction of the slope angle in the cases of the 50-keV argon and the 80-keV arsenic implantations for the chemically etched samples as well as the plasma-etched layers. The minimum slope angle achieved with plasma-etched samples was 9° (1×10^{16} As cm^{-2}, 80 keV), whereas with the chemically

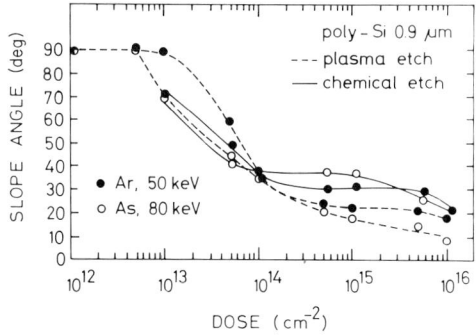

FIG. 4.10. Slope angles of window walls in polysilicon versus ion dose after implantation with 50-keV argon and 80-keV arsenic; chemical and plasma etching.

etched samples the angle could only be decreased to 23° (1 × 10¹⁶ cm⁻², 80-keV arsenic and 50-keV argon). There is also a different behavior with respect to the formation of a step in the wall profile between the two sets of samples. The chemically etched samples showed no step, whereas for the plasma-etched samples, a step was present with a height of 0.1 μm (Ar, 50 keV) or 0.2 μm (As, 80 keV) for doses greater than 5×10^{14} cm⁻². The implantation of argon with 200 keV in polysilicon layers only had an influence on the slope angle for ion doses greater than 1×10^{14} cm⁻².

The implantation of boron ions with 60 keV in SiO₂ and polysilicon layers gave no reliable results. At the higher ion doses (10¹⁵–10¹⁶ cm⁻²), a taper angle of about 40° could be observed, but the reproducibility was only ±10°. Most of the samples also showed a high step of about 0.3 μm in the wall profile due to the large range of the ions.

These results show that in SiO₂, Si₃N₄, PSG, and polysilicon layers, the slope angle of window walls can be tapered very precisely by ion implantation, allowing for a good step coverage. The slope angle depends on ion species, ion dose, and ion energy. The minimum angles achieved were about 10°, e.g., the etch rate of the damaged surface region showed a maximum enhancement of about six by ion implantation. Only low energies should be used to avoid a step on top of the desired slope resulting from enhanced etching of the thick damaged layer.

This method can also be used to avoid the underetching of windows that always occurs with chemical etches and plasma etching. The damage implantation is done with a relatively high energy through a resist window. The energy has to be adjusted to enhance the etch rate throughout the volume of the insulator. A schematic description of this technique is shown in Fig. 4.11. In this way very fine lines can be realized. The enhancement of the etching rate is larger up to a factor of 10 for SiO₂, PSG, Si₃N₄, and polysilicon after implantation.

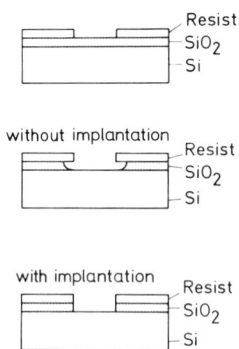

FIG. 4.11. Etching of narrow lines using implantation damage.

4.3. LOCOS Process Using Nitrogen Implantation

A very important process for LSI and VLSI circuits is the local oxidation of silicon (LOCOS) in order to obtain the necessary insulation between different devices (78). Usually this is performed using a CVD Si_3N_4 layer on top of a thin SiO_2 layer. The shortcomings of LOCOS using deposited Si_3N_4 are stress at the Si_3N_4–Si interface, lateral oxidation, anomalous oxidation at the edge of the nitride layer (white ribbon), and pinholes in the Si_3N_4 layer (78, 79). A thin SiO_2 layer beneath the deposited Si_3N_4 is required to reduce problems with stress between the Si_3N_4 and the silicon, but this results in severe problems for small structures: formation of the so-called bird's beak, due to the lateral underoxidation of the nitride. The effect is shown schematically in Fig. 4.12. The reduction in spatial resolution due to the bird's beak, and all other detrimental effects mentioned, can be circumvented using nitrogen implantation instead of CVD Si_3N_4.

Silicon nitride and silicon dioxide layers have been prepared by implanting large doses (10^{16}–2×10^{18} cm^{-2}) of nitrogen and oxygen (80–84). Wada et al. (83) could show that a strong retardation of the oxidation of crystalline and polycrystalline silicon can be obtained by using moderate doses ($\sim 10^{17}$ cm^{-2}) of nitrogen. Most investigations, however, have been devoted to the MOS properties of such synthesized layers (80, 81). It has been found that stable and uniform insulators are achievable, but capacitance measurements show that no simple nitride (or oxide) is formed; rather, a very complex structure with a doped and damaged layer underneath (81) that makes it useless as a MOS dielectric results.

Ramin et al. (84) investigated the technological properties of nitrogen-implanted silicon layers with regard to the masking behavior against thermal oxidation and etching. They used nitrogen doses of 5×10^{15}–5×10^{16} cm^{-2} with a current density of approximately 2 μA cm^{-2} at an energy of 200 keV, and doses of 6×10^{16} and 2.4×10^{17} cm^{-2} at 30 keV (obtained by im-

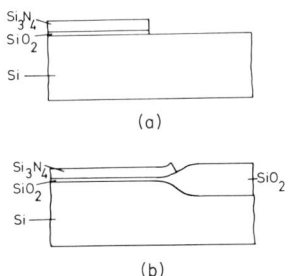

FIG. 4.12. Formation of the bird's beak. (a) Structure masked with a thin nitride layer on top of a thin oxide layer; (b) after oxidation, the bird's beak penetrates deep below the nitride.

planting 60-keV N_2^+ ions). The current density was approximately 4 μA cm^{-2} in the latter case. The annealing of the implanted layers prior to oxidation was done at 1000°C for 15 min–16 hr in dry nitrogen. The oxidation was performed in wet oxygen at 900, 950, and 1000°C for the 30-keV implantations.

In Fig. 4.13, the oxide thicknesses obtained after wet oxidation at 900°C for 15–600 min are shown for different nitrogen doses at 200 keV in ⟨100⟩-oriented silicon. Starting from a dose of 5×10^{15} cm^{-2}, the oxidation of silicon is markedly reduced.

To obtain a stoichiometric nitride in an area of $\pm \Delta R_p$ around the peak of the nitrogen distribution, a dose of 3×10^{18} cm^{-2} would be required, assuming a range R_p of 468 nm and a range straggling ΔR_p of 107 nm (23). The results indicate that a dose far below this stoichiometric dose is sufficient to suppress oxidation; the highest suppression, however, is obtained for the highest dose used, which was 5×10^{16} cm^{-2} in this case.

At 30 keV, much higher doses are required to obtain the same reduction of the oxidation. This is due to the smaller range straggling, which results in a thinner Si_3N_4 layer (for 30 keV, ΔR_p is only 32 nm). An out-diffusion or a diffusion broadening of the nitrogen does not take place because of the low diffusion coefficient of nitrogen in silicon.

A dose of 2.4×10^{17} cm^{-2} is sufficient to form a Si_3N_4 layer with a width of approximately 64 nm as measured by ellipsometry. In Fig. 4.14, the results for an oxidation at 950°C are given. After an oxidation at 950°C

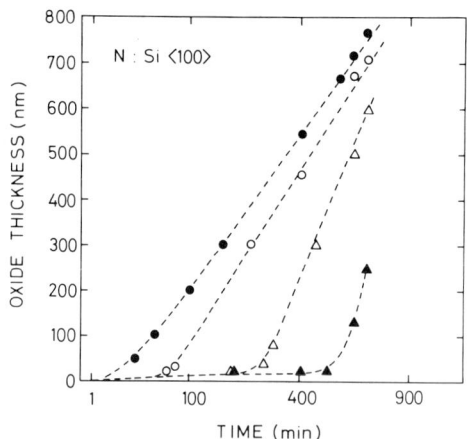

FIG. 4.13. Time dependence of oxide thickness of a 200-keV nitrogen-implanted ⟨100⟩-oriented silicon wafer with different nitrogen doses: $E = 200$ keV; $T_{ox} = 900°C$; ●, unimplanted; ○, 5×10^{15} cm^{-2}; △, 1×10^{16} cm^{-2}; ▲, 5×16^{16} cm^{-2}.

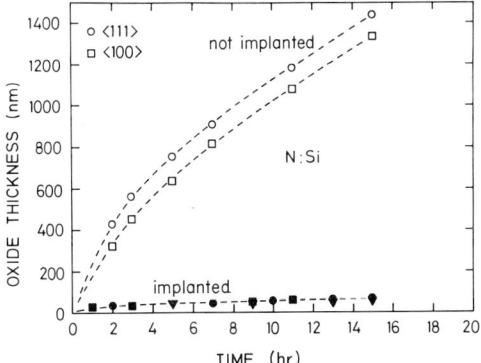

FIG. 4.14. Time dependence of oxide thickness of a 30-keV nitrogen implantation with a dose of 2.4×10^{17} cm^{-2} into $\langle 100 \rangle$- and $\langle 111 \rangle$-oriented silicon; preannealing at $T_A = 1000°C$ for (●) 15 min, (■) 4 hr, and (▼) 16 hr; $T_{ox} = 950°C$.

for 15 hr the implantation-synthesized Si_3N_4 still masks the silicon against oxidation. The oxide thickness is 1.3 μm in the case of $\langle 100 \rangle$ silicon and more than 1.4 μm in the case of $\langle 111 \rangle$ silicon, whereas the oxidized nitride has a thickness of 600–700 Å, depending on the preoxidation heat treatment. It was found by IR spectroscopy that no heat treatment is necessary to form the Si_3N_4 before the thermal oxidation starts.

The oxidation of implantation-synthesized Si_3N_4 is proportional to the time t. This conclusion is in contrast to the results of Enomoto (104), who found that the oxidation of CVD Si_3N_4 is proportional to $t^{2/3}$.

The oxidation rates for the implantation-synthesized Si_3N_4 layers are between 10 and 0.12 Å min^{-1}, depending on dose and orientation. For the 30-keV implant with a dose of 2.4×10^{17} cm^{-2}, the oxidation rate of Si_3N_4 increases by a factor of approximately two when the oxidation temperature is changed from 900 to 1000°C, respectively. The largest reduction in oxidation rate was obtained for the highest dose.

The etching behavior in H_3PO_4 and buffered HF is similar to that for CVD Si_3N_4, and is given in Table 4.1.

No indication of pinholes could be found in any of the experiments. This is obviously due to the very homogeneous distribution of nitrogen obtained by the scanned implantation of the ions. This assumption is not in conflict with the assumed existence of silicon islands in the implantation-synthesized Si_3N_4. Due to their microscopic nature, they cannot influence the oxidation-inhibiting properties of the Si_3N_4.

In Fig. 4.15 a comparison for the Si_3N_4–SiO_2 edge is given between CVD Si_3N_4 and implantation-synthesized Si_3N_4. In the case of the CVD Si_3N_4, the formation of the bird's beak is clearly seen, whereas in the case of

TABLE 4.1

ETCHING RATES OF IMPLANTATION-SYNTHESIZED
Si_3N_4 AND CVD Si_3N_4

Layer	Etch	Etching rate (Å min^{-1})
CVD Si_3N_4	H_3PO_4	45
	HF (4%)	10
$I^2Si_3N_4$ (6 × 10^{16} cm^{-2})	H_3PO_4	Not measured
	HF (4%)	50
$I^2Si_3N_4$ (2.4 × 10^{17} cm^{-2})	H_3PO_4	30
	HF (4%)	20
Thermal oxide	H_3PO_4	<1
	HF (4%)	800

the implanted Si_3N_4 no indication of it exists. Only a small underoxidation without the typical beak is seen. This can be explained by the absence of the SiO_2 layer required with CVD Si_3N_4 to reduce stress.

4.4. Ion-Beam Lithography

One of the oldest ideas of ion-implantation specialists is the direct implantation of structures by means of ion-beam writing (105, 110). In this technique no masking layers would be required since the diodes and transistors are directly written with a focused ion beam into a bare or (preferably) a thin oxidized wafer. The main disadvantage of this technique has been the time required to write a circuit. But by using a shaped beam similar to that used in electron-beam exposure the time needed can be considerably reduced. The second problem with this method has been that at the high current densities required for a reasonable implant speed, the silicon would melt. But now we possess the technique of laser annealing. With focused current densities of up to 10^6 A cm^{-2} and doses of 10^{15} cm^{-2}, one can obtain the temperatures required for solid-state epitaxy. Recently, several new and very bright ion sources have been developed, capable of delivering large current densities (106–108).

A much better and more immediate application for ion-beam writing, however, is the technique of ion-beam lithography. The advantages of ions over electrons, which are nowadays used for advanced lithography, are the following:

(1) Electrons produce a huge amount of fast backscattered electrons. They form an electron cloud that limits the possible resolution of the resist, depending on its thickness, to about 0.1–0.3 μm. In contrast, ions produce

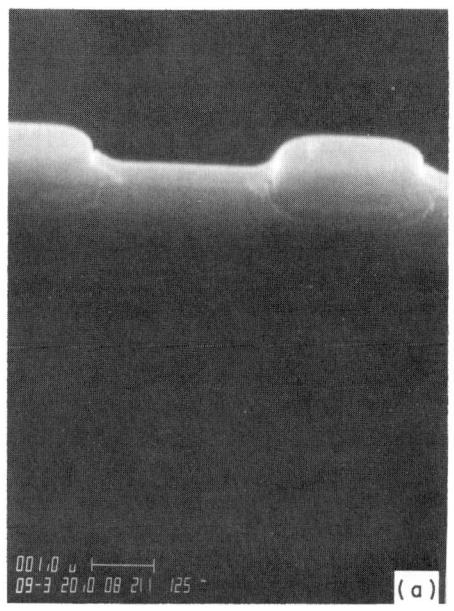

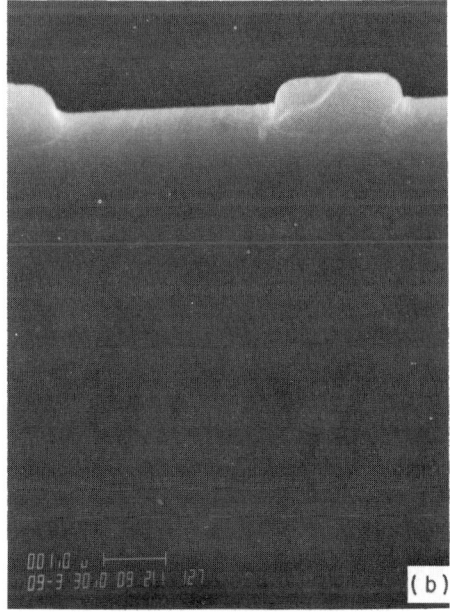

FIG. 4.15. LOCOS-oxide of 1-μm thickness. Masking during oxidation by (a) CVD nitride; (b) nitrogen-implanted silicon with a dose of 2.4×10^{17} cm^{-2} and an energy of 30 keV.

only low-energetic secondary electrons (in the range of several dozen electron volts) due to the inefficient energy transfer process. Therefore, the proximity effect, which requires difficult correction in electron-beam lithography, is circumvented.

(2) The second great advantage of ion beams is the much higher sensitivity of the resists to ion exposure, as compared to electron exposure. This depends somewhat on the ion mass, but is up to a factor of 100 higher for hydrogen ions in the case of positive resists and up to a factor of 10 for negative resists. For heavy ions such as argon, sensitivity enhancement up to a factor of 300 could be found (109). Some characteristic exposure curves for different resists are given in Fig. 4.16. EPMF 1, Elvacite 2008, and XXL-15 are PMMA resists, whereas OEBR 1000 and OEBR 1010 are mixtures of PMMA and ECA, and of PMIBK, anone, and cyclohexanol, respectively.

(3) Ions deposit their energy nearly constantly along their path, resulting in a very homogeneous exposure. In Fig. 4.17 an example of a vertical step obtained in a 1-μm-thick resist layer (PMMA Elvacite 2041) exposed with 3×10^{12} cm^{-2} hydrogen at 120 keV is shown. The proximity distance between mask and wafer was ~ 25 μm. The developing time is not a critical parameter for obtaining vertical walls, as is the case with electron-beam exposure.

Two different methods are possible in ion-beam lithography: With a focused beam, either the exposure of masks or direct exposure of the wafer is possible, just like in electron-beam exposure but avoiding the latter's previously mentioned shortcomings. Using a suitable mask, however, pro-

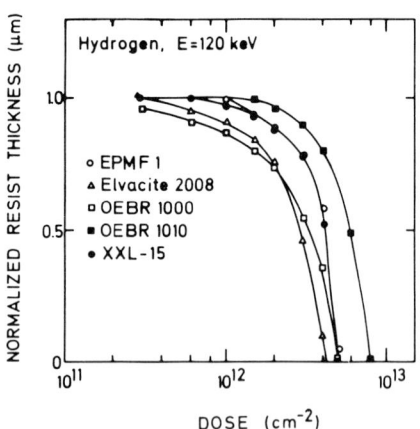

FIG. 4.16. Exposure characteristics of different positive resists irradiated with hydrogen ions at 120 keV (109).

FIG. 4.17. Profile of 1-μm PMMA exposed through a shadow mask with a dose of 3×10^{12} cm^{-2} hydrogen at 120 keV.

jection printing is also feasible (*111–113*). How this can be done using the channeling effect is shown in Fig. 4.18. A thin silicon foil supported by a thick rim is used to carry a masking layer made from gold or other materials. The silicon foil is oriented so as to allow a channeling of the ions and the masking metal must be thick enough to stop the ions.

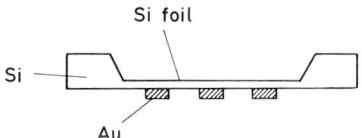

FIG. 4.18. Ion projection mask.

4.5. Various New Applications

In this part of the article some possible future applications will be treated that are worth mentioning but are still in a very premature state. Therefore no firm statements concerning their future can be made.

Buried conducting layers were proposed for device applications long ago (132), but until recently the depth required was too large and the accelerators needed were too costly. But for VLSI with high doping levels, in the collector as well as with shallow structures, this now seems realistic. Recently, a fully implanted transistor using high-energy ion implantation was developed at IBM (133). Details will be given in Section 5.4.

The insulation of integrated circuits is usually accomplished using reverse-biased junctions or a thick thermal oxide. For shallow structures the implantation of buried insulating layers also seems to be realistic. Several papers have been published on SiO_2 and Si_3N_4 production by ion implantation (81). The properties, however, were poor. For buried layers used for insulation only the conductivity is of major importance. The implantation of 3×10^{17} cm^{-2} ions with an energy of 1.2 MeV resulted in a 0.5-μm-thick insulating layer 2.2 μm deep, with a resistivity greater than 4×10^{12} Ω cm (134), and with breakdown voltages greater than 5×10^6 V cm^{-1}. We could realize diodes and transistors in the 2.2-μm-thick top layer of such structures, but further investigations are required to clarify the possibilities of this technique. The method for producing completely insulated devices using this technique is shown in Fig. 4.19. After implantation, the silicon is etched halfway down to the Si_3N_4 layer and reoxidized, yielding a planar surface and completely isolated islands in which devices can be built using standard techniques. With this method fully insulated structures similar to those obtainable using the silicon-on-saphire (SOS) technique seem to be possible.

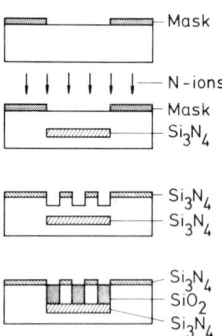

FIG. 4.19. Fully insulated structures by high-energy implantation of nitrogen.

An alternative technique uses only medium energies, with a subsequent epitaxy, to form a conducting layer within which devices can be formed (135). In Fig. 4.20, the process is schematically given. Nitrogen molecular ions of 10^{18} cm^{-2} were implanted at 180 keV in the substrate, and then annealed for 14 hr at 1000°C. At this stage, the nitrogen-implanted layer became highly resistive silicon nitride, with a breakdown voltage of 160 V. The thickness of the insulating layer was about 2 μm. Silicon single crystals remained on top of this. An epitaxial layer about 1 μm thick was then grown on the surface, and MOSFETs were fabricated in the epilayer. The electrical characteristics of these FETs showed a surface carrier mobility comparable to conventionally processed ones.

A further technique uses multiple neon implants to produce insulating silicon layers (136). Sheet resistivities up to 3×10^5 Ω cm and an insulation depth of 1.1 μm could be achieved. These layers were stable up to temperatures of 455°C.

A completely different approach to obtain layers for VLSI circuits uses low-energy ion deposition. With this technique, ions from an ion source are extracted, mass-separated, and after deceleration directed onto the target. The final energy is about 100 eV, and is therefore much higher than with molecular beam epitaxy. Using this technique, multilayer structures—as in the case of molecular beam epitaxy—can be fabricated, but as yet no devices have been realized.

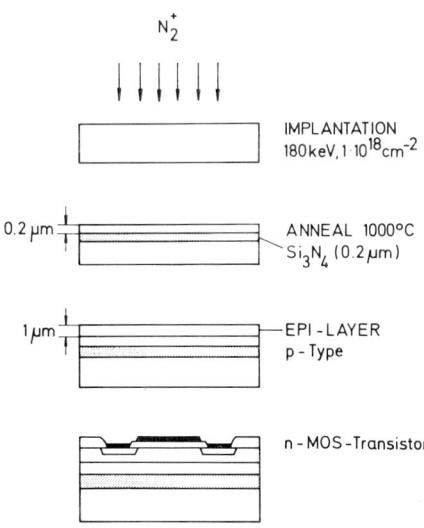

FIG. 4.20. Buried-nitride MOS fabrication process (135).

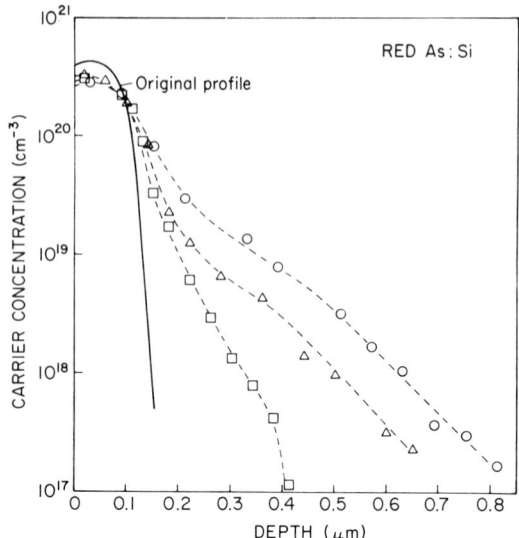

Fig. 4.21. Radiation-enhanced diffusion (RED) of arsenic at 900°C; □, 3 min; △, 7 min; ○, 30 min; $R_{PH_{eff}} = 1.2$ μm; $E_H = 200$ keV; $t_{ox} = 0.55$ μm.

Radiation-enhanced diffusion (RED) is caused by an irradiation of silicon at relatively low temperatures with light ions that produce vacancies. These vacancies enhance the diffusion, since all standard dopants diffuse via a vacancy effect (*114, 115, 137*).

So far we have conducted experiments using hydrogen, argon, and helium. Good results are only possible using hydrogen, since the other elements produce bubbles and damage the silicon permanently. An example of a RED arsenic profile is given in Fig. 4.21.

An application of this technique is possible for selective diffusion, using a suitable mask or a focused beam. In this way deep contacts or subcollector pedestals are possible, as shown in Fig. 4.22.

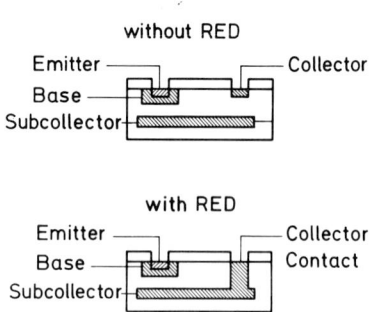

Fig. 4.22. Fabrication of a collector contact using radiation-enhanced diffusion (RED).

5. Application of Implantation to Devices

In this section, no attempt is made to cover fully all possible applications of ion implantation in modern device technology or its impact on future VLSI circuits. Too much is changing in too short a time. Therefore, only some typical device applications where ion implantation is used extensively, or where the devices are only possible using implantation, are given to show the possibilities and the power potentials of ion-implantation doping. Not mentioned again are the general features of superior homogeneity and reproducibility that are inherent when using this doping technique.

5.1. MOS Devices

The first applications of ion implantation for devices were in the field of MOS circuits. The oldest application is the self-aligned gate. At that time, drain and source were diffused, and only the area between them and the gate was implanted, using the gate as a mask to reduce the gate–drain overlap

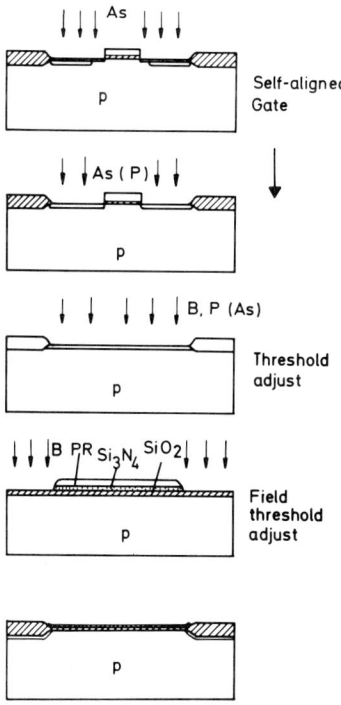

FIG. 5.1. Applications of implantation to MOS transistors.

(Miller capacitance) (6). Nowadays, self-aligned structures using polysilicon gates completely avoid the diffusion process. The implantation is usually done through the gate oxide, which also covers the source and drain area. The threshold of the transistors is also adjusted using ion implantation (7, 8). In the case of n-channel devices, which are the most prevalent MOS circuits, boron is used for enhancement, and phosphorus or arsenic for depletion devices. There is no way other than ion implantation for a free adjustment of the threshold voltages. A second deep boron implantation can be used to raise the punch-through voltage between source and drain. In Fig. 5.1 these applications are shown schematically. Another implantation step is used to increase the field threshold beneath the thick oxide. For this purpose boron is implanted before the field oxide is grown. In Fig. 5.2 examples of the dependence of the field threshold and the enhancement gate threshold voltages are given as a function of the implanted dose and the substrate voltage (116). Figure 5.3 shows a cross section through a complete structure consisting of depletion and enhancement devices.

For complementary MOS (CMOS) devices, all the above-mentioned techniques apply as well as the doping of the p-well (or sometimes the n-well). Doses of about $2-10 \times 10^{13}$ cm^{-2} boron are implanted and driven into a depth of several microns at 1200°C. After this step, complementary MOS transistors are fabricated using standard techniques (Fig. 5.4). The great advantage of ion implantation for this application is the exact control of the carrier concentration in the well.

In the near future all diffusion steps in the manufacture of MOS integrated circuit devices will be eliminated due to the greatly superior reproducibility obtainable by implantation. For VLSI circuits, boron and arsenic will be used for threshold adjustment and source–drain implantation, as well as phosphorus and arsenic for depletion load devices in n-type-channel MOS (NMOS) technology. For CMOS, p-wells or n-wells can be implanted using boron or phosphorus, respectively.

In the following, some special applications of ion implantation to MOS circuits will be given.

An advanced process for fabricating fully implanted 1-μm MOS circuits using six implantation steps is summarized in Table 5.1 (117). Electron-beam exposure is used throughout. Back-side ion implantation of boron and argon, the first step in wafer processing, is used both for gettering and ohmic back-side contacts. The third implant is the field implant to avoid the formation of inversion channels. The threshold of enhancement transistors is adjusted using a boron implant, and that of the depletion transistors using an arsenic implant. Source and drain are implanted using the gate oxide as a screen. An etch-back procedure of the thick oxide is used to avoid problems with low thick-oxide parasitic thresholds, believed to be caused by trapped

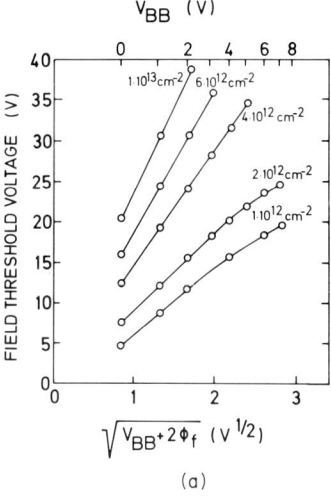

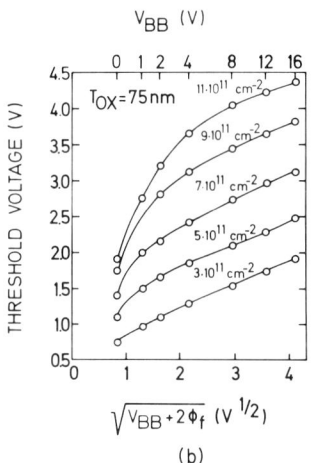

FIG. 5.2. Dependence of field threshold (a) and enhancement threshold (b) on dose and substrate bias (116). (Copyright © 1975, IEEE.)

holes in the arsenic-implanted field oxide. This has been done using the etch-rate ratio of 10:1 for implanted unannealed SiO_2 to thermal SiO_2, to minimize undercut of gate and field oxide beneath polysilicon. This process involves no diffusion step and is an example of a typical advanced VLSI technology for NMOS. In the future the LOCOS process might be modified using ion implantation of nitrogen (Section 4.3) to further increase dimensional control and thus decrease the dimensions.

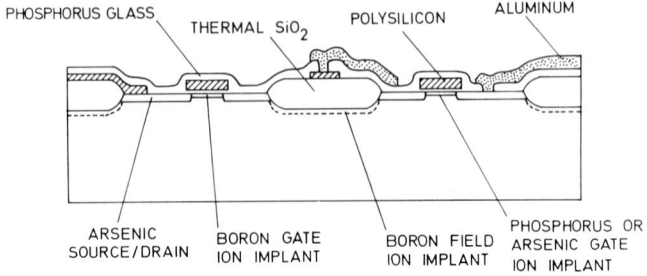

FIG. 5.3. Cross section of a high-speed n-channel silicon-gate MOS structure.

Another very interesting application of ion implantation is the double implanted MOS (DIMOS) transistor (*118*). In Fig. 5.5 the structure of the standard double diffused MOS (DMOS) in comparison to the ramp-gate DIMOS transistor is given. Here, only the latter technique will be discussed. The ramp shape of the polysilicon stripes is obtained by damaging the polysilicon layer using arsenic ion implantation (1×10^{15} cm^{-2}, 30 keV) prior to photolithography. The polysilicon thickness is about 0.5 μm.

The source–drain arsenic implantation is made without additional masking, whereas for the channel boron implantation the drain regions are covered by photoresist. In this example three implantation steps are used, and the device is not possible without implantation. The slope of the polysilicon gate can be adjusted by the dose of the damaging ions, therefore allowing a very precise control of the channel length.

The application of fluorine implantation to obtain polysilicon growth on SiO$_2$ and epitaxial growth on single-crystalline silicon for the manufacture

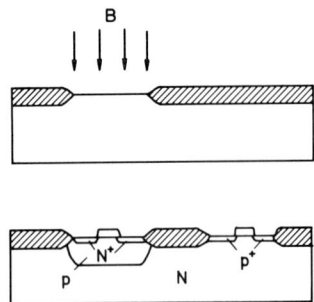

FIG. 5.4. CMOS manufacture by ion implantation.

TABLE 5.1

Step-by Step 1-μm MOS Fabrication Process[a]

Step 1. Starting substrate—p, 5 Ω cm, (100)
Step 2. Back-side I/I—^{11}B, 200 keV, 3×10^{15} cm^{-2}
Step 3. Back-side I/I—^{40}Ar, 350 keV, 5×10^{15} cm^{-2}
Step 4. Mask 1: defines alignment marks
Step 5. Mask 2: defines active area
Step 6. I/I field—^{11}B, 65 keV, 5×10^{12} cm^{-2}
Step 7. Growth of semirecessed field oxide
Step 8. I/I enhancement channel—^{11}B, 20 keV, 9.3×10^{11} cm^{-2}
Step 9. Growth of gate oxide
Step 10. Mask 3: defines depletion channel I/I mask
Step 11. I/I depletion channel—^{75}As, 72 keV, 1.9×10^{12} cm^{-2}
Step 12. Mask 4: Defines polysilicon gate
Step 13. I/I source–drain—^{75}As, 100 keV, 1×10^{16} cm^{-2}
Step 14. Mask 5: defines contact hole
Step 15. Mask 6: defines metallization

Nominal vertical dimensions (nm) in finished devices are:
Gate oxide	25
Field oxide under polysilicon	305
Polysilicon	320
Contact hole stack over source–drain and polysilicon	320
Field oxide under aluminum	435
Junction depth	350
Aluminum	500

[a] After (117). (Copyright © 1978, IEEE.)

of high-density MOS circuits (BOMOS) is a further example (119). The process is shown in Fig. 5.6. Several ion species such as boron, BF$_2$, phosphorus, arsenic, and fluorine were tried at various doses and energies. No significant differences were observed between species, which are equally effective on the enhanced nucleation. Among them, only the fluorine ion

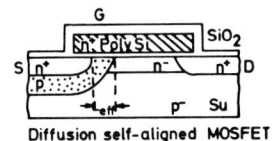
Diffusion self-aligned MOSFET

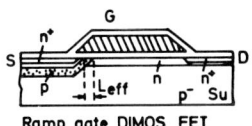
Ramp gate DIMOS FET

FIG. 5.5. Comparison of diffusion self-aligned and DIMOS field-effect transistors.

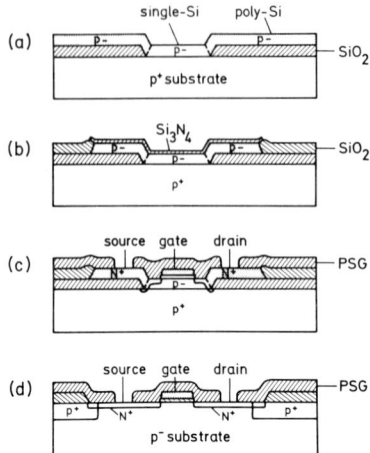

FIG. 5.6. Schematic diagram of various stages in the fabrication process of BOMOS (a–c) and bulk MOS (d) (*119*). (Copyright © 1978, IEEE.)

keeps the deposited silicon film neutral in polarity after successive annealing, and fluorine is fortunately available from boron trifluoride (BF_3), which is routinely used as a source gas for BF_2 implantation in MOS production. The physical reason for the enhanced nucleation effect is not yet well understood, but the ion-induced damage or internal stress near the oxide surface seems to produce strong nucleation sites or traps for migrating Si atoms on the surface. Another effect of ion implantation is the taper-etch of the oxide window by $HF-HNO_3$ etchant, which becomes significant above 1×10^{14} cm^{-2}—an excess tapering is not favorable for fine patterns. Fluorine implants of 5×10^{13}–1×10^{14} cm^{-2} give the best result for a moderate taper-etch of the oxide and the nucleation effect. The threshold is adjusted using boron implantation after local oxidation and gate oxidation. In this case, source and drain were produced by diffusion from doped oxide through the polysilicon. In the same way, RAM cells are also possible (*120*). The process could be further improved by also using implantation for source and drain.

A growing number of additional new applications of ion implantation may be found in the specialized literature, especially in the proceedings of the annual IEDM meetings. Among these are (without judging their technological future): implantation under different angles to obtain extremely small channel length, double implants into the channel to adjust the threshold and also the source–drain punch-through voltage, and adjusting the barrier voltage of Schottky diodes.

5.2. Schottky and Junction FET Devices

Junction and Schottky field-effect transistors (FETs) can be fabricated in silicon using implantation (junction formation, adjustment of barrier height), but they are used only for special applications, and presently play no role in VLSI technology. The situation is different with GaAs, where only Schottky or junction FETs are used.

Integrated circuits in GaAs are an especially good example for the application of ion implantation. Without implantation, integrated circuits were fabricated using mesa-etching of epitaxial layers on semiinsulating silicon, with all the well-known shortcomings of this technique. Now, with selective ion implantation instead of full-wafer epitaxy, planar structures insulated against each other without etching are possible. Moreover, these implanted devices show a better stability than epitaxial ones.

Integrated circuits in GaAs will gain a part of the VLSI market due to their inherent speed. Especially for on-line data handling and management, they will make gains against silicon. This includes not only military applications such as fast radar and other target-recognition techniques, but also optical communications, TV tuner, and satellite TV applications. Presently GaAs ICs are beginning to enter the field of LSI, but progress is rapid, and soon VLSI circuits will be available.

In Fig. 5.7, a process using two implantation steps to fabricate Schottky FETs is depicted (121). Initially, the GaAs is coated with a thin Si_3N_4 layer that remains on the wafer throughout all subsequent processing steps. A selenium implant (400 keV, 2.2×10^{12} cm^{-2}) provides for source and drain. Simultaneously, diodes are made. The sulfur implant results in a much deeper profile (400 Å versus 1500 Å with selenium), ideally suited for high-speed switching diodes, whereas both implants are used for level-shifting diodes and enhancement of the doping under all ohmic contact regions. During these implants, the photoresist (~ 1.5 μm) is used as the ion-beam mask. After each implant, a shallow step is plasma-etched into the Si_3N_4 for registration of the implanted regions. Following the implants, an additional dielectric layer is added prior to the annealing step.

A process for fabricating enhancement junction FETs (JFETs) is shown in Fig. 5.8 (122). The implantation of selenium is used to form source, drain, and other contact areas, whereas silicon ions are used for the n-channel and resistor load areas. Subsequently an annealing step is performed at 775°C for 15 min using a sputtered Si_3N_4 layer. Magnesium ions are implanted at a specific dose so that the resulting junction depth yields the desired threshold voltage. After the gate implant, the masking resist is stripped, and the wafer is capped and annealed at 700°C.

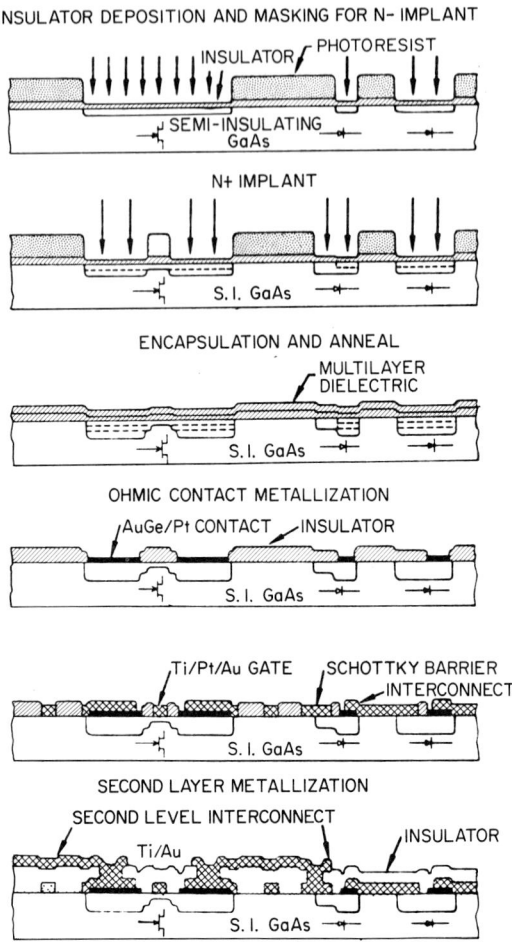

FIG. 5.7. Planar GaAs integrated circuit process steps (*121*). (Copyright © 1979, IEEE.)

5.3. Charge-Coupled Devices

Charge-coupled devices (CCDs) are related to MOS devices. They use the same technology, but their operation depends on the storage of minority carriers in potential wells, instead of the conduction of majority carriers. The first application of implantation to CCDs was the manufacture of drain electrodes to avoid blooming (*123*). CCDs are used for the detection of optical signals (line and area images), for memories, or for analog signal processing (*124*).

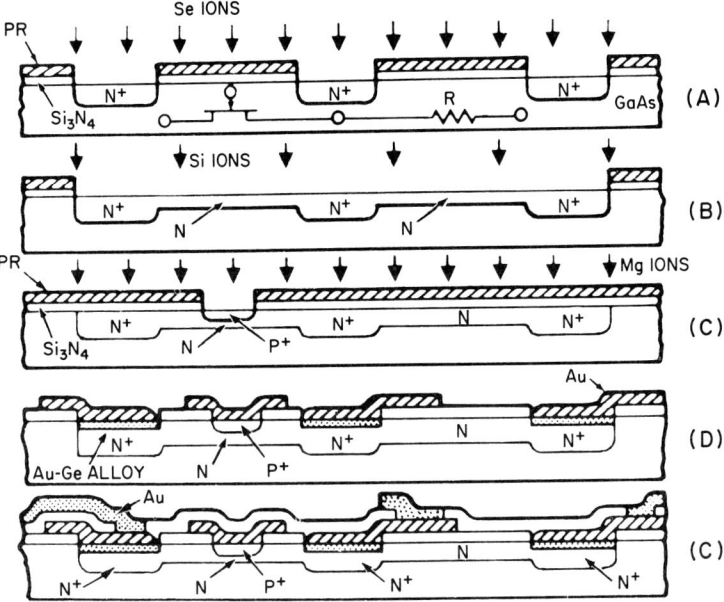

FIG. 5.8. GaAs planar enhancement-JFET process (*122*). (a) N⁺ ion implantation; (b) N ion implantation; (c) P⁺ ion implantation; (d) Ge–Au–Ni ohmic contacts and second metal; (e) dielectric insulation and final metal.

The applications of ion implantation are mainly the formation of buried channels to reduce the influence of the surface (trapping, etc.) by implanting low doses, e.g., 5×10^{12} cm^{-2}, of boron ions for *p*-channel devices (*125*) at fairly high energies. The charge transfer is enhanced due to fringing fields, but the area consumption is slightly larger than with surface channel CCDs. At the surface, a channel stop implant using phosphorus or arsenic can also be performed. From the standpoint of density considerations, the two-phase CCD is the most important for storage applications. A cross section of a high-density two-phase coplanar electrode structure is given in Fig. 5.9 (*126*). This structure consists of two levels of polysilicon gates. A potential well is formed under one-half of each phase electrode by selectively implanting arsenic ions in the vicinity of the SiO$_2$–Si interface, but is completely contained in the SiO$_2$ through a photoresist mask. The polysilicon is deposited and etched in order to be offset from the implanted arsenic. After etching the gate oxide to self-align the arsenic implant to the first gate level, the gate oxide is regrown, and a similar process is used to define a potential well under the second polysilicon gate electrodes. After this, a diffusion step is used to drive the arsenic into the silicon. To fabricate a buried-channel

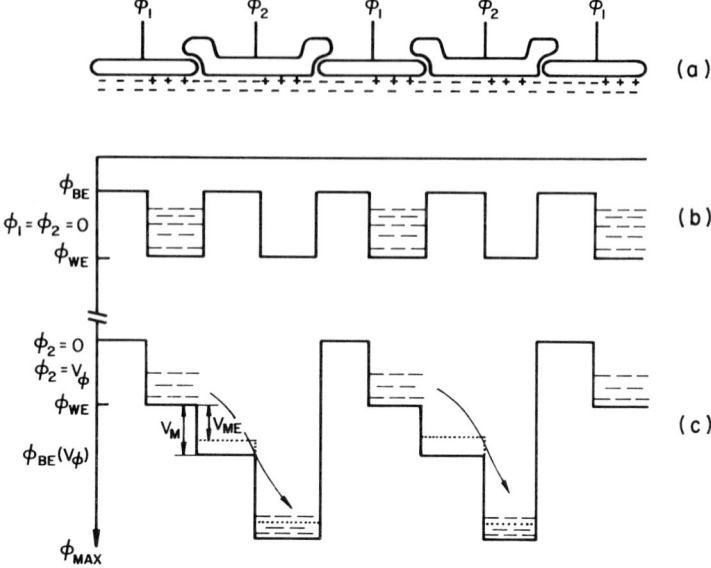

FIG. 5.9. (a) Schematic cross section of two-phase coplanar electrode CCD. (b) Lateral potential distribution for the "store" condition; both ϕ_1 and ϕ_2 clocks are low. (c) Lateral potential distribution for the "transfer" condition, to transfer charge from ϕ_1 to ϕ_2; V_M is the clock voltage overdrive; V_{ME} represents the reduced overdrive for the enhanced capacity structure.

CCD, a phosphorus implant is performed prior to the formation of the CCD structure (126).

5.4. Bipolar Devices

The first experiments in the fabrication of bipolar transistors using implantation were not very successful. The breakthrough came with the use of arsenic as an emitter dopant (9). Since that time, much work has been done on bipolar transistors, implanting boron for the base and arsenic for the emitter. Very often, two implants are used for the base. A low-dose implant is employed for the active (intrinsic) base, which determines the current gain and the cut-off frequency, and a higher dose implant is used for the extrinsic base as a low-series resistance contact. An excellent control of current gain is possible by varying the intrinsic base dose (see Figs. 5.10 and 5.11) and the energy (127).

In the near future, all fast VLSI bipolar circuits will use implanted arsenic emitters, boron bases and, for special applications, perhaps phosphorus as

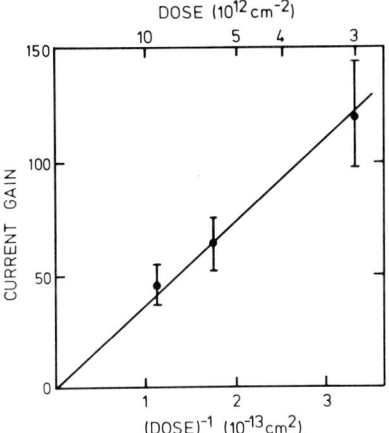

FIG. 5.10. Graph of h_{FE} versus the inverse of the deep-base implantation dose ($N_\square^{-1}$) for 5-GHz all-implanted transistors (*127*). (Copyright © 1974, IEEE.)

well. Low-frequency transistors will still be fabricated using phosphorus and boron diffusion.

In the following, some brief examples of bipolar devices using implantation are given.

An example for an advanced 1-μm I²L technique was presented by Evans *et al.* at the 1979 IEDM Meeting (*128*). They used 10 electron-beam patterning steps, and made full use of ion implantation. The process sequence is given in Table 5.2. The first implant is arsenic for the collector; it is partially driven before resistor definition to minimize loss of arsenic during the plasma-oxide etching that defines the resistors. After the resistor implant and annealing, the base contact is made using a double-energy implantation

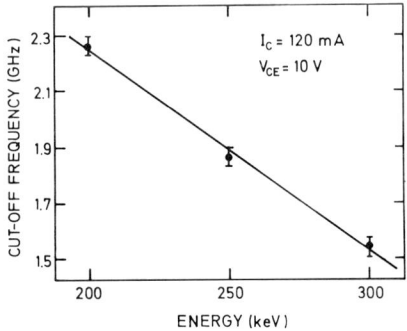

FIG. 5.11. Graph of f_r versus the energy of the deep-base implantation for 2-GHz all-implanted transistors (*127*). (Copyright © 1974, IEEE.)

TABLE 5.2

STEP-BY-STEP 1-μm BIPOLAR VLSI FABRICATION PROCESS[a]

Process step; minimum geometry (μm)	Conditions
1. Alignment marks; 10.0	5-μm Si etch (CF_4, O_2)
2. Isolation; 1.6	5500-Å anisotropic plasma Si etch 11,000-Å oxidation at 950°C
3. Collector; 3.8	As; 80 keV; 3×10^{15} cm^{-2} (50 Ω/□) anneal at 1000°C
4. Resistor; 2.0	As; 100 keV; 2.3×10^{13} cm^{-2} (500 Ω/□); anneal at 1000°C
5. p^+ extrinsic base; 1.4	Implant mask—6300-Å plasma oxide; B; 50, 70 keV; 1.5×10^{15} cm^{-2} (57 Ω/□)
6. p^- intrinsic base; 8.1	Implant mask—16,000-Å resist; B; 300 keV; 2×10^{12} cm^{-2} (~20 kΩ/□); anneals and passivation at 900°C
7. Contacts; 1.0	
8. First-level metal; 3.0	Pt (200 Å)/TiW (1750 Å)/Al + Cu (4000 Å)
9. Bias; 1.1	6000-Å plasma oxide dielectric
10. Second-level metal; 3.8	TiW (1750-Å)/Al + Cu (6000 Å)

[a] After (*128*). (Copyright © 1979, IEEE.)

at 50 and 70 keV to obtain a flatter profile. The active base is made by implanting boron with 300 keV. For comparison, I^{2}L circuits with Schottky outputs were also made. In that case no collector implantation was performed, but rather a 150-keV phosphorus implant to increase down beta and a 60-keV arsenic implant to lower the Pt–Si barrier height.

High-voltage transistors can be produced on the same chip as I^{2}L circuits, if the I^{2}L emitter is doped using an additional phosphorus implant

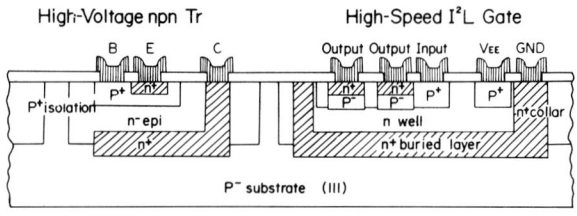

FIG. 5.12. Schematic cross section of a high-speed I^{2}L gate and a high-voltage linear n–p–n transistor (*129*). (Copyright © 1979, IEEE.)

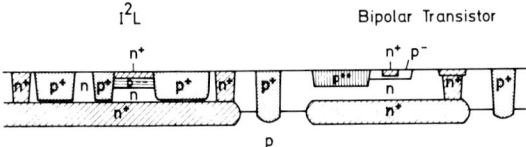
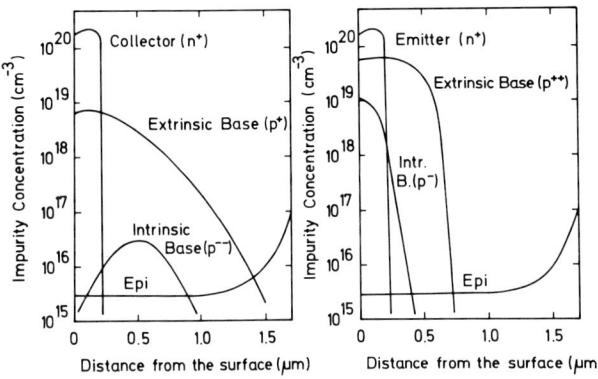

FIG. 5.13. Cross section and doping profiles of I²L and linear transistor (130). (Copyright © 1979, IEEE.)

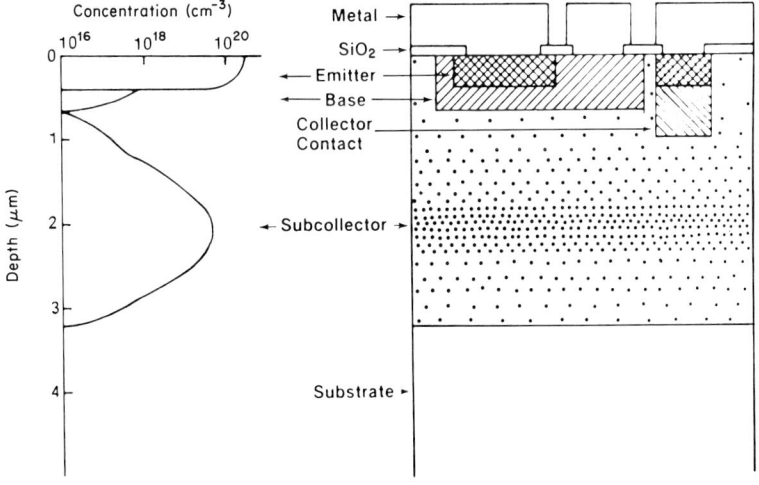

FIG. 5.14. Schematic of a nonepitaxial bipolar structure produced by high-energy ion implantation (114).

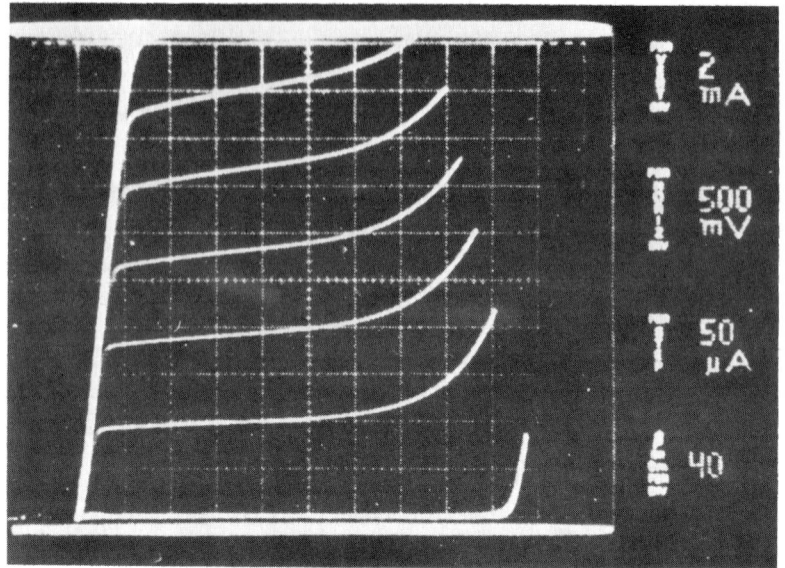

FIG. 5.15. Common emitter transfer characteristics of a nonepitaxial bipolar structure (*114*).

(*129*), as shown in Fig. 5.12. A different approach to obtain high-speed emitter-coupled logic (ECL) and I²L circuits on a chip is given in Fig. 5.13 (*130*). The high upward gain for I²L inverters and moderate downward gain for linear transistors were obtained using four different implants for the active and inactive base regions of the two structures. The implants were 260 keV with 10^{12} cm^{-2} and 60 keV with 3.4×19^{14} cm^{-2} for I²L, as well as 40 keV with 1.7×10^{14} cm^{-2} and 40 keV with 10^{15} cm^{-2} for ECL circuits, respectively.

A transistor completely fabricated using high-energy implantation was developed at IBM (*133*). A cross section of this transistor, together with a doping profile, is given in Fig. 5.14. Phosphorus with 2.9 MeV and a dose of 3.5×10^{15} cm^{-2} was implanted into p^+-silicon, forming the buried layer and the collector. Subsequently boron and arsenic were implanted to form the base and emitter. For the collector, an additional phosphorus reach-through implant was used. A single high-temperature step at 1000°C was employed. In Fig. 5.15 a typical characteristic is shown. The transistors showed a current gain of 120. The sheet resistivity of the subcollector was 22 Ω/□. Such buried layers could be produced selectively, and with higher yield than extremely thin epitaxial layers.

6. Conclusions

For VLSI circuits, extremely shallow highly doped layers with a low lateral spread are required. Implanted profiles always have the maximum of the distribution in the bulk, leaving a low doping concentration toward the surface. This low-doped region has to be surpassed by the current in devices by tunneling; otherwise it would add a series resistance. To circumvent this, a short drive-in diffusion has to be performed, modifying the doping profile to give a high concentration at the surface, but simultaneously reducing process control. In the case of bases for bipolar transistors, however, the reduced concentration in the emitter region is advantageous. Profiles with a maximum at the surface can be obtained by the procedure for sputter-saturated profiles, but this is probably possible only for special applications, or by knock-on implantation out of surface layers. In this case, however, process control is not perfect due to possible thickness variations of the layer, and the problem of damage caused by the primaries has to be considered. Sputtering itself does not limit the maximum obtainable doping concentration with standard dopants, since this effect only affects the silicon at concentrations above the solubility of the dopants.

An alternative is implantation through an oxide or nitride of appropriate thickness, so that the maximum of the distribution is at the silicon surface. Of course, this also reduces process control slightly due to possible thickness variations, but probably to no more than a tolerable extent. The recoil-implanted atoms lie in a more shallow region of the profile and usually may be tolerated.

The annealing of implanted layers raises no special problems for VLSI circuits too, but oxiding treatments will be critical. Process modeling has to be used for all technological steps to allow for a precise device and circuit modeling. For this purpose, many basic parameters concerning doping depending on diffusion coefficients, segregation coefficients, and enhanced oxidation still have to be determined.

Many new, mostly nondoping applications of ion implantation, such as damage gettering, damage-enhanced etching, and the LOCOS technique using nitrogen implantation, will find application in VLSI technology. Ion-beam lithography will allow the reduction of the size of devices to the possible physical limits.

In the last part of this article some examples of present-day applications of ion implantation to advanced devices were given. However, no attempt was made to fully cover all possible applications, since too much is presently changing in device technology as it progresses toward VLSI.

Acknowledgments

This work was partly assisted by the "Deutsche Forschungsgemeinschaft" and the "Bundesministerium für Forschung und Technologie" of West Germany. I am indebted to my coworkers J. Götzlich, K. Haberger, K. Hoffmann, H. Kranz, and G. Prinke for their contributions. The technical assistance of B. Schmiedt, J. Bosch, M. Bleier, E. Traumüller, and H. Zeller is very much appreciated.

References

1. W. Shockley, U.S. Patent No. 2,787,564 (1957).
2. J. W. Mayer, L. Erikson, and J. A. Davies. "Ion Implantation in Semiconductors." Academic Press, New York, 1970.
3. G. Dearnaley, J. H. Freeman, R. Nelson, and J. Stephen, "Ion Implantation." North-Holland Publ. Amsterdam, 1973.
4. H. Ryssel and I. Ruge, "Ionenimplantation." Teubner, Stuttgart, 1978.
5. S. Namba and K. Masuda, *Adv. Electron. Electron Phys.* **37**, 263 (1975).
6. R. W. Bower and H. G. Dill, *IEDM Washington* (1966).
7. K. G. Aubuchon, *Int. Conf. Prop. Use MIS Struct., Proc., Grenoble, 1969.*
8. J. Macdougall, K. Manchester, and R. B. Palmer, *Electronics*, June 22, p. 86 (1970).
9. R. S. Payne and R. J. Scavuzzo, *IEDM Washington* (1971).
10. V. G. R. Reddi and A. Y. C. Yu, *Solid State Technol* **15**, 35 (1972).
11. E. I. Shtyrkov, I. B. Khaibullin, M. M. Zaripov, M. F. Galyatudinov, and R. M. Bayazitov, *Sov. Phys.—Semicond. (Engl. Transl.)* **9**, 1309 (1976).
12. O. G. Kutukova and L. N. Strel'tsov, *Sov. Phys.—Semicond. (Engl. Transl.)* **10**, 265 (1976).
13. G. K. Celler, J. M. Poate, and L. S. Kimerling, *Appl. Phys. Lett.* **32**, 464 (1978).
14. R. H. Dennard, F. H. Gaensslen, L. Kuhn, and H. N. Yu, *IEDM Washington* (1972).
15. B. Hoeneisen and C. A. Mead, *Solid State Electron.* **15**, 891 (1972).
16. D. Widmann, in "Process and Device Modeling for Integrated Circuit Design" (F. van de Wiele, W. L. Engl, and P. G. Jespers, eds.). Noordhoff, Leyden, 1977.
17. J. Lindhard, M. Scharff, and H. E. Schiøtt *K. Dan. Vidensk. Selsk., Mat.-Fys. Medd.* **33**, No. 14 (1963).
18. W. K. Hofker, *Philips Res. Rep., Suppl.* No. 8 (1975).
19. H. Ryssel, G. Prinke, K. Haberger, K. Hoffmann, K. Müller, and R. Henkelmann, *Appl. Phys.* **24**, 39 (1981).
20. J. F. Gibbons and S. Mylroie, *Appl. Phys. Lett.* **22**, 568 (1973).
21. N. L. Johnson and S. Kotz, "Continuous Univeriate Distributions," Vol. II. Wiley, New York, 1970.
22. V. G. K. Reddi and A. Y. C. Yu. *Solid State Technol.* **15**, 35 (1972).
23. J. F. Gibbons, W. S. Johnson, and S. W. Mylroie, "Projected Range Statistics." Halstead Press, Stroudsburg, Penna. 1975.
24. B. L. Crowder, R. S. Title, M. H. Brodsky, and G. D. Pettit, *Appl. Phys. Lett.* **16**, 205 (1970).
25. J. P. Biersack, *Nucl. Inst. Meth.* **182/183**, 199 (1981).
26. F. Jahnel, H. Ryssel, J. Biersack, K. Haberger, K. Müller, and R. Henkelmann, *Nucl. Inst. Meth.* **182/183**, 223 (1981).
27. H. Matsumura and S. Furukawa, *Jpn. J. Appl. Phys.* **14**, 1983 (1976).

28. H. Runge, *Phys. Status Solidi A* **39**, 595 (1977).
29. H. Ryssel, K. Haberger, K. Hoffmann, G. Prinke, R. Dümke, and A. Sachs, *IEEE Trans. Electron Devices* **ed-27**, 1484 (1980).
30. H. H. Andersen and H. L. Bay, *J. Appl. Phys.* **46**, 1919 (1975).
31. P. Sigmund, *Phys. Rev.* **184**, 383 (1969).
32. T. C. Cass and V. G. Reddi, *Appl. Phys. Lett.* **23**, 268 (1973).
33. R. A. Moline and A. G. Cullis, *Appl. Phys. Lett.* **26**, 551 (1975).
34. G. Fischer, G. Carter, and R. Webb, *Radiat. Eff.* **38**, 41 (1978).
35. P. Sigmund, *J. Appl. Phys.* **50**, 726 (1979).
36. T. Hirao, K. Inoue, and T. Takayanagi, *J. Appl. Phys.* **50**, 193 (1979).
37. A. Goetzberger, *IEDM Washington* (1975).
38. R. A. Moline, G. W. Reutlinger, and J. C. North, *in* "Atomic Collisions in Solids," Vol. 1 (J. Datz, B. R. Appleton, C. D. Moak, eds.). New York, 1975.
39. H. Ishiwara and S. Furukawa, *Jpn. J. Appl. Phys.* **16**, 53 (1977).
40. W. K. Chu, M. J. Sullivan, S. M. Ku, and M. Shatzkes, *Proc. Int. Conf. Ion Beam Mod. of Matls.*, Budapest, 1978.
41. H. Ishiwara and S. Furukawa, *in* "Ion Implantation" (F. H. Eisen and L. T. Chadderton, eds.), p. 231. Gordon and Breach, New York, 1971.
42. J. M. Shannon, *Conf. Ser.—Inst. Phys.* **28**, 37 (1976).
43. T. Ito, S. Hijya, H. Nishi, M. Shinoda, and T. Furuya, *Jpn. J. Appl. Phys.* **17**, 201 (1978).
44. H. Müller, H. Ryssel, and K. Schmid, *J. Appl. Phys.* **43**, 2006 (1972).
45. F. F. Morehead, B. L. Crowder, and R. J. Title, *J. Appl. Phys.* **43**, 1112 (1972).
46. L. Csepregi, J. W. Mayer, and T. W. Sigmon, *Appl. Phys. Lett.* **29**, 92 (1976).
47. L. Csepregi, W. K. Chu, H. Müller, and J. W. Mayer, *Radiat. Eff.* **28**, 227 (1976).
48. L. Csepregi, E. F. Kennedy, and J. W. Mayer, *J. Appl. Phys.* **49**, 3906 (1978).
49. L. Csepregi, E. F. Kennedy, T. J. Gallagher, and J. W. Mayer, *J. Appl. Phys.* **48**, 4234 (1977).
50. E. F. Kennedy, L. Csepregi, J. W. Mayer, and T. W. Sigmon, *J. Appl. Phys.* **48**, 4241 (1977).
51. S. Prussin and A. M. Fern, *J. Electrochem. Soc.* **122**, 830 (1975).
52. S. Prussin, *in* "Ion-Implantation in Semiconductors" (S. Namba, ed.), p. 499. Plenum Press, New York, 1975.
53. J. J. Comer and S. A. Roosild, *Radiat. Eff.* **25**, 275 (1975).
54. S. Hasegawa, J. E. Forward, and H. Hartnagel, *Electron. Lett.* **11**, 53 (1975).
55. S. Prussin, *J. Appl. Phys.* **45**, 1635 (1974).
56. T. E. Seidel, R. S. Payne, R. A. Moline, W. R. Costello, J. C. C. Tsai, and K. R. Gardner, *Tech. Dig.—IEDM Washington*, p. 581 (1975).
57. S. Mader and A. Michel, *J. Vac. Sci. Technol.* **13**, 391 (1976).
58. R. A. Moline and A. G. Cullis, *Appl. Phys. Lett.* **26**, 551 (1975).
59. M. Tamura, *Appl. Phys. Lett.* **23**, 51 (1975).
60. R. B. Fair, *Semiconductor Silicon 1977* (H. R. Huff and E. Sintl, eds.), *Proc. Electrochem. Soc.* **77-2**, 51 (1975).
61. D. A. Antoniadis, A. G. Gonzalez, and R. W. Dutton, *J. Electrochem. Soc.* **125**, 813 (1978).
62. R. B. Fair and J. C. C. Tsai, *J. Electrochem. Soc.* **125**, 2050 (1978).
63. J. W. Colby and L. E. Katz, *J. Electrochem. Soc.* **123**, 409 (1976).
64. T. M. Buck, J. M. Poate, and K. A. Pickar, *Surf. Sci.* **35**, 362 (1973).
65. B. J. Masters, J. M. Fairfield, and B. L. Crowder, *in* "Ion Implantation" (F. H. Eisen and L. T. Chadderton, eds.), p. 81. Gordon and Breach, New York, 1971.
66. T. E. Seidel, R. L. Meek, and A. G. Cullis, *Conf. Ser.—Inst. Phys.* **23**, 494 (1975).
67. C. M. Hsieh, J. R. Mathews, H. D. Seidel, K. A. Pickar, and C. M. Drum, *Appl. Phys. Lett.* **22**, No. 238 (1973).
68. A. G. Nassibian and B. Golja, *IEEE Trans. Electron Dev.* **ed-26**, 245 (1979).

69. H. Ryssel, H. Kranz, P. Bayerl, and B. Schmiedt, *Radiat. Eff.* **48,** 125 (1980).
70. B. Golja and A. G. Nassibiam, *Solid-State Electron. Devices* **3,** 127 (1979).
71. W. Kern and R. S. Rosler, *J. Vac. Sci. Technol.* **14,** 1082 (1977).
72. R. A. Moline, R. R. Buckley, S. E. Haszko, and A. U. Mac Rae, *IEEE Trans. Electron Dev.* **ed-20,** 840 (1973).
73. J. C. North, T. E. McGahan, D. W. Rice, and A. C. Adams, *IEEE Trans. Electron Dev.* **ed-25,** 809 (1978).
74. G. Bell and J. Hoepfner, *in* "Proc. Symp. Etching for Pattern Definition" (H. G. Hughes and M. J. Rand, eds.), p. 47. The Electrochemical Society, Princeton, NJ (1976).
75. Y. Akasaka, K. Horie, K. Nomura, and S. Kawazu, *J. Jpn. Soc. Appl. Phys.* **43,** 493 (1974).
76. J. Götzlich and H. Ryssel, *J. Electrochem. Soc.* **128,** 617 (1981).
77. P. D. Parry and S. P. Bristol, *J. Vac. Sci. Technol.* **15,** 664 (1978).
78. J. A. Appels, E. Kooi, M. M. Paffen, J. J. H. Schatorje, and W. H. C. G. Verkuylen, *Philips Res. Rep.* **25,** 118 (1970).
79. J. A. Appels and M. M. Paffen, *Philips Res. Rep.* **26,** 157 (1971).
80. M. Watanabe and A. Tooi, *Jpn. J. Appl. Phys.* **5,** 737 (1966).
81. J. H. Freemann, G. A. Gard, D. J. Mazey, J. H. Stephen, and F. B. Whiting, *Proc. Eur. Conf. Ion Impl., 7–9 Sept., 1970, Reading*, p. 74.
82. C. R. Fritsche and W. Rothemund, *J. Electrochem. Soc.* **120,** 1603 (1973).
83. Y. Wada and M. Ashikawa, *Jpn. J. Appl. Phys.* **15,** 1725 (1976).
84. M. Ramin, H. Ryssel, and H. Kranz, *Appl. Phys.* **22,** 393 (1980).
85. J. M. Poate and T. C. Tisone, *Appl. Phys. Lett.* **24,** 391 (1974).
86. N. Nishi, T. Sakurai, T. Akamats, and T. Furuya, *Appl. Phys. Lett.* **26,** 337 (1974).
87. Z. L. Liau and J. W. Mayer, *in* "Treatise on Materials Science and Technol." (J. K. Hirvonen, ed.). Academic Press, New York, 1980.
88. B. Y. Tsaur, Z. L. Liau, and J. W. Mayer, *Appl. Phys. Lett.* **34,** 167 (1979).
89. J. L. Prince and F. N. Schwettmann, *J. Electrochem. Soc.* **121,** 705 (1974).
90. H. Ryssel, H. Kranz, J. Biersack, K. Müller, and R. A. Henkelmann, *in* "Ion Implantation in Semiconductors" (F. Chernov, J. A. Borders, and D. K. Brice, eds.), p. 727. Plenum Press, New York, 1977.
91. C. P. Ho, J. D. Plumer, J. D. Meindl, and B. E. Deal, *J. Electrochem. Soc.* **125,** 665 (1978).
92. D. A. Antoniadis, S. E. Hansen, and R. W. Dutton, *Tech. Rep.* 5019-2, Stanford Univ. (1978).
93. K. Lehovec and, A. Slobodskoy, *Solid State Electron.* **3,** 45 (1961).
94. C. L. Anderson, G. K. Celler, and G. A. Rozgonyi (eds.), "Laser and Electr. Beam Processing of Electronic Materials." The Electrochem. Soc., Princeton, NJ, 1980.
95. S. D. Ferris, H. J. Leamy, and J. M. Poate (eds.), "Laser–Solid Interactions and Laser Processing, 1978." American Inst. of Physics, New York, 1979.
96. E. Rimini (ed.), *Proc. Laser Effects Ion-Implanted Semicond., Catania, 1978.*
97. E. Rimini, *in* Ref. *94*, p. 270.
98. P. H. Tsien, H. Ryssel, D. Röschenthaler, and I. Ruge, *J. Appl. Phys.* **52,** 4775 (1981).
99. A. Lietoila, J. F. Gibbons, J. L. Regolini, T. W. Sigmon, T. J. Magee, J. Peng, and J. D. Hong, in Ref. *94*, p. 350.
100. A. Gat and J. F. Gibbons, *Appl. Phys. Lett.* **32,** 142 (1978).
101. J. F. Gibbons, *in* Ref. *94*, p. 1.
102. S. D. Allen, M. von Almen, and M. Wittmer, *in* Ref. *94*, p. 514.
103. M. Wittmer, *in* Ref. *94*, p. 514.
104. T. Enomoto, R. Ando, H. Morita, and H. Nakamura, *Jpn. J. Appl. Phys.* **17,** 1049 (1978).
105. R. L. Seliger and W. P. Fleming, *J. Appl. Phys.* **45,** 1416 (1974).

106. R. L. Seliger, R. L. Kubena, R. D. Olney, J. W. Ward, and V. Wang, *J. Vac. Sci. Technol.* **16**, 1610 (1979).
107. G. R. Hanson and B. M. Siegel, *J. Vac. Sci. Technol.* **16**, 1875 (1979).
108. R. Clampitt and D. K. Jefferies, *Conf. Ser.—Inst. Phys.* **38**, 12 (1978).
109. H. Ryssel, H. Kranz, K. Haberger, and J. Bosch, "Microcircuit Engineering 80" (R. P. Kramer, ed.). Delft Univ. Press, Amsterdam, 1981.
110. K. Haberger, H. Ryssel, L. Träger, and H. Kranz, *Int. Conf. Ion Implantation Equipment, Trento, Italy, 1978.*
111. L. Csepregi, F. Iberl, and P. Eichinger, *in* "Microcircuit Engineering 80" (R. P. Kramer, ed.). Delft Univ. Press, Amsterdam, 1981.
112. D. B. Rensch, R. L. Seliger, G. Csanky, R. D. Olney, and H. L. Stover, *J. Vac. Sci. Technol.* **16**, 1897 (1979).
113. G. Stengl, R. Kaitna, H. Loeschner, P. Wolf, and R. Sacher, *Aacher Technik Wien*, June Report (1979).
114. J. C. Bourgoin and J. W. Corbett, *Radiat. Res.* **36**, 157 (1978).
115. B. J. Masters and E. F. Gorey, *J. Appl. Phys.* **49**, 2717 (1978).
116. J. T. Clemens, R. H. Doklan, and J. J. Nolen, *IEDM*, p. 299 (1975).
117. W. R. Hunter, L. M. Ephrath, W. Grobman, C. M. Osburn, B. L. Crowder, A. Cramer, and H. E. Luhn, *Tech. Dig.—IEDM*, 54 (1978).
118. J. Tihanyi and D. Widmann, *Tech. Dig.—IEDM*, p. 399 (1977).
119. J. Sakurai, *IEEE J. Solid-State Circuits* **sc-13**, 468 (1978).
120. J. Sakurai, *IEDM*, p. 197 (1978).
121. B. M. Welch, R. C. Eden, Y. D. Shen, and R. Zucca, *Tech. Dig.—IEDM*, p. 493 (1979).
122. G. L. Troeger, A. F. Behle, P. E. Friebertshauser, K. L. Hu, and S. H. Watanabe, *Tech. Dig.—IEDM*, p. 497 (1979).
123. C. H. Sequin, *Bell Syst. Tech. J.* **51**, 1923 (1972).
124. W. F. Kosonocky, *Wescon 1974*, Tech. Session 2 (1974).
125. S. Shimizu, S. Iwamatsu, and M. Ono, *Appl. Phys. Lett.* **22**, 286 (1973).
126. P. K. Chatterjee, C. W. Taylor, and A. F. Tasch, Jr., *IEEE Trans. Electron Dev.* **ed-26**, 871 (1979).
127. R. S. Payne, R. J. Scavuzzo, K. H. Olson, J. M. Nacci, and R. A. Moline, *IEEE Trans. Electron Dev.* **ed-21**, 273 (1974).
128. S. A. Evans, S. A. Morris, and J. Englade, *Tech. Dig.—IEDM*, 196 (1979).
129. O. Ozawa, S. Kameyama, Y. Sasaki, Y. Tokumaru, M. Nakai, and T. Tanji, *Tech. Dig.—IEDM*, 188 (1979).
130. K. Kanzaki, M. Taguchi, G. Sasaki, A. Furukawa, and K. Aoki, *Tech. Dig.—IEDM*, 328 (1979).
131. P. H. Tsien, H. Ryssel, D. Röschenthaler, and I. Ruge, *J. Appl. Phys.* **52**, 2987 (1981).
132. J. F. Ziegler, B. L. Crowder, and W. J. Kleinfelder, *IBM J. Res. Dev.* **15** (1971).
133. A. E. Michel and W. H. Dexter, to be published.
134. P. Bayerl, H. Ryssel, and M. Ramin, *Radiat. Eff.* **47**, 217 (1980).
135. T. Tokuyama, *in* "Ion Implantation in Semiconductors" (F. Chernow, J. A. Borders, and D. K. Brice, eds.). Plenum Press, New York, 1977.
136. J. A. Yasaitis, *Electron. Lett.* **14**, 460 (1978).
137. H. Ryssel, H. Kranz, and P. Eichinger, *Conf. Ser.—Inst. Phys.* **281** (1976).

Stimulated Čerenkov Radiation

JOHN E. WALSH

Department of Physics and Astronomy
Dartmouth College
Hanover, New Hampshire

I. Introduction 271
   A. Čerenkov Radiation 271
   B. Čerenkov Masers 273
II. Theory 275
   A. Čerenkov Gain on a Strongly Magnetized Beam 276
   B. Gain from an Unmagnetized Beam 283
   C. Bounded Structures 285
   D. The Effect of Beam Velocity Spread 294
   E. Comments on Nonlinear Behavior 300
III. Experiment 301
   A. The Electron Beam 301
   B. A Millimeter-Wavelength Experiment 304
   C. Čerenkov Devices in the Short-Wavelength Limit 307
IV. Conclusion 309
   References 310

I. INTRODUCTION

A. Čerenkov Radiation

The electromagnetic wave produced by a charged particle moving faster than light in a dielectric medium is known universally as Čerenkov (*1*) radiation. Čerenkov's experiments, which were performed independently during the 1930s, and the subsequent analysis of the phenomena by Frank and Tamm (*2*) did, however, have some precursors.

Heaviside (*3*), in 1889, analyzed the problem of the radiation produced by a charged particle when it moved with uniform velocity. This work was done prior to the development of the special theory of relativity, and Heaviside assumed that it was possible for a particle to move with a velocity greater than that of light in a vacuum. When it was so assumed, radiation was produced. In a formal sense, his results were similar to those of Frank and Tamm. Sommerfeld (*4*), in 1904, without apparent knowledge of Heaviside's results, performed a similar analysis. There were also some experimental precursors to Čerenkov's work. In 1911 M. Curie (*5*) observed that radiation produced in the walls of glass containers holding radioactive material was probably due in part to the penetration of the glass by fast charged particles. Some experiments performed by Mallet (6) in 1926 were, in part, observations of Čerenkov radiation. None of this early work, however, lessens the importance of the pioneering experiments of P. A. Čerenkov.

Following the initial experiments of Čerenkov and the theory of Frank and Tamm, a very great number of both theoretical and experimental contributions have appeared. General discussions, with hundreds of additional references, may be found in Jelley (7), Zrelov (8), and the review article by Bolotovski (9). The major interest of many contributors has been the potential use of the Čerenkov process as a practical radiation source. Notable among these contributions were the papers of Ginzburg (10), in which he considered a number of ways in which electrons could be coupled to dielectrics so as to produce radiation in the millimeter and submillimeter regions of the electromagnetic spectrum.

Much of the early work dealt with the radiation produced by single electrons. As we shall see, however, this spontaneous radiation is a relatively weak process for all wavelengths longer than the blue UV region of the spectrum. Hence, in order to produce useful amounts of radiation, it was natural to consider the radiation produced by a bunched electron beam. At wavelengths long compared to the length of the bunch the radiated power is proportional to the square of the number of electrons involved, and hence the power emitted rises dramatically. A number of experiments were designed to explore the properties of Čerenkov radiation produced by prebunched electron beams moving in close proximity to a dielectric surface. Important contributions were made by Coleman (11), Danos (12), Lashinsky (13), and Ulrich (14). In these experiments no provision was made for feeding back the emitted radiation on subsequent bunches, and hence these results could be categorized as observations of enhanced spontaneous emission.

Suggestions have also been made that Čerenkov radiation could be used as the basis of a microwave tube (15–17). In these experiments a dielectric tube was used as a slow-wave structure. The general configuration suggested was similar to that used in traveling-wave tubes. When electron beams in the energy and current range found in conventional microwave tubes are used, however, the resulting devices are unsatisfactory for several reasons. We develop this line of argument carefully in subsequent sections, since these difficulties must be surmounted in constructing a useful Čerenkov source.

A major difficulty in constructing a Čerenkov source capable of producing useful amounts of radiated power is the coupling of the electron beam to the dielectric. In elementary discussions it is usually assumed that the electron passes right through the dielectric. This can actually occur in the limiting case of very-high-energy particles and gaseous or liquid dielectrics. In this regime Čerenkov radiation actually finds wide practical application as a diagnostic tool (7). There have also been serious attempts (*18–20*) to observe stimulated Čerenkov radiation in the visible and UV regions from a high-energy beam–gaseous dielectric combination. In these latter experiments, momentum modulation (*18, 19*) by an applied electromagnetic field has been observed, but as yet there is no clear-cut evidence of true stimulated emission. An alternative to passing an electron beam directly through a dielectric is to let a beam propagate along a channel. Recent experiments (*21–23*) in which millimeter-wavelength stimulated Čerenkov radiation has been observed were of this type.

A primary purpose of the present article is to explore the potential of the latter option. We will establish criteria necessary for producing usable levels of stimulated Čerenkov radiation at wavelengths short compared to the characteristic scale length of both the transverse and longitudinal dimensions of a dielectric resonator.

B. *Čerenkov Masers*

The goal of the general area of research pertaining to devices now often called free-electron lasers is to produce coherent and tunable moderate- or high-power radiation in those parts of the electromagnetic spectrum where such a source is not now available. All of the devices suggested to date have much in common with microwave tubes, and hence the designation "maser" or "laser" could be the subject of debate. It is possible, but not necessary, to formulate the equations of motion quantum mechanically. The electron transitions are between continuum states. The recoil due to single-photon emission is negligible, and thus Planck's constant does not appear in any final working formula. A classical analysis based on either fluid or kinetic equations will lead to the same expressions. Therefore, much of what is known about microwave tubes will apply also to free-electron lasers. Microwave tubes, however, operate at wavelengths comparable to or greater than the device, whereas the opposite will be the case for any free-electron laser or maser. This difference, although minor from some viewpoints, accounts for many of the difficulties encountered in attempting to build short-wavelength beam-driven radiation sources.

A Čerenkov maser (Fig. 1) is a device consisting of a dielectric resonator, an electron beam, and an output coupling structure.[1] The device is, in essence, a traveling-wave tube with the dielectric resonator serving as the slow-wave structure. When low-relative-dielectric-constant materials are used for the resonator and at least mildly relativistic electron beams are used

---

[1] The name *Čerenkov* in the designation follows from the fact that it is the Čerenkov criterion that the beam velocity must satisfy if gain is to be obtained.

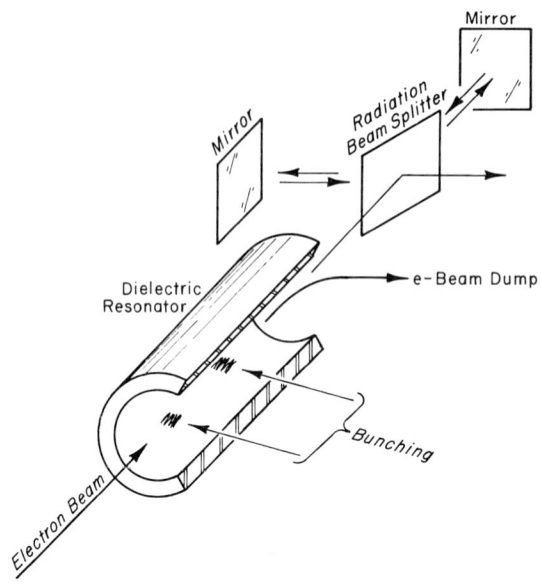

FIG. 1. Čerenkov maser components.

for the drive, gain can be obtained at wavelengths comparable to and less than the transverse dimension of the resonator. We shall see from the subsequent analysis that a device such as the one shown in Fig. 1 may be expected to work in the lower millimeter, submillimeter, and far-IR portions of the spectrum.

In the device shown in the sketch, the resonator supports a wave going slower than the speed of light in vacuum. The electron beam propagates slightly faster than the wave, and hence it will bunch in the region of the retarding field. Work is done and the wave grows. This process is analyzed in detail in Section II.

Shown in Fig. 2 are two other possible configurations for a Čerenkov source. In the first the beam runs over the top of a slab of dielectric, and in the second it is assumed to pass through the dielectric. The first form may be used as shown, or it may be the limiting form of a thin cylindrical resonator–hollow beam configuration. The second form is convenient for analysis since the boundary-value problem implied in the first version is much simplified. We shall use Fig. 2a for simplification. When extremely relativistic electron beams and gaseous dielectrics are used, the second sketch might also serve as the basis of a practical device. The fundamental problems of practical implementation of the direct device, which are the production and propagation of a sufficiently monoenergetic electron beam, are beyond the scope of

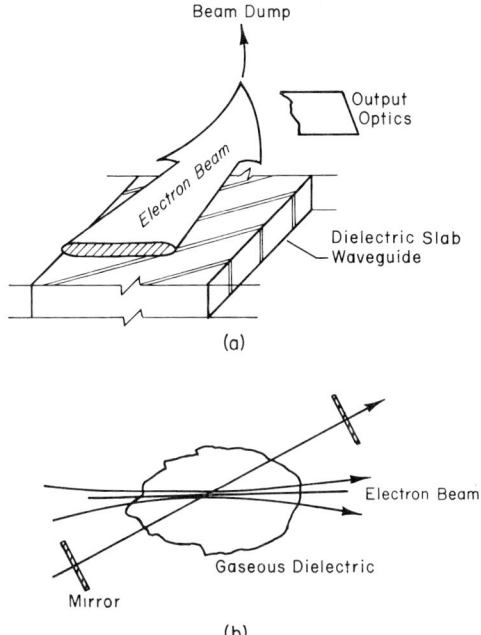

FIG. 2. Alternative configurations of the basic device: (a) slab waveguide; (b) possible Čerenkov gas laser.

the present analysis. Hence, we do not speculate seriously about experimentally realistic devices where the beam propagates through the dielectric.

Emphasis throughout the analysis and discussion is on resonators that are separate from the beam. Furthermore, we always assume that the devices operating at lower-millimeter wavelengths or less are the ones of interest. In Section II we establish conditions that must be obtained in this wavelength region. Discussion in Section III is devoted to experimental matters. Then, finally, some general conclusions are given in Section IV.

II. Theory

A series of calculations aimed at establishing the beam energy, current, and velocity spread required to obtain growth of stimulated Čerenkov radiation is presented in this section. The analysis will proceed along classical lines similar to those used in traveling-wave tube and beam plasma theory. In Sections II,A and B we examine the exponential gain of stimulated Čerenkov radiation obtained when it is assumed that either a strongly magnetized or a completely unmagnetized monoenergetic electron beam passes directly through a dielectric medium. The limit implied by the assumption that the beam is monoenergetic is examined in Section II,C, and modified gain formulas are derived. Section II,D is devoted to some resonator configurations that are more practical for the present application. Emphasis will be on the slab geometry, since in this case it is possible to present a reasonably compact analytic result, but the results obtained from other geometries are similar. A few brief comments and calculations related to nonlinear effects are outlined in Section II,E.

A. Čerenkov Gain on a Strongly Magnetized Beam

We consider first the case of a plane wave propagating at an angle to a strongly magnetized electron beam. The geometry is shown in Fig. 3.

1. Current Modulation

When the beam is strongly magnetized, the beam density and modulation are one dimensional and lie along the beam and magnetic axis. In this limit, the linearized equation for the velocity modulation has only one component, v_z, where

$$dv_z/dt = -eE_z/m\gamma^3 \qquad (1)$$

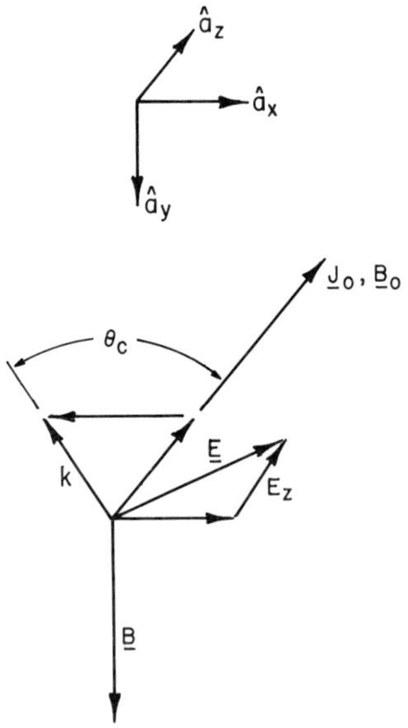

FIG. 3. Geometry for the plane wave–electron beam interaction in a dielectric medium.

The solution of this equation for the assumed E_z is readily found:

$$v_z = -(ie/m\gamma^3)E_z/(\omega - kv_0) \qquad (2)$$

This result, together with a linearized equation of continuity, gives for the density modulation n:

$$n = (n_0 k_0 v)/(\omega - kv)$$
$$= -[(in_0 e)/m\gamma^3][kE_z/(\omega - kv_0)^2] \qquad (3a)$$
$$= -[in_0 e/(m\gamma^3)]kE_z/(\omega - kv_0)^2 \qquad (3b)$$

Thus the current produced by the wave is

$$J_z = -n_0 ev - nev_0 \qquad (4a)$$
$$= (i\omega_p^2/4\pi\gamma^3)\omega E_z/(\omega - kv_0)^2 \qquad (4b)$$

where $\omega_p^2 = 4\pi n_0 e^2/m$ is the beam plasma frequency.

2. The Wave Equation

The current given by Eq. (4b) appears in Maxwell's equations as a source term. These are

$$\nabla \times \mathbf{E} = -c^{-1}(\partial \mathbf{B}/\partial t) \qquad (5a)$$
$$\nabla \times \mathbf{B} = (4\pi/c)\mathbf{J} + c^{-1}(\partial \mathbf{D}/\partial t) \qquad (5b)$$

In writing the second of these, we assume that the wave and the beam are in a dielectric medium, where

$$\mathbf{D} = \epsilon \mathbf{E} \qquad (6)$$

Taking the time derivative of the second Maxwell equation and substituting into the first gives a single wave equation:

$$\nabla \times \nabla \times \mathbf{E} + \frac{\epsilon}{c^2}\frac{\partial^2 \mathbf{E}}{\partial t^2} = -\frac{4}{c^2}\frac{\partial \mathbf{J}}{\partial t} \qquad (7)$$

There is no current component in the direction perpendicular to the beam and thus the perpendicular component of Eq. (7) may be used to express E_x in the terms of E_z. Doing this, substituting into the longitudinal component of Eq. (7), and making use of the assumed time and z dependence, we obtain a single wave equation for E_z:

$$\frac{\partial^2 E_z}{\partial x} + \left(\frac{(\omega^2\epsilon/c^2) - k^2}{c^2}\right)E_z = \frac{4\pi i}{\omega\epsilon}\left(\frac{(\omega^2\epsilon/c^2) - k^2}{c^2}\right)J \qquad (8)$$

Since we have also assumed a plane-wave dependence in the perpendicular as well as the longitudinal direction, we also obtain immediately:

$$\left[\frac{\omega^2\epsilon}{c^2} - k^2 - p^2 - \frac{\omega_p^2}{\epsilon\gamma^3}\frac{(\omega^2\epsilon/c^2) - k^2}{(\omega - kv_0)^2}\right]E_z = 0 \qquad (9)$$

where p is the perpendicular component of the wavenumber and ω_p is the plasma frequency.

3. The Dispersion Relation

We are obviously interested in the case $E_z \neq 0$ and, hence, the coefficient of Eq. (9) is the dispersion relation:

$$\frac{\omega^2\epsilon}{c^2} - k^2 - p^2 - \frac{\omega_p^2}{\epsilon\gamma^3}\frac{(\omega^2\epsilon/c^2) - k^2}{(\omega - kv_0)^2} = 0 \qquad (10)$$

for a plane wave propagating through a dielectric medium at an angle to a strongly magnetized electron beam.

Equation (10) is a quartic in both ω and k, and hence it has four roots. When $v_0 < c/\sqrt{\epsilon}$, all four roots are real, whereas if $v_0 > c/\sqrt{\epsilon}$ it has two real roots and a complex conjugate pair. One of the real roots is related to a wave propagating in the direction opposite to that of the beam (in the negative z direction). The other three result from the coupling of an electromagnetic wave propagating in the positive z direction and two beam space-charge waves. The latter two (fast and slow) space-charge waves, would be normal modes of the free beam. In the presence of the dielectric, however, they become coupled to the electromagnetic wave. When the velocity threshold $v_0/c = 1/\sqrt{\epsilon}$ is exceeded, the beam–wave dielectric system becomes unstable.

4. Čerenkov Gain

The presence of the beam is obviously felt most strongly for waves near "synchronism," i.e., when

$$\omega \simeq kv_0 \qquad (11a)$$

$$kv_0 \simeq \omega_{0k} \qquad (11b)$$

where we define ω_{0k}^2 as

$$\omega_{0k}^2 = c^2(k^2 + p^2)/\epsilon \qquad (11c)$$

the dispersion relation of the electromagnetic waves in the absence of the electron beam.

In the region where Eqs. (11a) and (11b) are valid, the dispersion relation Eq. (10) becomes an approximate cubic:

$$(\omega - kv_0)^3 - (\omega_p^2/2\gamma^3\epsilon)\omega(1 - 1/\beta^2\epsilon) = 0 \qquad (12)$$

Equation (12) follows from (10) when kv_0 is set equal to ω in those terms where the substitution does not give zero. This is a valid assumption provided that ω_p^2 is small in a sense that we shall define shortly.

When $\beta^2\epsilon < 1$, Eq. (12) has three real roots, whereas in the reverse case the roots are

$$\omega - kv_0 = [(\omega_p^2\omega/2\epsilon\gamma^3)(1 - 1/\beta^2\epsilon)]^{1/3} \qquad (13a)$$

$$\omega - kv_0 = (\omega_p^2\omega/2\gamma^3\epsilon)^{1/3}(1 - 1/\beta^2\epsilon)^{1/3}\tfrac{1}{2}(1 \pm i\sqrt{3}) \qquad (13b,c)$$

The root corresponding to Eq. (13b) is an exponentially growing wave either in time, Im $\omega \neq 0$, or space, Im $k \neq 0$. The choice between these will be determined by initial and boundary conditions.

We assume, for the moment, that the spatial growth is of interest and we let Im $k = \alpha$; then

$$\alpha = \frac{\sqrt{3}}{2}\left(\frac{\omega_p^2\omega}{2\epsilon\gamma^3}\right)^{1/3}\frac{(1 - 1/\beta^2\epsilon)^{1/3}}{c\beta} \qquad (14)$$

Examination of Eq. (14) shows that the spatial gain increases with the two-thirds power of the beam density and the one-third power of the frequency. It vanishes as the beam energy approaches the Čerenkov threshold and decreases as ϵ and γ become large.

Shown in Fig. 4 are sketches of free-wave dispersion curves for two different perpendicular wave numbers p_1 and p_2. The curves leave the $k = 0$ axis at the point $\omega/c = p/\sqrt{\epsilon}$, cross the speed of light at $\omega/c = p/(\epsilon - 1)^{1/2}$, and then asymptotically approach a wave propagating in the z direction. Along this curve the angle of propagation varies from $\theta = \pi/2$ to $\theta = 0$. Also shown in Fig. 4 is a beam "velocity" line, $\omega = ck\beta$. The points at which this line intercepts the dispersion curves are points at which the beam velocity and the phase velocity of the free waves are the same; they are in "synchronism."

Consideration of Eqs. (11a) and (11b) shows that, at this point, the angle of propagation is the same as the Čerenkov angle $\theta_c = \cos^{-1}(1/\beta\sqrt{\epsilon})$. At this point the dispersion is modified by the beam and the wave will grow at a rate given by Eq. (14). If γ, ϵ, and the beam density are left unchanged, the rate of growth at the synchronous point on the p_1 curve will be greater than that on the p_2 curve by an amount equal to the frequency ratio to the one-

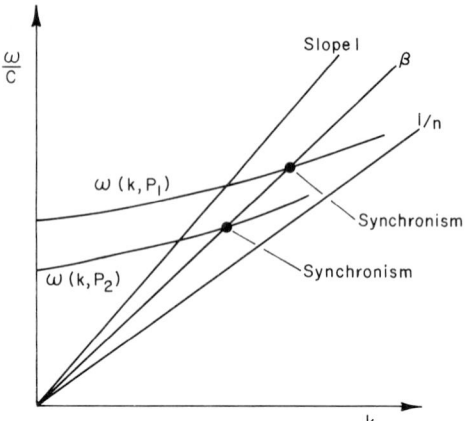

FIG. 4. The relations ω versus k for two different perpendicular components of wavenumber $(P_1 > P_2)$.

third power. Thus the stimulated Čerenkov process is a potential short-wavelength radiation source.

Growth will also occur at angles other than the Čerenkov angle. Shown in Fig. 5 is a numerical solution of the complete dispersion relation [Eq. (10)]. We see that there are three solutions in the positive, positive k quadrant of the $\omega - k$ plane. One is purely real, whereas the other two are a complex-conjugate pair in the region below and near synchronism and real above

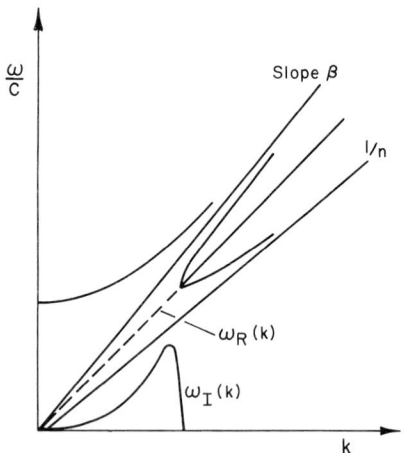

FIG. 5. The complete dispersion relation for the electron-beam dielectric system and the region of instability.

this point. The gain peaks just below synchronism [the shift is equal to $\mathrm{Re}(\omega - kv_0)$ given in Eq. (13)] and goes identically to zero at the point $\omega = kv_0$. On the small k side, $\mathrm{Im}(\omega/c)$ goes to zero more slowly. The exact shape of this curve will depend on γ, ϵ, and the beam density.

We have now established that by controlling the angle of propagation ϵ and the beam energy, the frequency as this maximum growth occurs increases as $\omega^{1/3}$. It is instructive to consider the magnitude of the gain as these parameters are manipulated. In order to do this, we again rewrite Eq. (14), now in this form:

$$\alpha = (\beta \omega_p^2 \omega/2c^3)^{1/3} G(\gamma_T) F(\gamma, \gamma_T)(\gamma_T/\gamma) \tag{15}$$

where

$$\gamma_T^2 = \epsilon/(\epsilon - 1) \tag{16a}$$

is the threshold energy ($\beta_T^2 = 1/\epsilon$),

$$G(\gamma_T) = (1 - 1/\gamma_T^2)^{1/3} \gamma_T^{5/3} \tag{16b}$$

$$F(\gamma, \gamma_T) = (1 - \gamma_T^2)^{1/3}/(1 - 1/\gamma^2) \tag{16c}$$

One power of β has been inserted in front of ω_p^2 so that we may subsequently express it in terms of the beam current, a form we find convenient in our numerical evaluation of the gain. But before we do this evaluation we examine the functions G, F, and γ_T/γ.

The function G depends only on the dielectric constant of the material. A sketch is shown in Fig. 6a. It shows a vertical rise at $\gamma_T = 1$ (the point where the dielectric constant of the material approaches infinity), reaches a maximum at $\gamma_T^2 = 7/5$ ($\epsilon = 7/2$), and finally decreases like $\gamma_T^{-5/3}$ as γ_T becomes large ($\epsilon \to 1$). Thus, in considering a practical Čerenkov source, one cannot move profitably in the direction of low-beam-energy optically dense materials ($\gamma \to 1$, $\gamma_T \to 1$, $\epsilon \to \infty$) since the gain vanishes rapidly in this limit. As a practical matter, one could not propagate a beam in this type of material in any event. In the opposite limit we should have gases ($\epsilon \to 1$). In this region the gain will also decrease, but conclusions as to the usefulness of this limit must also include consideration of the $\omega^{1/3}$ term. It is interesting (and perhaps important) for practical millimeter–submillimeter-wavelength devices that G peaks in the region of the dielectric constant of quartz.

The function F depends both on the threshold energy γ_T and the beam energy γ. It rises vertically from $\gamma = \gamma_T$ and asymptotically approaches unity from below. Sketches of F, γ_T/γ, and their product are shown in Fig. 6b. There is obviously a local maximum in the growth rate. The value of the product at this maximum is about 0.5.

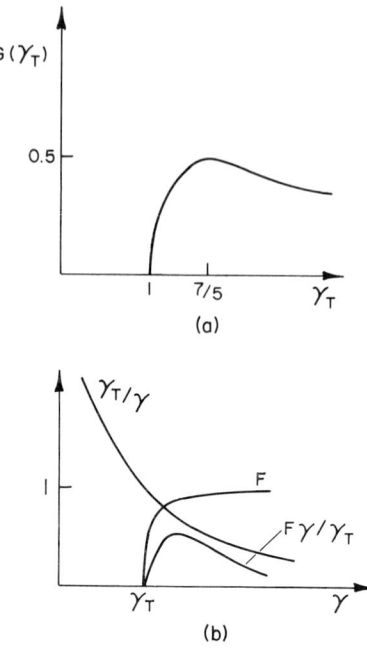

FIG. 6. The components of growth rate: (a) the dielectric contribution; (b) the beam threshhold function F.

Before we consider some actual numerical values for the growth rate, it will be useful to consider one further scaling, which will be that of the beam density. We assume for the present that the beam is now a rectangular slab of thickness a and that the variation of E in the x direction is still given by $\exp(ipx)$. The term

$$\frac{\beta \omega_p^2}{2c^3} = \frac{4\pi \beta n e^2}{2mc^2} \frac{\omega}{c} \tag{17a}$$

can be reexpressed as

$$\beta \omega_p^2 / 2c^3 = (2\pi I / I)(\omega a / c)(1/a^3) \tag{17b}$$

where

$$r_0 = e^2/mc^2 \tag{18a}$$

$$I_0 = ec/r_0 \tag{18b}$$

and I is the electron beam current. When I is measured in amperes, I_0 has the value 17 kA. Hence, the factors preceding the energy and material form factors in the expression for gain, Eq. (15), are given by

$$\left(\frac{\beta\omega_p^2\omega}{2c^3}\right)^{1/3} = \left(\frac{2\pi I}{I_0}\right)^{1/3}\left(\frac{\omega a}{c}\right)^{1/3}\frac{1}{a} \quad (19)$$

When I is approximately 3 A, the first of the three factors is approximately equal to 0.1, whereas if it is 3 mA it becomes 0.01. The second factor may, in principle, vary from zero to a moderately large number, and the characteristic scale length a may be anything from 0.01 to 1 cm. Hence, substantial gain is possible in principle. A discussion of ways in which this may be achieved in practical cases is deferred until after we have made some mention of wave-guiding structures.

B. Gain from an Unmagnetized Beam

The preceding analysis presumes that the current density modulation occurs only in the z direction. As we shall see in the following discussion, one class of Čerenkov device makes use of mildly relativistic electron beams; thus they somewhat resemble microwave tubes. The beams in these devices almost certainly propagate along a strong axial guide field, and in this limit the assumptions made in the last section will be at least approximately valid.

Another class of device, however, might make use of a more relativistic beam, such as that used in the injector of a linear accelerator, in a linear accelerator itself, or perhaps in some other type of accelerator. The beam in this case may very well not be magnetized. It will then have rapidly varying components in the transverse as well as the longitudinal direction, and the gain formulas will be modified.

When the beam is unmagnetized, the linearized equation for the perpendicular motion is

$$d\mathbf{v}_\perp/dt = (e/m\gamma)[\mathbf{E}_\perp + (\mathbf{v}_0/c) \times \mathbf{B}] \quad (20)$$

whereas the longitudinal motion is still governed by Eq. (1). Assuming the same geometry given in Fig. 1, the one nonvanishing component of this equation will lie in the x direction:

$$dv_x/dt = -(e/m\gamma)[E_x - (v_0 B_y/c)] \quad (21)$$

Equation (21), with the aid of Faraday's law, may be restated in the form

$$\frac{dv_x}{dt} = -\frac{e}{m\gamma}\left[\left(\frac{\omega - kv_0}{\omega}\right)E_x + \left(\frac{pv_0}{\omega}\right)E_z\right] \quad (22)$$

The $\mathbf{v} \times \mathbf{B}$ term gives rise to an E_z as well as E_x dependence for v_z.

The solution to Eq. (21), together with the linearized equation of continuity, may be used to construct expressions for the current density. These are

$$J_x = \frac{ip^2}{4\pi\gamma} \frac{1}{\omega}\left(E_x + \frac{pv_0}{\omega - kv_0} E_z\right) \tag{23a}$$

$$J_z = \frac{i\omega p^2}{4\pi\gamma} \frac{1}{\omega}\left[\frac{pv_0}{\omega - kv_0} E_x + \left(\frac{\omega^2}{\gamma(\omega - kv_0)^2} + \frac{p^2 v_0^2}{(\omega - kv_0)^2}\right) E_z\right] \tag{23b}$$

The current terms can now be substituted in Eq. (7). When this is done, we have as our new wave equation:

$$\begin{pmatrix} k^2 + \dfrac{\omega_p^2}{c^2} - \dfrac{\omega^2\epsilon}{c^2} & pk + \dfrac{pv_0}{\omega - kv_0}\dfrac{\omega_p^2}{c^2} \\ pk + \dfrac{pv_0}{\omega - kv_0}\dfrac{\omega_p^2}{\gamma c^2} & p^2 + \left(\dfrac{\omega_p^2}{\gamma c^2} \dfrac{\omega^2\gamma^2 + p^2 v_0^2}{\gamma^2(\omega - kv_0)^2}\right) - \dfrac{\omega^2\epsilon}{c^2} \end{pmatrix} \begin{pmatrix} E_x \\ E_z \end{pmatrix} = 0 \tag{24}$$

The determinational equation for Eq. (24) is now the dispersion relation for the unmagnetized beam–dielectric combination. It is

$$\left(k^2 + \frac{\omega_p^2}{\gamma c^2} - \frac{\omega^2\epsilon}{c^2}\right)\left(p^2 + \frac{\omega_p^2}{\gamma^3 c^2} \frac{\omega^2\gamma^2 p^2 v_0^2}{(\omega - kv_0)^2} - \frac{\omega^2\epsilon}{c^2}\right)$$

$$- \left(pk + \frac{pv_0}{\omega - kv_0}\frac{\omega_p^2}{\gamma c^2}\right)^2 = 0 \tag{25}$$

Equation (25), which appears quite cumbersome in comparison with Eq. (9), is still a quartic in either ω or k or both. All qualitative comments made about the strongly magnetized case apply here as well. However, the results are quantitatively somewhat different. Again, the strongest coupling region of the beam to the wave is in the velocity synchronism ($\omega/ck = \beta$).

If terms proportional to $1/(\omega - kv_0)^2$ are collected separately, we obtain for the dispersion relation:

$$\frac{\omega^2\epsilon}{c^2}\left(\frac{\omega^2\epsilon}{c^2} - k^2 - p^2 - \frac{\omega_p^2}{c^2}\right)$$

$$- \frac{\omega_p^2}{\gamma c^2}\frac{1}{(\omega - kv_0)^2}\left[\left(\frac{\omega^2\epsilon}{c^2} - k^2 - p^2\frac{\omega_p^2}{c^2}\right)p^2 v_0^2\right.$$

$$\left. + \frac{\omega^2}{\gamma^2}\left(\frac{\omega^2\epsilon}{c^2} - k^2\right) - p^2(\omega^2 - v_0^2 p^2 - v_0^2 k^2)\right] = 0 \tag{26}$$

Near synchronism, this reduces to

$$\left(\frac{\omega^2\epsilon}{c^2} - k^2 - p^2 - \frac{\omega_p^2}{\gamma c^2}\right) - \frac{\omega_p^2}{\gamma c^2} = \frac{\omega^2(\epsilon - 1)(\beta^2\epsilon - 1)}{(\omega - kv_0)^2} \tag{27a}$$

or

$$\left(\frac{\omega - kv_0}{v_0^3}\right)^3 = \frac{\omega_p^2 \omega}{2\gamma c^3} \frac{(1 - 1/\beta^2\epsilon)}{\beta^3 \gamma_T^2} \qquad (27b)$$

Once again, the dispersion relation is cubic, and the frequency and the dependence on the size of the Čerenkov angle $[\theta = \sin^{-1}(1 - 1/\beta^2\epsilon)]$ are the same. However, the beam energy and ϵ dependence are different. If we use the functions defined earlier, we have for the spatial growth rate:

$$\alpha = (\beta\omega_p^2 \omega/c^3)^{1/3} G(\gamma) F(\gamma, \gamma_T)(\gamma_T/\gamma)^{1/3} \qquad (28)$$

The energy dependence is now $\gamma^{-1/3}$ in the high-energy limit, as opposed to the more constrictive γ^{-1} dependence in the strongly magnetized limit. If all other factors are the same, the gain in the unmagnetized limit will be greater than that for the strongly magnetized beam. This is because the electrons in the beam can now do work on the wave in both the transverse and longitudinal directions.

C. Bounded Structures

Except for the possibly interesting limit of extremely relativistic beams and gaseous dielectrics, it is not practical to have the beam penetrating the dielectric. Hence, in assessing the practicality of Čerenkov sources, it is important to consider dielectric waveguides and resonators that have channels for the beam propagation. This complicates the analysis. Thus, before we take up the cases quantitatively, it will be useful to consider, at this point, the regime where the results of the preceding section are qualitatively useful.

First, we note that with minor changes, the results of the last section will apply exactly to a metal-bounded, cylindrical, dielectric waveguide through which an electron beam propagates. The perpendicular wave number p is now a root of a zero-order Bessel function and is no longer completely free. The only other change is that the factor π in the current term no longer appears, because the beam is now also cylindrical. The field symmetry is now transversely magnetic.

1. Cylindrical Guide with a Beam Channel

When the beam propagates in a hole in the dielectric, we have a situation such as that sketched in Fig. 7. If the diameter of the hole is sufficiently small (the condition for smallness will shortly be quantitatively specified), the results of the preceding section might be expected to apply more or less exactly.

It is obviously the relative size of the hole which is the fundamental difference. Fortunately, it is possible to attain considerable insight into its

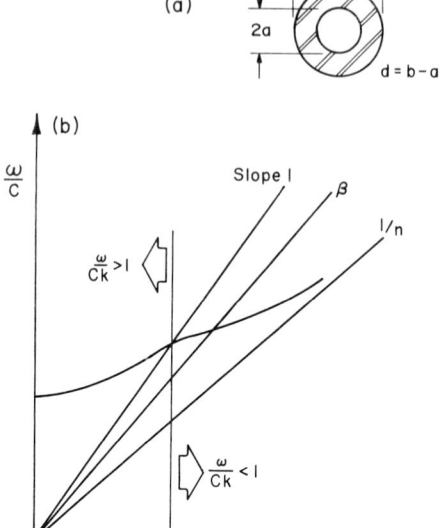

FIG. 7. Dispersion relation for the free waves of a partially filled guide.

effect with little analysis. We consider for the moment a metal-lined guide partially filled with dielectric. The dispersion curves sketched in Fig. 7 are similar to those shown in Fig. 4. The main difference is the shape near the light line $\omega = ck$. The point where the curve crosses this line is now controlled by the relative filling factor d/b as well as the dielectric constant of the material. As d/b and ϵ become small, the point where the partially filled guide becomes a slow-wave structure, $\omega/ck < 1$, can thus still be made to occur at an arbitrarily high frequency.

When $\omega/ck > 1$ (outside the light line), the field in the hole is proportional to $J_0(pr)$, an ordinary Bessel function. In this regime it peaks in the center of the hole. However, we must operate in the regime $\omega/ck < 1$, and in this case, the radial dependence is proportional to a modified Bessel function $I_0(qr)$. The field is now a minimum at $r = 0$, and the beam–wave coupling is obviously decreased.

A sketch of the field dependence in the two regimes is shown in Fig. 8. The wave number in the dielectric p is still given by

$$p^2 = (\omega^2/c^2) - k^2 \tag{29}$$

whereas the wave number in the hole when $\omega/ck < 1$ is now given by

$$q^2 = k^2 - (\omega^2/c^2) \tag{30}$$

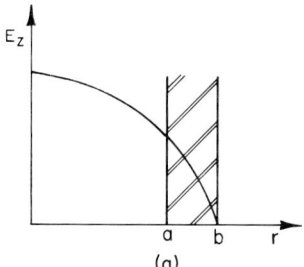

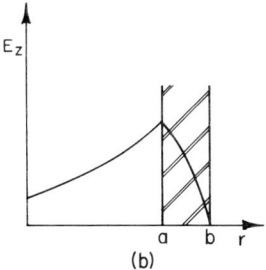

FIG. 8. Radial dependence of the axial electric field: (a) fast-wave region, $\omega/ck > 1$; (b) slow-wave region, $\omega/ck < 1$.

The latter is obviously one measure of the field depression in the hole. Since we operate near synchronism ($\omega = ck\beta$) we have for q:

$$q = k/\gamma \tag{31a}$$

$$q = \omega/c\beta\gamma \tag{31b}$$

or

$$q = 2\pi/\lambda\beta\gamma \tag{31c}$$

Hence, when nonrelativistic beams are used ($\beta\gamma \simeq v/c$) the field drops off away from the dielectric in a distance small compared to a wavelength. If, however, the beam is at least mildly relativistic ($\beta\gamma \gtrsim 1$) the opposite limit applies and we can operate with wavelengths that are small compared to the hole.

The latter considerations actually apply to any structure supporting a wave for which $\omega/ck < 1$. One might then ask about the relative advantages of a dielectric tube, since if $\beta\gamma \gtrsim 1$ then coupling would be improved at short wavelengths only for a slow-wave structure.

The advantages of the dielectric tube also lie in the short-wavelength range. In a conventional slow-wave structure, the periodicity must also be comparable to the wavelength. Structures of reasonable length are, therefore, a great many wavelengths long and they become very difficult to fabricate at relatively long wavelengths (a few millimeters). It is possible, but not easy, to build conventional structures with a fundamental period smaller than a few millimeters. The dielectric is, however, a smooth structure and easy to fabricate. When the beam is relativistic, the coupling impedance becomes comparable to that of other structures. Modifications of this basic structure, such as a dielectric tube with no metal boundary and multiple coupled tubes, may also be of practical use. Another basic structure, a dielectric slab bounded on one side by a conductor, also shows promise for application in the shorter wavelength region. This follows from the fact that a greater mode separation at small wavelengths can be obtained from this more open structure. Hence, it may well be easier to make single-mode devices with this type of structure, and for this reason we will analyze it in some detail.

The basic geometry is shown in Fig. 9. Assuming, for the moment, that no beam is present, we have for the TM modes of the guide

$$\mathbf{E} = (0, E_y, E_z) \tag{32a}$$

where

$$\{(d^2/dy^2) + [(\omega^2 \epsilon/c^2) - k^2]\} E_z = 0 \tag{32b}$$

$$E_y = \frac{ik}{(\omega^2 \epsilon - k^2)/c^2} \frac{\partial E_z}{\partial x} \tag{32c}$$

In the region $0 \leq y \leq d$ the dielectric constant ϵ appears, whereas in the region $y \leq d \leq \infty$ it is set equal to unity.

Anticipating the fact that we are concerned only with slow waves bound to the surface guide, we have for the electric fields:

$$E_z = A \sin py \tag{33a}$$

where

$$p^2 = (\omega^2 \epsilon/c^2) - k^2 \tag{33b}$$

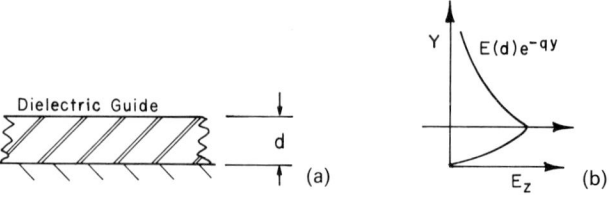

Fig. 9. Basic slab guide geometry (a) and axial field dependence (b).

in the region $0 \leq y \leq d$. Outside the dielectric, the field is

$$E_z = Be^{-qy} \tag{34a}$$

$$q^2 = k^2\omega^2/c^2 \tag{34b}$$

Matching of the tangential electric and magnetic fields may be used to eliminate the constants A and B. Thus we have for the dispersion relation of the dielectric slab waveguide

$$\epsilon q \cot pd = p \tag{35}$$

A plot of the roots of this function is given in Fig. 10. The lowest order mode has no cutoff. It comes up along the light line $\omega = ck$, until pd gets somewhat closer to the neighborhood of $\pi/2$. Thereafter, as ω becomes larger it asymptotically approaches the speed of light in the dielectric. In the region $\pi/2 \leq pd \leq \pi$ there are no solutions to Eq. (35), whereas, when $\pi \leq pd < 3\pi/2$, a second mode which has a finite-ω cutoff frequency can also propagate. At successively higher frequencies, more of these modes appear. Several are shown in Fig. 10.

2. Coupling of Beam to Bounded Resonator

Also shown in Fig. 10 is a beam speed line $\omega = ck\beta$. It is obvious that if the beam velocity satisfies the Čerenkov conditions $\beta > 1/\sqrt{\epsilon}$ phase synchronization between an electron beam and a wave can be maintained.

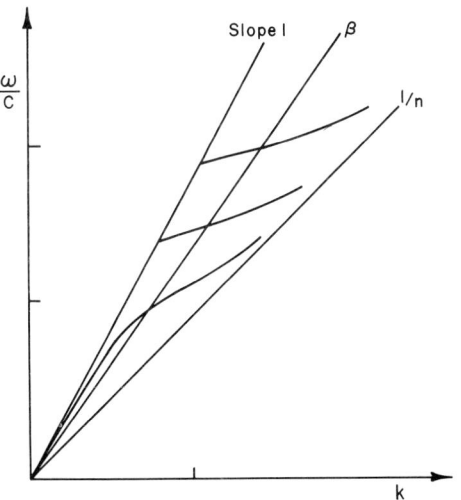

FIG. 10. Dispersion relation for the slab guide.

When the beam is added, the wave equation in the vacuum region becomes

$$\{(d^2/dy^2) + [(\omega^2/c^2) - k^2]\epsilon_\|\}E_z = 0 \tag{36a}$$

where

$$\epsilon_\| = 1 - \frac{\omega_p^2/\gamma^3}{(\omega - kv_0)^2} \tag{36b}$$

In arriving at Eqs. (36a) and (36b), it has been presumed that beam density modulation occurs only in the z direction, that the left edge of the beam is close to the dielectric, and that the beam extends indefinitely in the region $y > d$. The size of the actual gap between the beam and the dielectric is an important parameter in a short-wavelength device and its role is discussed separately.

3. *The Beam-Guide Dispersion for a Bounded Structure*

When a bounded structure is used to support the wave, as it must in almost any practical source, the dispersion relation becomes a transcendental (as opposed to an algebraic) function. It is more or less straightforward to obtain values for the roots by numerical means, but it is not immediately obvious how to obtain a good qualitative understanding of the roots.

One method appropriate for relatively weak beams is the following. Assume a relation of the form

$$D(\omega, k, \omega_p^2) = 0 \tag{37}$$

where the presence of ω_p^2 in Eq. (37) indicates the presence of the beam. If the beam is weak, we can write

$$D(\omega, k, \omega_p^2) = D^{(0)}(\omega, k) + \omega_p^2(\partial D/\partial \omega_p^2) \tag{38a}$$

where

$$D^{(0)}(\omega, k) = D(k, \omega, 0) \tag{38b}$$

is the dispersion relation for the waves supported by the structure when no beam is present. This function can, in a region near the solution $D^{(0)}(\omega, k) = 0$, be written as

$$D^{(0)}(\omega, k) = (\omega - \omega_k)(\partial D^0/\partial \omega) \tag{39}$$

where ω_k are the roots of Eq. (35).

The second term in Eq. (38a) can also be further reduced. The dependence of the dispersion relation on ω_p^2 always enters through ϵ, and hence the second term of Eq. (38a) takes the form

$$\omega_p^2 \frac{\partial D}{\partial \omega_p^2} = \frac{\omega_p^2 C(\omega, k)}{\gamma^3(\omega - kv_0)^3} \tag{40}$$

where $C(\omega, k)$ is a function which depends on the details of the structure. It may, for example, have zeros, but it will not have any poles near either $\omega = kv$ or $\omega = \omega_k$.

Thus, near synchronism ($\omega_k = kv_0$) and for beams that are not too strong ($I/I_0 \ll 1$), we again have a cubic dispersion relation:

$$0 = (\omega - \omega_k)\frac{\partial D^0}{\partial \omega} + \frac{\omega_p^2}{\gamma^3(\omega - kv_0)^2} C(\omega_k, k) \tag{41a}$$

or

$$(\omega - kv_0)^3 = \frac{\omega_p^2}{\gamma^3} \frac{C(\omega_k, k)}{[\partial D^0(\omega_k, k)/\partial \omega]} \tag{41b}$$

Thus, the qualitative nature of the roots is rather independent of the exact geometry of the wave-supporting structure.

When the wave-supporting structure is a dielectric slab and the assumptions made earlier apply, the dispersion relation becomes

$$q \cot pd = (p/\epsilon)\sqrt{\epsilon_\parallel} \tag{42}$$

The expansion procedure outlined in the preceding section then gives for Eq. (41b),

$$\left(\frac{\omega - kv_0}{v_0}\right)^3 = \frac{\beta\omega_p^2}{c^3} \frac{(1 - 1/\beta^2\epsilon)}{\epsilon\beta^4\gamma^3} \frac{\sin^2 pd}{(kd/\gamma) + (\gamma^2/\epsilon\gamma_T^2)\sin^2 pd} \tag{43}$$

The first two groups of factors on the right-hand side of Eq. (43) are identical to the results obtained when it was assumed that the beam propagated in the dielectric, and much of the discussion presented at that point applies here as well. The last group of factors contains the dependence on the geometry. It can be seen that in addition to the Čerenkov threshold dependence, the coupling also goes to zero as the thickness of the slab goes to zero; this is a result that could easily be anticipated.

The other trends in the gain can be understood as follows: On the fundamental mode, the value of pd varies from 0 at $\omega = 0$ up to $\pi/2$ as $\omega, k \to \infty$. On the higher branches, it varies from $n\pi$ at cutoff ($\omega = ck$) to $(2n + 1)\pi/2$ as the curve asymptotically approaches the speed of light in the dielectric. The value of $\sin^2 pd$ thus varies monotonically from zero to one. Assuming that the velocity synchronism is maintained along the dispersion curve, the gain will vanish at $\omega = ck$, because in this limit $\gamma \to \infty$ and it becomes increasingly difficult to modulate the beam. Furthermore, as $\beta \to 1/\sqrt{\epsilon}$, the gain also vanishes due to the factor $(1 - 1/\beta^2\epsilon)$ in the coefficient. The gain thus vanishes at both ends of the dispersion curve and peaks in between. A sketch of the general behavior is shown in Fig. 11.

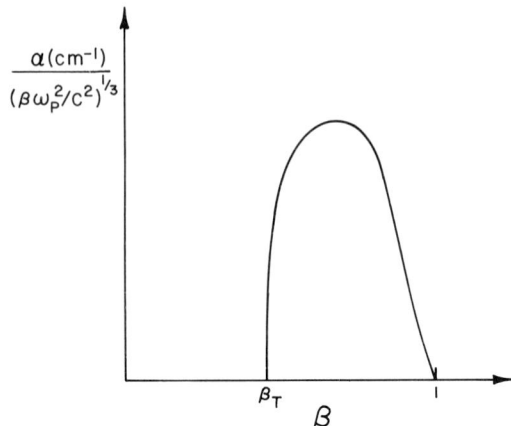

Fig. 11. Gain curve shape versus beam velocity for the beam-slab guide system.

The maximum value the gain can achieve is similar to that of the filled guide case. Some typical results for a thin-quartz-slab waveguide are shown in Fig. 12a,b. In these plots, the factor $(\beta\omega_p^2/c^2)^{1/3}$ has been omitted for convenience. The remaining factors contain all the relevant frequency and energy dependencies. Maximum values somewhat greater than unity are obtained for this particular set of parameters. The omitted term $(\beta\omega_p^2/c^2)^{1/3}$ is actually the beam current density in A cm^{-2} divided by I_0 ($=17$ kA), all to the one-third power. It is relatively easy to obtain values of 0.1 for this number; hence the plots shown in Fig. 12a,b demonstrate that with a quartz-slab guide it is possible in principle to have relatively large gain ($\alpha = 0.233$) gives 1 dB cm^{-1} well into the submillimeter part of the spectrum.

The gain plot in Fig. 12 also indicates that the gain is a bit higher on the higher order modes. This trend is a reflection of the $\omega^{1/3}$ factor in the gain. It is real, but it depends on two assumptions whose validity are also frequency-dependent. These are (1) that the beam is infinitesimally close to the dielectric, and (2) that the beam is monoenergetic. The first of these will be discussed now and the second point will be covered in Section II,D.

4. *Finite Gap between Beam and Resonator*

If we assume a small gap between the beam and the dielectric surface, we should have a situation such as that shown in Fig. 13. The analysis proceeds as before, but the resulting dispersion relation

$$[\omega_p^2 e^{-qd_2}]/[2(\omega - kv_0)^2] = (q\epsilon \cot pd - p)/p \qquad (44)$$

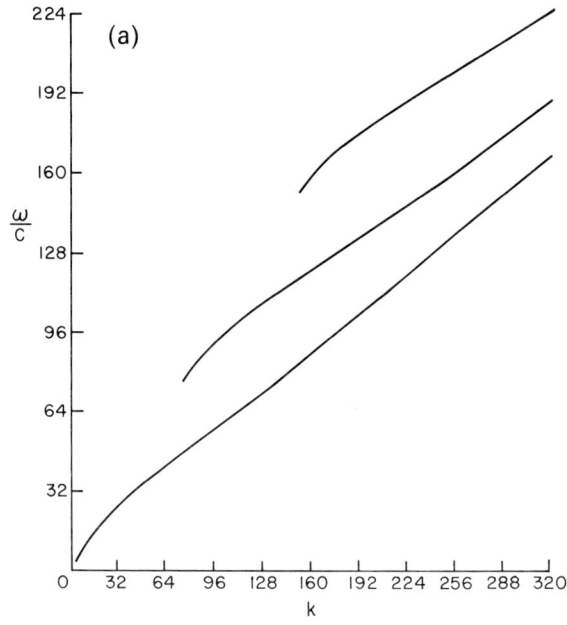

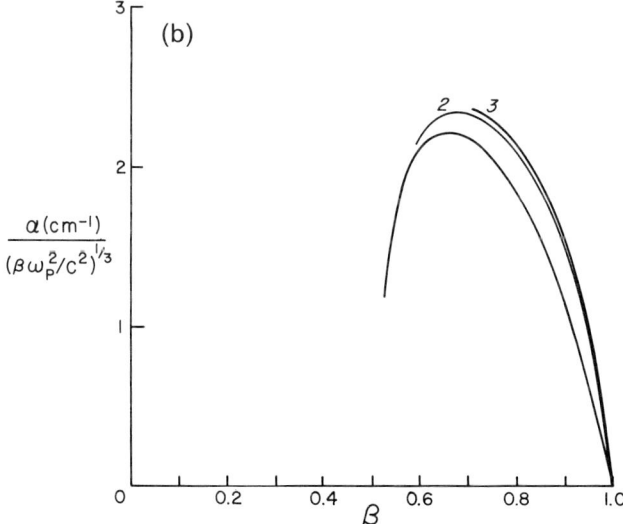

FIG. 12. Numerical values of dispersion (a) and spatial gain versus beam velocity (b) for a 0.025-cm-thick quartz guide.

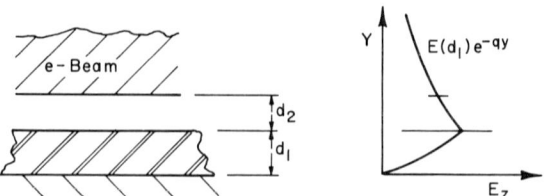

Fig. 13. Geometry of slab guide–beam system when there is a finite gap between the beam and the guide.

is, at first sight, much more complicated. However, if we again assume that the roots at synchronism lie along the dispersion curve for the free modes, the situation simplifies considerably; the end result is that the gain is modified by an exponential factor that depends on the size of the gap:

$$\alpha = \alpha(d) \exp(-kd_2/\gamma^3) \tag{45}$$

As long as kd_2/γ is small, the gain on the higher order modes will be comparable to or greater than the gain on the fundamental mode. Values of d_2 of about 1 mm would be conservative, and fractions of this are easily obtained. Hence, provided that ones uses $\beta\gamma \geq 1$, the quartz guide system discussed above will still be viable well into the submillimeter region of the spectrum.

We have been assuming that the beam extends indefinitely in the positive y direction. As long as the beam is at least a few e-foldings thick, this assumption does not affect the gain. Since we are primarily interested in high frequencies, this assumption will normally be valid.

The falloff of the field in the transverse direction may also be useful in obtaining some mode selection. If a relatively thin beam is used, the fields for the lower order modes may penetrate through to the other side. If a lossy material is placed above the beam, it may be possible to further reduce the gain on the lower order mode.

In the falloff of the electric field, operation at arbitrarily short wavelengths could be obtained if γ were allowed to become large, i.e., kd_2/γ remained small. This would involve a penalty in the maximum value of gain obtainable, but since it is relatively large to begin with, the resulting system would still be potentially useful. In this way, with more relativistic electron beams, it might be possible to operate well into the IR portion of the spectrum. This will be discussed further in Section III,C.

D. The Effect of Beam Velocity Spread

Prior to this point in our discussion, we have assumed that the electron beam was perfectly monoenergetic. It is intuitively plausible that this is a wavelength-dependent assumption, and we now examine its consequences.

The discussion is divided into three parts. First, we determine wavelength limit for a simple beam space-charge wave. Then, this result is compared with a similar criterion for a Čerenkov instability. Finally, having set the limiting wavelength for treating the beam as monoenergetic, we derive gain expressions valid in the region where the assumption is violated.

1. *Beam Space-Charge Waves*

The linearized equation of motion for a strongly magnetized electron beam is given by Eq. (1). If this is taken along with the equation of continuity Eq. (3a), Poisson's equation, and assumptions similar to those of Section II,A, the dispersion relation for space-charge waves

$$\omega = kv_0 \pm \omega_p/\gamma^{3/2} \qquad (46)$$

may be easily derived.

The upper (lower) sign in Eq. (46) corresponds to a fast (slow) space-charge wave (see Fig. 14a).

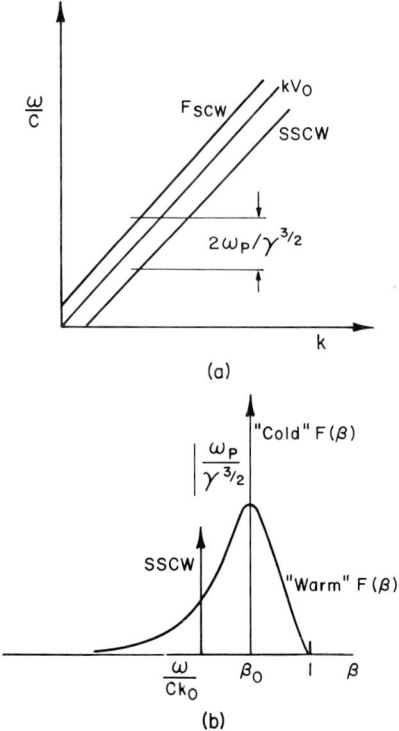

FIG. 14. (a) Relative positions of the phase velocity and (b) the beam velocity distribution function ($F\beta$) when the beam is "cold" and when it is "warm."

We concern ourselves with a slow space-charge wave. Shown in Fig. 14b is a sketch that illustrates the meaning of the statement: "The wave is resolved from the beam." The wave is clearly resolved when the beam may be regarded as a delta function in frequency space (the arrows located at ω and kv_0). If, on the other hand, the velocity spread of the beam Δv is such that the self-consistent frequency separation $\Delta\omega = \omega - kv_0$ (derived under the assumption that the beam was monoenergetic) is less that $k\,\Delta v$, the assumption is violated. A quantitative criterion for this critical k is

$$k_c\,\Delta v = \omega_p/\gamma^{3/2} \tag{47}$$

Equation (47) may be reexpressed in terms of physically more intuitive variables if we write: $k_c = \omega/c\beta = 2\pi/\lambda_c\beta$, Δv in terms of $\Delta\gamma$ and β, and ω_p in terms of the beam current density J_b. Then we have:

$$\lambda_c = \pi(\Delta\gamma/\gamma)(I_0/\beta\gamma J_b)^{1/2} \tag{48}$$

where I_0 is still $ec/r_0 = 17$ kA. If $\beta\gamma$ is of order unity, $\Delta\gamma/\gamma$ is of order 10^{-2}, and J_b is a reasonable fraction of an A mm^{-2}, than λ_c is a fraction of a millimeter. These are relatively modest requirements, and thus we predict that it should be possible to make effectively cold beams well into the submillimeter part of the spectrum.

The critical wavelengths given by Eqs. (47) and (48) are dependent on the assumption of a simple space-charge wave. When we are considering a Čerenkov instability, however, $\Delta\omega = \omega - kv_0$ is actually larger than $\omega_p/\gamma^{3/2}$, and hence the beam can be effectively colder at a given wavelength. The criterion for resolution is

$$k_c\,\Delta v = \omega_I/2 \tag{49}$$

where the right-hand side of Eq. (49) is the real part of the detuning [Eq. (13b)]. Substitution of the expressions for ω_I can be made for the appropriate case.

When the beam propagates through the dielectric Eq. (13b) applies directly and we have

$$\lambda_c = (2\pi)^{1/2}(\Delta\gamma/\beta\gamma)^{3/2}(I_0/\beta\gamma J_b)^{1/2}$$
$$\times (1 - 1/\gamma)^{-1/2}(\gamma^2/\gamma_T^2 - 1)^{-1/2} \tag{50}$$

The current-density dependence is similar to that of Eq. (48), but provided the beam is at least mildly relativistic ($\beta\gamma \geq 1$), the energy dependence is more favorable. Overall, presuming that γ_T and γ/γ_T are not excessively large, the value λ_c given by Eq. (50) will be at least as small as that given by Eq. (48). The addition of the form factor associated with a more practical resonator does not alter this essential conclusion.

2. Gain in the Warm-Beam Limit

When the criteria given in the preceding paragraphs are violated, the beam is to be regarded as "warm" at the wavelength in question. The gain does not vanish in this limit, but it does begin to drop as ω^{-1}, as opposed to the general $\omega^{1/3}$ trend in the cold-beam limit. This trend means that oscillators can probably be built in the warm-beam limit, but amplifiers will be impractical.

In calculating the gain we use the Vlasov equation as the basic equation of motion, and we will retain the assumption that the beam is strongly magnetized. In this case, the Vlasov equation is

$$(\partial f/\partial t) + v_z(\partial f/\partial z) + p_z(\partial f/\partial p_z) = 0 \tag{51}$$

If this is linearized ($f = f_0 + \delta f$) and Fourier-transformed, we have for the perturbed component distribution,

$$\delta f = -ieE_z \left(\frac{\partial f_0/\partial p_z}{\omega - kv_z}\right) \tag{52}$$

The current is now given by:

$$J_z = \int v_z \, \delta f \, dp \tag{53a}$$

$$= ieE_z \int v_z \left(\frac{\partial f_0/\partial p_z}{\omega - kv_0}\right) dp_z \tag{53b}$$

Substitution of this into Eq. (8) will then lead to a dispersion relation. If the beam distribution is a delta function, then the integral can be performed immediately and the current given by Eq. (4b) is recovered. However, we are now interested in the limit where the beam velocity spread is finite.

An exact solution of a dispersion relation containing a integral kernel, such as that of Eq. (54), can be found using numerical techniques. The results of such a procedure are discussed below. However, some insight into the general behavior can be obtained in the limit where k times the width of the beam distribution is broad in comparison with the gain that would be obtained from a calculation in which one assumed that the beam was cold (monoenergetic).

The dispersion relation obtained from the above procedure is

$$\frac{\omega^2 \epsilon}{c^2} - k^2 - p^2 + \frac{\omega_p^2}{\omega \epsilon}\left(\frac{\omega^2 \epsilon}{c^2} - k\right)\int_{-\infty}^{\infty}\left(\frac{mv}{\omega - kv}\right)\frac{\partial f}{\partial p} dp = 0 \tag{54}$$

In its present form the integral that appears in Eq. (54) is to be performed along the real momentum line, and hence the procedure for handling the singularity at synchronism is not yet defined. Borrowing from the theory of plasma physics, we handle it by formally extending the integral into the complex plane. First, we reexpress Eq. (54) as a velocity integral. Then

$$\frac{\omega^2\epsilon}{c^2} - k^2 - p^2 + \omega_p^2 \frac{\omega^2\epsilon/c^2 - k^2}{\omega\epsilon} \int \frac{v}{\gamma^3(\omega - kv)} \frac{\partial F}{\partial v} dv = 0 \quad (55)$$

we now let

$$(\omega - kv)^{-1} = P(\omega - kv)^{-1} - i\pi\delta(\omega - kv) \quad (56)$$

where P stands for the principal part of the integral. Multiplying through by c^2/ϵ and using the condition $v = \omega/k\omega\epsilon$, we find for the imaginary part of the dispersion relation:

$$D'' = -\frac{\pi\omega_p^2\beta^2(1 - 1/\beta^2\epsilon)}{\epsilon\gamma^3} \frac{\partial F(\omega/k)}{\partial(\omega/k)} \quad (57)$$

It is sufficient for the purpose of the present discussion to ignore the small correction to the real part of the dispersion represented by the principal part of the integral. In the limit, the real part of the dispersion is

$$D' = \omega^2 - \omega_k^2 \quad (58a)$$

and providing the growth is small, the imaginary part of the frequency is adequately represented by

$$\omega_I = \frac{D''}{\partial D'/\partial \omega}\bigg|_{\omega = \omega_k} \quad (58b)$$

or

$$\omega_I = \pi\omega_p^2 \frac{\beta^2(1 - 1/\beta^2\epsilon)}{2\epsilon\omega_k\gamma^3} \frac{\partial F(\omega_k/k)}{\partial(\omega_k/k)} \quad (58c)$$

Obviously, there will be wave growth (inverse Landau damping) in the region of velocity space where $\omega_k \leq kcv_0$). Sketches of the unperturbed dispersion relation $\omega = \omega_k$ and ω_I are shown in Fig. 15. The region of positive ω_I lies on the larger k side of the synchronous wave number $k_s = \omega/v_0$, and peaks at a velocity which is below v_0 by an amount approximately equal to the half-width of the velocity distribution.

Thus, the wave growth in the warm-beam limit has a shape which is complimentary to that obtained in the cold-beam limit. This result may seem to imply that the growth due to inverse Landau damping is fundamentally different from the growth obtained in the warm-beam limit. However, this is not the case. If the roots of Eq. (55) are followed as the beam

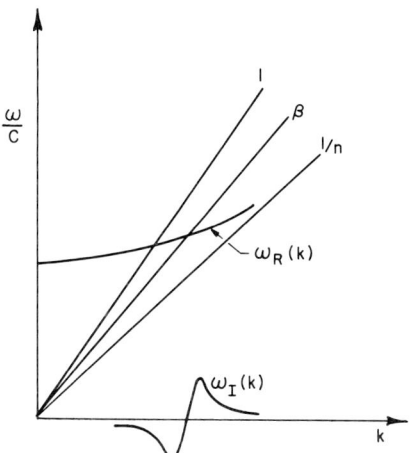

FIG. 15. Qualitative shape of the dispersion and gain in the "warm"-beam limit.

width is varied from a value of zero up to $k\,\Delta v\omega_I$ (cold), we find that one regime passes smoothly into the other. The peak absorption shown in Fig. 15 is always somewhat less than the gain. It is a composite of the long-wavelength cold-beam gain and Landau damping caused by that part of the beam having $\partial F/\partial v < 0$. The peak gain obtainable in the warm-beam limit will always be less than ω_I (cold). This occurs because, as the beam distribution becomes arbitrarily sharp, the self-consistent beam-density-dependent frequency shift moves the phase velocity of the wave downward relative to the position of maximum $\partial F/\partial v$.

In spite of this, the warm-beam limit may very well be of interest. An assessment of this requires that we estimate Eq. (58c), which, in general, depends on the detailed shape of the beam distribution. We can, however, proceed without undue complication if we recognize that if $\sim 1/\Delta v$, and thus $\partial F/\partial v$ at its maximum is $\sim 1/\Delta v^2$. Then, in terms of $\Delta\gamma$, an approximate expression for ω_I becomes

$$\omega_I \simeq \left(\frac{\pi\omega_p^2}{2\epsilon}\right)\left(\frac{\gamma}{\omega}\right)\left(\frac{\beta\gamma^2}{\Delta\gamma}\right)\left(1 - \frac{1}{\beta^2\epsilon}\right) \tag{59a}$$

or

$$\alpha \simeq \left(\frac{\lambda}{\epsilon}\right)\left(\frac{J_b}{I_0}\right)\left(\frac{\gamma^2}{\Delta\gamma}\right)\left(\frac{\gamma^2/\gamma_T^2 - 1}{\beta^2\gamma}\right) \tag{59b}$$

It is interesting to evaluate the possible gain in the 10-μm range. In this case, $\lambda = 10^{-3}$ cm. If we can achieve $J_b/I_0 \sim 10^{-3}$ and $\Delta\gamma/\gamma > 10^{-2}$, it appears that $\alpha \sim 0.1$ cm^{-1} is within the realm of possibility. Thus, it might be possible to construct Čerenkov lasers down to wavelengths comparable to those achieved in stimulated Compton devices.

It is also interesting to evaluate Eq. (59b) when λ is equal to λ_c. Upon substitution, we obtain

$$\alpha = (2\pi)^{1/2}(\gamma/\Delta\gamma)^{1/2}(J_b/\gamma I_0)^{1/2}(\gamma^2/\gamma_T^2 - 1)^{1/2} \qquad (60)$$

Examination of Eq. (60) supports the conjecture that values of $\alpha \geq 0.1$ cm^{-1} are attainable in the cold-to-warm crossover region ($\lambda \simeq \lambda_c$).

E. Comments on Nonlinear Behavior

It would be possible at this point to develop a reasonably complete nonlinear theory for single-mode Čerenkov devices. This would, in part, follow lines of argument originally established to explain microwave tubes, beam plasma interactions, and more recently, free-electron lasers. In the wavelength regime which is of primary interest, however, the ratio L/λ is large and there will be, at the very least, a few axial modes within the halfwidth of the gain curve. There may also be a mixture of transverse modes, although if the device can be made to operate on a single transverse mode, it will be advantageous to do so. Relatively less is known about electron beam devices in the multimode region, and the development of a complete nonlinear theory, analogous to that developed by Lamb (24) for the gas laser, should take account of multimode operations. This would be a substantial undertaking. It could be productive, however, since we generally expect that Čerenkov devices will exhibit many phenomena intrinsic to all multimode oscillators, and that some of these may be useful in applications (e.g., mode locking). Hence, because the development would at this point omit some of the most interesting parts of the problem (due to its length and because the motivation for experimental development rests primarily on the prediction of the linear gain which would be expected in specific wavelength regions), we will restrict discussion of nonlinear problems to a few simple scaling arguments.

Nonlinear Scaling Arguments

In the operating regime where the beam velocity distribution can be regarded as cold and the motion one dimensional, the relative density modulation is given by

$$\delta n/n_0 = kv/(\omega - kv_0) \qquad (61)$$

Now, the change from electron orbits that move progressively forward relative to the phase of the wave (untrapped) to orbits that are winding up (trapped), occurs in the vicinity of $|\delta n/n_0| \simeq 1$. Thus it occurs where

$$k\,\delta v \simeq |\omega - kv_0| \qquad (62a)$$

or

$$k\,\delta v \simeq \omega_I \qquad (62b)$$

This can be converted to a prediction of the magnitude of the axial component of the electric field at which saturation of the linear growth is in progress. We find from Eqs. (1) and (61a), (61b):

$$|E_z| \simeq (m\gamma^3/e)(\omega_I^2/k) \qquad (63)$$

The Poynting flux, and hence also the total power carried to the wave, is proportional to E_z^2 and thus up to ω_I^4. The latter is in turn proportional to $(I/I_0)^{4/3}$. Hence, up to form factors (which can vary between small numbers and unity) the overall power at the separatrix crossing value of E_z [Eq. (62)] becomes

$$P_{\text{wave}} \sim (I/I_0)^{1/3} P_{\text{beam}} \qquad (64)$$

This is a conversion factor that may be expected in any traveling-wave device. Enhanced conversion could be obtained by tapering the phase velocity and thus "deepening" the trapping well. More energy still could be recovered if the beam were collected at high potential (depressed collector operation). An estimate of the overall efficiency obtainable from a Čerenkov device is thus a subtle and complex matter. In the final analysis, however, tubelike efficiencies of perhaps 50% could be attainable.

III. Experiment

A. The Electron Beam

The single most important component of an electron-beam-driven radiation source is the beam itself. Its parameters will, in large part, determine the performance of the system. In order to examine the potential of Čerenkov sources in the millimeter and submillimeter parts of the spectrum, it is useful to examine the parameters of some typical electron-beam generators suitable for this application.

A few general types of electron-beam generator and their parameters are listed in Table I. All of the beams are at least mildly relativistic; a choice is dictated by the coupling considerations developed in earlier sections. However, the values of the beam current and the modes of operation vary widely. We begin our discussion of the table entries by considering the role of beam energy.

The beam energy helps to determine the operating wavelength in several interrelated ways: (1) by synchronism; (2) by the magnitude of the gap between the beam and the resonator, which must be present in any real system; and (3) by its entry into the equations that determine the beam modulation. The first of these alone does not place any stringent limit on the attainable wavelength. This is because the design of dielectric resonators that will support a wave of any reasonable phase velocity does not present a problem. The second and third aspects of the energy dependence will therefore be more important in setting the short-wavelength limit to a device. The rate at which the electric field decreases as the distance from the resonator increases is given by $q = 2\pi/\lambda\beta\gamma$. The gain, however, increases with frequency and hence we would expect it to peak somewhere in the vicinity of $qa \simeq 1$, where a is a characteristic distance between the beam and resonator. A modest value for a would be approximately 1 mm, and a more difficult but attainable value would be about one-tenth of this. Taking this range and assuming $qa \simeq 1$, we can determine the limiting wavelength for good beam-to-resonator coupling. The range of this wavelength is shown in the table. Examination of the table will show that relatively compact machines could ultimately be expected to work well into the submillimeter region of the spectrum, and if one extends the range of beam generator complexity, operation in the IR is perhaps possible.

TABLE I

Electron Beam Generators of Potential Use in Čerenkov Sources

Type	Energy (MeV)	Current (A)	Pulse length	Peak power	"Good" coupling wavelength (mm)
1-Stage thermionic	0.1–0.2	0.1–0.2	∞	10–50 kW	1 µm
	0.04–5	10–100	µsec	1–100 MW	1
n-Stage thermionic	0.1–1	0.1	∞	10–100 kW	0.3
	1	kA	µsec	1 GW	0.5
Field-emission diode	0.5–5	multi-kA	0.01–0.1 sec	multi-GW	0.5
LINAC	2–100	0.01–1	10 psec	50 MW	1 µm

While the beam-to-resonator coupling decreases away from the resonator due to the falloff of the electric field with distance, an arbitrary increase in beam energy will lower the minimum wavelength by a corresponding amount. However, the beam energy also enters into consideration through the equations of motion for the electrons, and increasing γ furthers the difficulty of attaining good beam modulation. Hence these two effects must be traded off against each other. When the beam is strongly magnetized, the energy dependence of the growth is approximately $\alpha \sim (\gamma - 1)$, and when it is not magnetized $\alpha \sim (\gamma - 1/3)$. Since the criterion for strong magnetization becomes more difficult to satisfy as λ decreases, short-wavelength operation probably requires unmagnetized, or at least weakly magnetized, electron beams.

The current available from the variety of generators listed in Table I also covers a wide range. When the beam is cold, the gain will scale as $(J_b/I_0)^{1/3}$, and thus a beam with 17 A cm^{-2} will make this parameter 0.1. We have seen earlier that if this is used with reasonable values of the other parameters, the gain will be in the 0.1–0.5 cm^{-1} range. The first and third entries in the table can probably achieve values in this vicinity, while the second and fourth entries could do so without question. The greater current available from the second and fourth type of generator might also make up for difficiencies in another parameter.

Field-emission diode generators produce very large currents. Hence, they are in principle capable of producing a lot of gain. It was partly for this reason that a generator of this type was used in early experiments designed to demonstrate the utility of stimulated Čerenkov radiation. They also have the ability to produce beams whose energy is sufficient to couple well into the submillimeter region. Their drawback for short-wavelength operation may, ironically, be the fact that the current is large. This is because the self-fields (which have been neglected in our analysis of gain) may lead to a larger $\Delta\gamma/\gamma$, and hence to a limit on the usable range of wavelength.

The accelerators listed in the table are also capable of operating at current densities that give usable gain. The peak current will be low, but the focusing could be better. The current in a linear accelerator will also have a complicated time structure (the typical value of I is the peak in the micro-pulse), and this complicates the gain calculations. If tried, however, such a generator will be designed to work at short wavelengths and the micro-structure may be a minor feature. Both this complication and the role of self-fields are worth further analysis.

The pulse length and peak power entries in the table are largely self-explanatory, although one consequence of the pulse versus continuous operation is worthy of comment. We shall see below that when the Q of a cavity is reasonable, the current density required to initiate oscillation is quite modest. Thus systems with relatively low gain ($\alpha = 0.01$–0.1) may be very usable as oscillators well into the IR, whereas the same beam generator would not be a suitable source for an amplifier. Overall, it is to be expected that the pulsed-beam generators, due primarily to the larger current densities available, could be used as both oscillators and amplifiers, whereas the steady-state generators would be largely restricted to application as oscillators.

In addition to its relation to the gain of the device, the current density plays a role in determining the wavelength at which the beam may no longer be regarded as "cold." Clearly, if all parameters other than current remain fixed, increasing the current decreases the wavelength at which the beam must be regarded as warm. The current density, the energy, and the energy spread are not truly independent, but we shall, for the purpose of discussion, treat them as though they were.

Restating the expression $k\,\Delta v = \omega_I/2$ in terms of the spatial gain α (cm^{-1}) and a beam energy spread $\Delta\gamma$, we have for the wavelength at which the cold–warm transition occurs:

$$\lambda_c = [4\pi/(\beta\gamma)^2](\Delta\gamma/\beta\gamma)\alpha^{-1} \tag{65}$$

It is clear that the beam generator in the lower energy end of the range considered must achieve relatively better energy collimation if it is to operate in the "cold" regime at any given wavelength. The lower energy pulsed electrostatic devices with their larger currents can be expected to operate in this mode for wavelengths in the upper submillimeter to one-millimeter range. As the beam energy rises, the crossover wavelength becomes smaller, and hence with accelerator-produced beams, one would be encouraged to attempt experiments with IR Čerenkov sources. This is quite an extrapolation from any present stimulated Čerenkov experiment. But other modes of electron-beam use have achieved success in this range. Beam-dependent limitations are largely governed by the particular interaction, however, and thus comparisons of Čerenkov and, for example, stimulated Compton sources depend primarily on the gain. In the former device we must couple to a suitable resonator, but in the latter a relatively large transverse velocity modulation must be induced on the beam. Thus the two methods of electron-beam use may produce at least comparable results.

B. A Millimeter-Wavelength Experiment

Shown in Fig. 16a is the schematic diagram of an apparatus that has been constructed to test a number of the concepts discussed in Section II. It consists of a high-voltage pulse transformer, a thermionic cathode, a single-gap accelerating stage, a static magnetic field to guide the beam, and a tubular quartz–dielectric resonator. The beam is guided along the axis of the resonator and is collected on the surface of a plane mirror placed at a 45° angle to the end of the copper tube that supports the quartz. The output is then focused into a plane-parallel Fabry–Perot resonator.

The transformer has a peak capacity of 400 kV at 200 A for 10 μsec at a 300 Hz repetition rate. However, typical operating conditions for the present work are 100–300 kV at 0.5–20 A for approximately 1 μsec. At the present time the pulse repetition frequency is a few hertz. The primary of the transformer is driven by a hydrogen-thyratron-triggered single-stage pulse-forming network. A characteristic high-voltage trace and an accompanying millimeter-wavelength radiation pulse is shown in Fig. 16b. The voltage range of the generator, which provides adequate power for stimulated Čerenkov radiation experiments, is such that millimeter–submillimeter operation can be expected.

Coherent output radiation has been obtained at wavelengths extending from about 1 cm to below 1.5 mm. The wavelength of the radiation depends

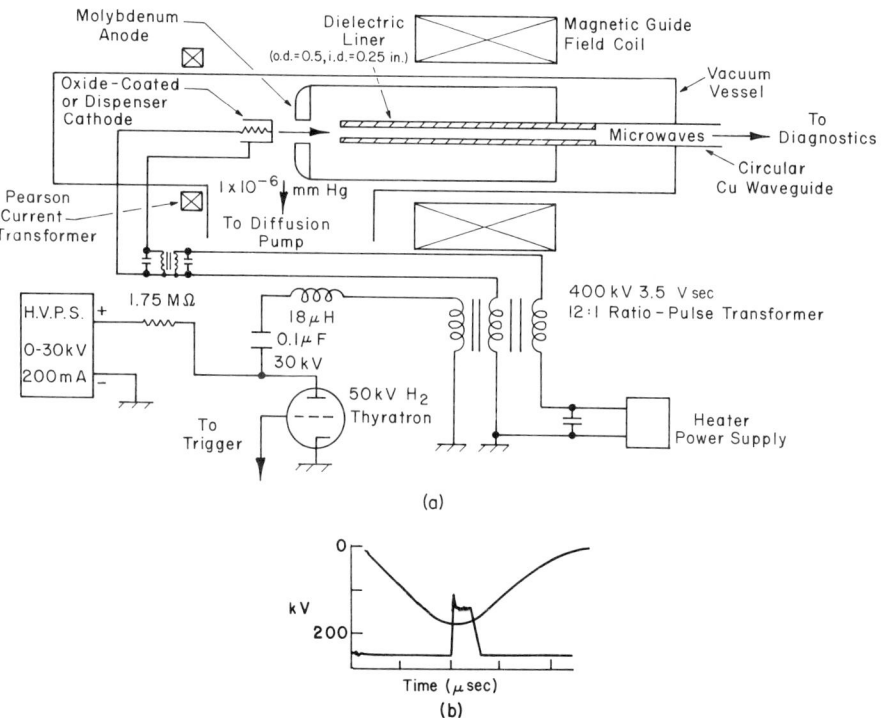

FIG. 16. Configuration of an experimental device designed to produce millimeter-wavelength stimulated Čerenkov radiation: (a) the device; (b) typical voltage and radiation pulses.

on the guide radius, the relative amount of dielectric and its dielectric constant, and the beam voltage. Configurations which should work in the fundamental mode over the 1 cm–3 mm range have been studied, and reasonable agreement with the theory of Section II is found. The diameter of the copper guide that supports the quartz tube is approximately 1.5 cm, and tubes with 1–3 mm wall thicknesses have been used most recently. Thus output wavelengths smaller than the transverse dimension of the waveguide have already been obtained.

At the longer wavelengths, the frequency has been determined with the Fabry–Perot interferometer, and some typical data illustrating the behavior are shown in Fig. 17. In Fig. 17a, a calibration trace made with a 35-Hz Gunn diode source is displayed, and in Fig. 17b experimental data with approximately the same wavelength are shown. The trace shown in Fig. 17b consists of many repetitions of the electron-beam pulser, and it shows that the average spectral width, which is itself quite narrow (0.1–1%), is primarily due to shot-to-shot reproducibility. The output of a single pulse is apparently very coherent. An interferometer for shorter wavelengths is under construction. The shorter wavelengths have been measured with cutoff filters.

The output is monitored by ordinary microwave diodes, either IN23s, IN26s, or IN53s, with the latter being used down to wavelengths below 1.5 mm. Attenuation levels of 30–60 dB are required in order to ensure that the output levels are below the burnout levels of the diodes. The absolute output power levels are not yet precisely determined except within an experimental uncertainty that becomes greater at shorter wavelengths and varies between 0.1 and 5% of the beam power. All of the factors which control this conversion efficiency are not yet well understood. The loaded Q of the resonator is modest, and the system is probably operating as a superradiant oscillator. If this is so, the previously stated conversion efficiencies are plausible, but as is the case with the theory, nonlinear behavior of the experimental device is practically unexplored.

The experiment described above is in a relatively early stage of development. It does appear, though, that millimeter–submillimeter Čerenkov devices are a realistic possibility.

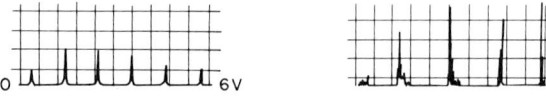

FIG. 17. Fabry–Perot interferometer output: (a) calibration (35-GHz source); (b) stimulated Čerenkov radiation. (TSSS–0.2 V cm^{-1}; k155 μVm–100-mV range; 15 × 18 mesh; IN53 detector.)

C. Čerenkov Devices in the Short-Wavelength Limit

The bulk of the analysis presented in the preceding sections was devoted to systems designed to work in a way similar to microwave devices of the traveling-wave type. We emphasized the region that was collective where there was an interaction of fields with the beam. Recently (25) however, radiation with a wavelength of about 3 μm was obtained from an electron beam produced by a linear accelerator and a coupler—which was a helical static magnetic field. This device operated in the so-called single-particle limit:

$$\omega_p L/c\beta\gamma^{3/2} \ll 1 \qquad (66)$$

In this expression, L is the length of the interaction region, and the other quantities are defined in earlier sections.

The interaction in this free-electron laser depends on the coupling between the transverse velocity modulation imparted to the beam by the helical pump, and the magnetic field component of the wave in question. The beat wave produced, sometimes called a ponderomotive wave, travels at a phase velocity. Hence it, too, is a slow-wave interaction.

There is no reason why the slow wave supported by a thin-film dielectric waveguide could not also be used in the 1 μm range. Furthermore, the beam quality requirements are the same if

$$kb/\gamma \simeq 1 \qquad (67)$$

where b is the beam thickness. The electric field associated with the slow-wave structure can be greater than that of the ponderomotive wave. A detailed comparison of the two modes of interaction (dielectric versus static B pump) would be too lengthy to present here, but in view of the above comment, it follows that there are beam–resonator combinations for which the Čerenkov device's gain will exceed the gain of the ponderomotive wave-based free-electron laser (FEL). In this case, the behavior of a Čerenkov device will be more "forgiving" than the FEL. An example of this is the relaxation of the energy collimation required if a beam in the single-particle limit is to be considered monoenergetic. Simple consideration of the phase shift accumulated in a length L leads to the requirement:

$$ck\,\Delta\beta L/v < \pi \qquad (68a)$$

or

$$\lambda/L > 2\delta\gamma/(\beta\gamma)^3 \qquad (68b)$$

Higher gain would allow a shorter length, and hence a system that was in the "cold" region at a shorter wavelength. Because of this and other con-

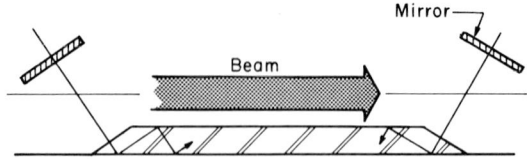

Fig. 18. Slab guide resonator configuration for a possible far-IR device.

siderations, it is of interest to make a detailed comparison of these two systems. We conclude this section with a brief discussion of two possible configurations that might be used in the 1 μm range. The first is shown in Fig. 18.

Shown in the figure is a thin dielectric guide with tapered ends. An electron beam passes over the guide and couples to the evanescent field. When the beam has a high enough energy, the coupling will be good. The tapered ends face mirrors that, together with the guide, form an optical resonator.

It might also be useful to taper the thickness of the guide near its ends. By doing this, the field energy in the guide can be increased at the expense of that stored above the guide. If the taper is adiabatic on the scale of k^{-1}, the dispersion relation will vary smoothly, as will the field distribution. Shown in Fig. 19 are the results of one possible experimental configuration including a taper. As the guide thickens, the field distribution is pulled down into the guide and formed into a half-sinusoid. The latter form, which is close to the normal mode distribution of a guide covered top and bottom, should give better control of the input–output coupling.

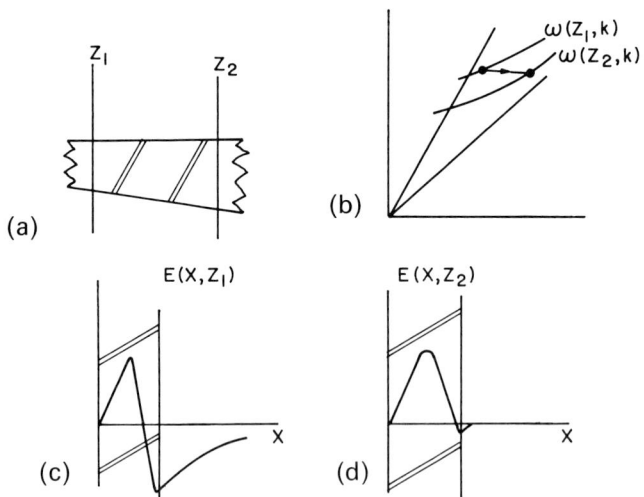

Fig. 19. (a) The effect of tapering the slab guide thickness; (b) the change in dispersion; (c) the field at Z_1; (d) the field at Z_2.

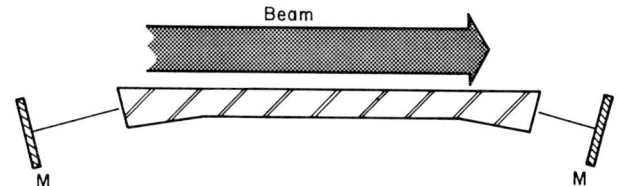

Fig. 20. Another possible configuration for short-wavelength operation.

Another possible version of a short-wavelength system is shown in Fig. 20. In this system, the output optics are placed below the beam. Incident radiation comes through the end face of the guide and is normal to the face, but the angle formed with the top surface is greater than that required for total internal reflection. This guide is also tapered. As it thins out, the field is pushed out to increase the coupling. It is then pulled in again and passed to a second mirror.

The two configurations just discussed are highly schematic. The beam-to-field coupling will be a straightforward matter, but the overall optical system may be quite complex. Self-reproducing patterns of the general type discussed should exist, however, and well-collimated relativistic beams, together with the resonator, might form the basis of far IR Čerenkov devices.

IV. Conclusion

The theme of the preceding three sections is the use of the combination of the electron beam and the dielectric resonator to generate coherent radiation in the millimeter through IR portions of the electromagnetic spectrum. At the present time, there is experimental confirmation of some of the basic theoretical assumptions in the 10–1 mm region. The theoretical advantages of smooth-bore slow-wave structures at shorter wavelengths indicate that submillimeter–IR operation is also a realistic possibility. Beams with the monoenergicity and current sufficient to achieve oscillation for moderate-Q resonators $[(\omega_I Q/\omega) > 1]$ currently exist. The long-term interest in these sources will not depend on whether they work at all, but on how they compare with other possible sources.

Acknowledgments

This is a report of work in progress. I owe a debt to those who have already contributed and to those who are currently engaged in the research. I would most especially thank Robert Layman for continuing to carry a major portion of the burden of the experimental work, of which only a small part has been reported herein. In addition, Kenneth Busby and Kevin Felch helped to build the first apparatus and took the first measurements. John Bagger and Geoffrey Crew performed calculations during the early stages of the work; portions of their senior theses appear in Section II. To be thanked also are John Branscum, John Golub, David Kapilow, James Murphy, David Speer, and Douglas Wise, who currently bear a great part of the responsibility for the experimental and theoretical program. Finally, my thanks to Desirée Rastrom for research and administrative assistance.

Support for this work has been provided by Dartmouth College, by the Air Force Office of Scientific Research, by the Army Research Office, and by the Office of Naval Research.

References

1. P. A. Čerenkov, *Dokl. Akad. Nauk SSSR* **2**, 451 (1934).
2. I. M. Frank and I. Tamm, *Dokl. Akad. Nauk. SSSR* **14**, 109 (1937).
3. O. Heaviside, *Philos. Mag.* Feb., p. 130; Mar., p. 202; May, p. 379; Oct., p. 360; Nov., p. 434; Dec., p. 488 (1888).
4. A. Sommerfeld, *Göttinger. Nachrichten* **99**, 363 (1904).
5. E. Curie, "Madame Curie." Heinemann, London, 1941.
6. L. Mallet, *C. R. Acad. Sci.* **183**, 274; **187**, 222; **188**, 445 (1926).
7. J. V. Jelley, "Čerenkov Radiation and Its Applications." Pergamon, London, 1958.
8. Z. P. Zrelov, "Čerenkov Radiation in High Energy Physics." Israeli Program for Scientific Translations, 1970.
9. B. M. Bolotovski, *Usp. Fiz. Nauk.* **75**, 295 (1961); *Sov. Phys. Usp. (Engl. Transl.)* **4**, 781 (1962).
10. V. L. Ginzburg, *Dokl. Akad. Nauk. SSSR* **3**, 253 (1947).
11. P. Coleman and C. Enderby, *J. Appl. Phys.* **31**, 1695 (1960).
12. M. Danos, *J. Appl. Phys.* **26**, 2 (1953).
13. H. Lashinsky, *J. Appl. Phys.* **27**, 631 (1956).
14. R. Ulrich, *Z. Phys.* **199**, 171 (1967).
15. A. Ahiezer, *Nuovo Cimento, Ser. 10* **3**, 591 (1956); M. Abele, *Nuovo Cimento, Ser. 9 3207* (1952).
16. G. Mourier, in "Microwave Tubes," Vol. 2, p. 132. Elsevier, Amsterdam, 1958.
17. J. G. Linhart in "Microwave Tubes," Vol. 2, p. 136. Elsevier, Amsterdam, 1958.
18. M. A. Piestrup, R. A. Powell, G. B. Rothbart, C. K. Chen, and R. H. Pantell, *Appl. Phys. Lett.* **28**, 92 (1976).
19. A. N. Chu, M. A. Piestrup, T. W. Barbee, and R. H. Pantell, *Proc. Int. Conf. Lasers*, p. 744–749 (1978).
20. M. Stockton and J. E. Walsh, *J. Opt. Soc. Am.* **68**, 1629 (1978).
21. J. E. Walsh, T. C. Marshall, and S. P. Schlesinger, *Phys. Fluids* **20**, 709 (1977).
22. K. L. Felch, K. O. Busby, R. W. Layman, D. Kapilow, and J. E. Walsh, in Free Electron generators of coherent radiation. "Physics of Quantum Electronics" (S. Jacobs, H. Pilloff, M. Sargent, M. Scully, and R. Spitzer, eds.), Vol. 7, Addison-Wesley, Reading, Massachusetts (1980).
23. K. L. Felch, K. O. Busby, R. W. Layman, D. Kapilow, and J. E. Walsh, *Appl. Phys. Lett.* **38**, 60 (1981).
24. W. Lamb, *Phys. Rev. A* **134**, 1429 (1964).
25. D. A. Deacon, L. R. Elias, J. Madey, G. Ramian, H. Schwettman, and T. Smith, *Phys. Rev. Lett.* **38**, 892 (1977).

Materials Considerations for Advances in Submicron Very Large Scale Integration*

D. K. FERRY

Department of Electrical Engineering
Colorado State University
Fort Collins, Colorado

I. Introduction 312
II. Submicron MOSFETs 314
   A. Introduction to MOSFETs 315
   B. The Inverter Circuit 320
   C. Field-Dependent Mobility 322
   D. Scaling the MOSFET 325
   E. Short-Channel Effects 327
III. Submicron MESFETs 332
   A. The JFET 333
   B. The Inverter Circuit 335
   C. Field-Dependent Mobility 336
   D. Scaling the MESFET 338
   E. Short-Channel Effects 339
IV. Switching of High-Speed Logic 340
   A. Transient Switching 341
   B. Drive Currents 345
   C. Velocity Response 347
   D. Effective Saturation Velocity 355
   E. Interconnection Capacitance 358
V. Band Structure Considerations 361
   A. Silicon 364
   B. The III–V Compounds 366
   C. Pseudobinary Compounds 367
   D. Alloy Ordering 373
VI. Comparisons and Limitations of Logic for Ultra-Large-Scale Integration 375
   A. Device Parameters 376
   B. Thermal Considerations 380
   C. Further Considerations 383
   D. Summary 386
   References 387

---

* The study leading to this article was supported by the U. S. Army Electronics Technology and Device Laboratory, Fort Monmouth, New Jersey.

I. Introduction

In recent years, the semiconductor electronics industry has entered the very large scale integration (VLSI) era, in which individual integrated circuits are characterized by increasingly complex circuitry and technology. Indeed, it may be said that it is actually the drive to provide less expensive, but more complex and sophisticated, integrated systems that has caused the growth of VLSI, a growth that is truly phenomenal in its range and impact. The increase of complexity in these circuits has followed the well-known Moore's law—an effective doubling of complexity approximately every two years. There are, of course, several factors which contribute to this increase in complexity, including major effects arising from increased die size, increased circuit cleverness, and reduced device size. This latter factor, the reduction of the individual feature size in a device, is of paramount importance, and dimensions of laboratory systems are currently down into the submicron range.

Based upon the currently utilized design technology, the development cost of a silicon VLSI chip can be approximately expressed as an exponential function of complexity, whereas the assembly costs per function decrease as the chip becomes more complex and includes more functions. At any rate, there is an optimum chip size based upon the tradeoff of these two costs (Russo, 1980). This optimum size increases as developments in technology and smaller device sizes become available. Indeed, it can be easily shown that power considerations require ever lower power-delay products as higher speed and more complex systems are developed. The argument may be stated simply, as the power dissipation in a chip will be approximately (see, e.g., Eden et al., 1979)

$$P \geq 2N_g f_c (P_D \tau_D) \tag{1}$$

where N_g is the number of gates, f_c is the gate-level clocking frequency, and $P_D \tau_D$ is the gate energy (average) required to switch—the power-delay or speed–power product. The power level in Eq. (1) will be restricted by the available ability of the chip to dissipate heat, a quantity related to the thermal conductivity of the semiconductor material, so that as N_g or f_c is increased, $P_D \tau_D$ must be reduced. This has been achieved to date primarily by reducing the size of a single gate or device. The average transistor area in microprocessors was reduced from more than 10 mil^2 to 1.8 mil^2 during the late 1970s (Russo, 1980), usually by using more regular structures. Indeed, fully regular structures such as random-access memories have been developed in the laboratory with less than 0.2 mil^2 chip area per transistor.

Progress in the microelectronics area will continue to be tied to the ability to continue to put ever increasing numbers of smaller devices on a chip. Today, VLSI involves gate lengths and linewidths in the 1–3-μm range.

It is apparent that extrapolation of current technology will produce individual devices whose dimensions are of the order of 0.1–0.5 μm (Hoeneisen and Mead, 1972; Chatterjee et al., 1980) and will usher in the era of ultralarge-scale integrated (ULSI) circuits. Although there are laboratory devices with gate lengths already at the 0.1-μm level (Hunter et al., 1981), there are serious concerns about the future of silicon ULSI in this dimensional range. As a consequence, many designers have turned to GaAs integrated circuits for high-speed applications (see, e.g., Van Tuyl and Liechti, 1977). The guiding light of these designers has been the supposedly superior electronic properties of GaAs, which should allow one to achieve speed and power advantages in VLSI.

Needless to say, the silicon semiconductor industry has not welcomed the onset of GaAs technology with open arms. The progress made in silicon large-scale integration (LSI) and VLSI, as noted above, has been nothing short of phenomenal, and this has generated enormous confidence in the future of this technology among the faithful. The arguments between the two have been boisterous and often not well founded on valid comparisons. GaAs technology is based upon the junction field-effect transistor (JFET) and metal–semiconductor FET (MESFET), whereas silicon technology is predominantly based upon the metal-oxide–semiconductor FET (or MOSFET) (we will not concern ourselves with bipolar technology here). In the submicron device size range, it is expected that transport will be limited by the saturation velocity, thought to be roughly comparable in the two materials. But this is only true for large devices: How do the saturation velocities compare in small devices, where one must account for overshoot-velocity effects? Is semiinsulating GaAs advantageous in reducing interconnection capacitance, or does an all-implantation technology in silicon allow the use of relatively high-resistivity silicon (Hyltin, 1965) and thus negate the advantage? Is thermal conductivity a limit, since that of silicon is four times that of GaAs? These questions are seldom easily answered. The principal requirements of a digital technology for high-speed VLSI are (Eden et al., 1979): (1) very high density, and thus low chip area per gate; (2) low power dissipation per gate; (3) extremely low delay-power product; (4) high speed and thus low gate delay [points (2)–(4) are thus interrelated]; and (5) a high-yield process. Factors (1) and (5) will be by-products of the technological push to VLSI–ULSI, and so it is really the requirements of low $P_D \tau_D$ and τ_D (and thus high f_c) that are of concern in comparing various technologies. Finally, one can ask the obvious question: Should we be limited to silicon and GaAs, or are there other materials that hold promise in the submicron device area?

In this work, an attempt is made to analyze the above questions and to provide some possible answers. We hope to go beyond the comparisons made earlier by Liechti (1977), and treat the prospects for devices in the submicron region. Any evolution of high-speed VLSI to this region of gate length is inextricably linked to the synthesis, control, and understanding of the properties of silicon and the III–V compounds, including the binary, ternary, and quaternary alloys. This aspect will not be pursued, as it is the subject of many other reviews. Rather, we shall seek the material properties required for a viable high-speed VLSI technology. To do so requires first an understanding of the operation of the pertinent devices, particularly in the submicron range. To this end, the operation of submicron MOSFETs and JFET–MESFETs is examined in Sections II and III, respectively. In Section IV, the requirements for switching high-speed logic are examined. The band structure and properties of silicon and III–V compounds are examined in Section V. Finally, in Section VI, the comparison between silicon and various III–V technologies is drawn for devices in the extreme submicron region, i.e., devices of 0.1–0.2-μm gate length.

II. Submicron MOSFETs

The concept of space-charge effects at the surface of a semiconductor has been well known for a considerable length of time. Such effects were in fact first studied in connection with metal–semiconductor contacts (Braun, 1874; Bose, 1904), although the suggestion of stable space-charge layers in these systems followed much later (Schottky, 1938). However, this latter result probably contributed to the suggestions at that time for a surface field-effect device in the patents of Lilienfeld (1930) and Heil (1935). Actual use of field-effect devices, and the oxide–semiconductor system for these devices, followed later (Shockley and Pearson, 1948; Moll, 1948). Advances in the field were rapid from that point onward, and the MOSFET occupies a major, if not totally dominant, role in VLSI of electronic circuits, a role that is becoming increasingly pronounced as device dimensions continue to decrease.

The MOS, or metal–insulator–semiconductor (MIS), structure was first proposed as a voltage-variable capacitor (Moll, 1948; Pfann and Garrett, 1959). The characteristics of the structure were subsequently analyzed by Frankl (1961) and Lindner (1962) and used to study the silicon surface by Terman (1962) and by Lehovec et al. (1963). Following the initial studies of the properties of the surface, Khang and Attala (1960) fabricated a thermally oxidized silicon FET. The basic device characteristics have been generally developed and many models formulated.

In the past several years, the integration density of silicon circuits has increased steadily and dramatically. A significant part of this steady increase lies in the reduction of channel lengths of the individual devices, and this reduction has been supported by several technological developments such as more accurate process control and fine-pattern lithography. However, as the channel length is reduced, many effects which heretofore were of second-order importance, become of primary importance and dominate device and circuit performance. The reduction of device size in order to achieve greater performance has followed a scaling principle (Dennard *et al.*, 1974), but this approach is limited by physical and practical problems.

In order to gain the complete insight and understanding of device behavior for the submicron device, more general and accurate two-dimensional (at a minimum) solutions are required which are generally only obtainable by numerical techniques (Reiser, 1973). The solution to the two-dimensional Poisson equation represents no conceptual difficulties and the major physical effects are dominantly tied up in the manner in which the charge fluctuations and current response are coupled to the local electric field, formally related through the continuity equation. However, these computations often give only detailed results that verify our more direct intuitive ideas on the principal physics of operation of the device. Between our intuition and these detailed calculations, a rather good *qualitative* (and often quite good quantitative as well) understanding of the physics germane to submicron devices has been obtained, and it is this qualitative understanding that we wish to utilize here. While the quantitative details may not be exact, the qualitative understanding is fully adequate for our purposes.

In the following, we treat the MOSFET itself, covering not only the gradual channel approximation and saturation, but also physical limits and scaling, as well as the basic enhancement–depletion inverter circuit. Then effects that arise from the two-dimensional nature of the device and the necessary charge sharing in small devices are considered.

A. Introduction to MOSFETs

A typical n-channel MOSFET structure is shown in Fig. 1. Two n-type regions are introduced (by diffusion or implantation) into the p-type substrate. These regions form the drain and source contacts. The gate structure is essentially combined with the p-type substrate to form an MOS diode. If the gate is biased positively, a negative surface space charge appears at the semiconductor surface next to the oxide. For a sufficiently large forward bias, an n-type inversion layer forms at the surface. This inversion layer forms a narrow channel between the source and the drain contacts, and this inversion channel conducts current from the source to the drain. The MOS structure modulates this current by varying the surface charge. The depth

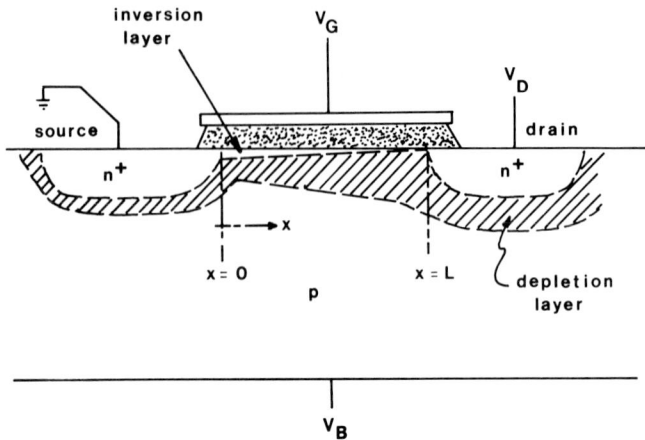

Fig. 1. The insulated-gate field-effect transistor. The conduction channel is formed by a surface inversion layer under the MOS capacitor.

of the channel into the p-region is determined by the gate voltage and the drain voltage, since it is the difference in voltage across the region $V_G - V(x)$, where $V(x)$ is the surface potential at x, which determines the surface charge density at that point. From this, it is apparent that the voltage across the oxide layer and the channel is a decreasing function of x, as x increases from the source to the drain. Hence, the channel width decreases slightly as we move from the source to the drain. Pinch-off occurs first at the drain end of the channel, and it occurs when the voltage is no longer able to maintain the inversion layer at that point. At this value of drain voltage, the drain current saturates.

The basic mode of operation, the enhancement mode, utilizes the approach discussed above. It is possible, however, that the work function difference and the oxide charge, or even an implanted n-layer in the channel region, can lead to a surface channel existing when no gate voltage is applied. In this case, the device is termed a depletion-mode device since gate voltage is used to turn the device off, rather than on. The depletion-mode device is regularly used as an active load device in integrated logic circuits and is traditionally fabricated by ion implantation techniques.

In order to model the MOS transistor in a relatively exact manner, it is necessary to account for the charge variation along the channel by writing differential equations. The incremental voltage drop along the channel is represented as a function of the current through the differential impedance. Integration of this equation leads to a relationship for the drain current in terms of the applied voltages. We assume that the gradual channel approximation is valid; that is, the fields in the direction of current flow are much smaller than the fields normal to the silicon surface. This assumption validates the use of a one-dimensional MOS analysis to find the carrier concentrations and the dimensions of the depletion region under the gate. This is equivalent to ignoring that portion of Poisson's equation arising from $\partial\Psi/\partial x$ (along the channel), allowing a one-dimensional Poisson equation to be retained. In practice the fields normal to the surface are generally at least an order of magnitude larger than fields along the channel, except in the pinched-off portion of the channel adjacent to the drain where the current is saturated and our analysis is not germane in any case. For the present, we also assume that the channel length L is large compared to the depletion width at the source and drain junctions, but will remove this limitation presently.

Strong inversion at the surface will occur when the minority carriers at the surface become equal to the majority carrier density in the bulk, or for p-type material, $n_s = p_0$. This occurs when

$$\Psi_s = -2\phi_b = 2\ln(N_a/n_i) \tag{2}$$

where N_a is the acceptor concentration in the p-type substrate, the reduced bulk potential is

$$\phi_b = (E_F - E_{Fi})/k_B T \tag{3}$$

($\phi_b < 0$ for p-type substrates), and

$$\Psi_s = \phi_s - \phi_b = eV_s/k_B T \tag{4}$$

is the reduced surface potential. Thus, the critical turn-on voltage for the channel is

$$V_T = k_B T \Psi_s/e = (2k_B T/e)\ln(N_a/n_i) \tag{5}$$

Actually, we shall include the possible effects of difference in the work functions and surface states through a flat-band voltage V_{FB}, and this will modify the turn-on voltage somewhat. Pinch-off occurs when V_D increases to a value such that

$$V_G - V_D \simeq V_T \tag{6}$$

at the drain end of the channel. If we have a silicon substrate with $N_a = 10^{16}$ cm^{-3}, we find that $V_T - V_{FB} = 0.72$ V, so that this would be the turn-on voltage in the absence of any surface states or work-function differences.

We now follow a procedure in which the current flow through the channel is calculated as a function of drain voltage. For FET operation it may be assumed that $V_G > V_T$. Then an increment of resistance along the channel is

$$dR = dx/Z\mu_e\rho_{s,n}(x) \tag{7}$$

where $\rho_{s,n}(x)$ is the inversion-layer charge density at the surface, Z is the lateral extent of the channel (gate width), and μ_e is an effective electron mobility. The total charge at the surface is composed of contributions from the depletion region and the inversion layer, so that

$$\rho_{s,n} = Q_T - Q_{depl} = \rho_s - eN_aW \tag{8}$$

where W is the width of the depleted region. At the onset of inversion W becomes nearly constant, but it will still vary somewhat and is a critical factor in short-channel effects. The applied gate voltage V_G divides between the capacitance of the oxide layer and the surface potential V_s as

$$V_G - V_{FB} = \rho_s/C_0 + V_s \tag{9}$$

where we have added V_{FB}, the flat-band voltage. We can combine Eqs. (8) and (9) to give $\rho_{s,n}$ as

$$\rho_{s,n} = C_0(V_G - V_{FB} - V_s) - eN_aW \tag{10}$$

The depletion width is just

$$W = [(2\epsilon/N_ae)(V_s - V_B - 2\phi_b k_B T/e)]^{1/2} \tag{11}$$

where V_s is the surface potential $V(x)$ in the channel and V_B is the substrate bias. For simplicity, we write

$$V_{TB} = -2\phi_b k_B T/e - V_B \tag{12}$$

as the bulk and substrate contributions to the turn-on voltage. In the following, we also replace V_s with $V(x)$, and

$$W = \{(2\epsilon/N_ae)[V(x) + V_{TB}]\}^{1/2} \tag{13}$$

Combining Eqs. (10)–(13) gives

$$\rho_{s,n} = C_0[V_G - V_T' - V(x)] - \{2\epsilon N_ae[V(x) + V_{TB}]\}^{1/2} \tag{14}$$

where

$$V_T' = V_T + V_{FB} \tag{15}$$

The increment of channel resistance is then just

$$dR = \{C_0[V_G - V_T' - V(x)] - \{(2\epsilon N_ae)[V(x) + V_{TB}]\}^{1/2}\}^{-1} dx/Z\mu_e \tag{16}$$

and $V(x = 0) = 0$ defines the source end of the channel.

The voltage drop along the increment of channel arises from the resistance drop in the increment of length dx, as

$$dV = I_D \, dR \tag{17}$$

and we can integrate this over the channel length. Thus

$$Z\mu_e C_0 V_D [V_G - V_T' - V_D/2]$$
$$- \tfrac{2}{3} Z\mu_e (2\epsilon N_a e)^{1/2} [(V_D + V_{TB})^{3/2} - V_{TB}^{3/2}] = I_D L \tag{18}$$

$$I_D = (Z\mu_e C_0/L)[V_G - V_T' - V_D/2 - \bar{Q}_B/C_0]V_D \tag{19}$$

where

$$\bar{Q}_B = (1/V_D) \int Q_b(x) \, dV = \tfrac{2}{3}(2\epsilon N_a e)^{1/2}[(V_D + V_{TB})^{3/2} - V_{TB}^{3/2}]/V_D \tag{20}$$

is an average of the depletion charge along the channel, and Eq. (19) is valid for $V_D < V_{Dsat}$, the pinch-off voltage.

When the drain voltage is increased to a level such that the charge at the drain end of the inversion layer $\rho_{s,n}(L) = 0$, pinch-off occurs. At this point, the current saturates at I_{Dsat} and this occurs for $V_D = V_{Dsat}$. The value of V_{Dsat} is obtained from the condition $\rho_{s,n}(L) = 0$, or from Eq. (14)

$$C_0(V_G - V_T' - V_{Dsat}) = [2\epsilon N_a e(V_{Dsat} + V_{TB})]^{1/2} \tag{21}$$

or

$$V_{Dsat} = V_G - V_T' + K^2\{1 - [1 + 2(V_G + V_{TB} - V_T')/K^2]^{1/2}\} \tag{22}$$

where $K = (\epsilon N_a e)^{1/2}/C_0$ is related to the average depletion charge along the channel. The saturation current is found by using Eq. (22) in Eq. (19). Thus

$$I_{Dsat} = G_0 V_{Dsat}(V_G - V_T' - V_{Dsat}/2 - \bar{Q}_{Bsat}/C_0)$$
$$\simeq (G_0/2)[V_G - V_T' - (\bar{Q}_{Bsat}/C_0)]^2 \tag{23}$$

where $\bar{Q}_{Bsat}$ is $\bar{Q}_B(V_{Dsat})$. For a doping level of 10^{17} cm^{-3} in silicon, and an oxide thickness of 200 Å, $K \simeq 0.25$.

In Eq. (23), we note that I_D is dependent upon the actual length of the channel L (through $G_0 = Z\mu_e C_0/L$). However, this length is a function of the drain voltage due to depletion region widening at the drain end of the channel. For increases of V_D above V_{Dsat}, the end of the ohmic channel (the pinch-off point) moves away from the drain. This causes the effective length L', the distance from the source to the end of the channel, to decrease. It is this L', rather than L, that should appear in the equations for the saturated current and transconductance. An accurate theory for the effective channel length is complicated by the two-dimensional nature of the potential and the interaction of the gate electrode in this region. A simple model, however, can take this into effect, provided that the channel is not too small. For V_D close to V_{Dsat}, the effective channel length can be approximated by a one-dimensional space-charge spreading analysis of the depletion region around the drain. The distance between the pinch-off point and the drain contact edge is then

$$L - L' = [2\epsilon(V_D - V_{Dsat})/eN_a]^{1/2} \tag{24}$$

and for $I_D > I_{Dsat}$,

$$I_D = \left(\frac{L}{L'}\right) I_{Dsat} = \frac{Z\mu_e C_0 [V_G - V_T'']^2/2}{L - [2\epsilon(V_D - V_{Dsat})/eN_a]^{1/2}} \tag{25}$$

B. The Inverter Circuit

Equation (23) or (25) provides the basic current equation that can be used to evaluate the performance of a simple MOSFET circuit. If we approximate $V_{Dsat} \simeq V_G - V_T''$, where V_T'' includes the bulk charge effect, then Eq. (23) can be written as

$$I_{Dsat} = k(V_G - V_T'')^2 \tag{26}$$

where $k = (G_0/2) = (G_0 L/2L')$. In Fig. 2, the basic enhancement–depletion inverter is depicted. Here, an enhancement-mode MOSFET T_1 is used as a pull-down transistor and a depletion-mode MOSFET T_2 is used as a pull-up or active load transistor. The active load is preferred over a resistive load for its generally lower power dissipation. The devices are interconnected in such a manner that T_1 is almost entirely saturated throughout the logic swing. When Eq. (26) is applied to the depletion-mode device, however, the effective threshold voltage V_T'' is negative, so that a negative gate voltage is required

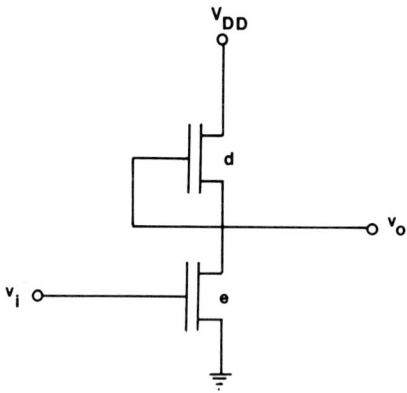

Fig. 2. The basic enhancement–depletion inverter circuit used in logic circuits.

to turn off the device. For convenience of logic levels though, the value of V_T'' in the pull-down transistor is usually set at about $0.2V_{DD}$. The value of $V_{po} = -V_T''$ in the pull-up transistor may be set to provide a convenient property of the circuit, such as equal currents or equal delay times in rise and fall of the output voltage. As we will see, this choice sets limits on the relative sizes of the pull-down and pull-up transistors.

We can write the currents for the two transistors as (using the properly defined node voltages and assuming that the pull-up transistor is nearly saturated),

$$I_{DS1} = k_1(v_i - V_T'')^2, \quad v_i > V_T''; \quad I_{DS2} = k_2 V_{po}^2 \tag{27}$$

where we have used the above definition for the pinch-off voltage of the depletion transistor. For $v_i < V_T''$, $I_{DS1} = 0$ and $v_0 = V_{DD}$. In this state, the output voltage is high, but little actual current flows, depending upon the load circuitry, as T_1 is turned off. For $v_i = V_{DD}$, T_1 is turned on, but is no longer saturated. Again, little current flows through T_1 as T_2 is turned off ($v_0 = 0$). Thus, the logic swing is approximately $[0, V_{DD}]$. Over much of this swing the transistors are in fact saturated, so that little error is introduced by using Eq. (27). Since T_1 does not begin conducting until $v_i = V_T''$, we require this same margin on the high state. Thus the symmetry point of the logic lies at about $0.5V_{DD}$. If we require equal currents at this point, then

$$k_1(0.4V_{DD})^2 = k_2(V_{po})^2 \tag{28}$$

or

$$k_1 = 4k_2 \tag{29}$$

for $V_{po} = 0.8V_{DD}$. Thus, the on-state conductance of T_1 must be four times that of T_2 to ensure equal currents for the two transistors. This is a common requirement in digital logic systems and is usually accomplished by having $L_2 = 4L_1$ (four times the gate length).

The variation in k between the two devices is significant in that it affects the drive capabilities of the circuit during switching. as T_1 turns on, it must discharge the following stages, whereas during the turn-off transient T_2 must provide the charging current to the following stages. Although $k_1 = 4k_2$, T_2 operates at a higher effective gate voltage. Thus Eq. (28) allows effectively equal current in the two transistors.

In many applications of fast logic, it is preferable to have the pull-up current $I_{DS2} = I_{DS1}/2$, as this provides equal drive current during both transient swings. In this case

$$k_1 = 2k_2 \tag{29a}$$

results. However, in this case the logic symmetry point is at $0.7V_{DD}$ for the above parameters, an unacceptably high value as it does not provide much margin at the upper voltage level. Thus there results a situation in which a compromise must be made. A consistent set of parameters results if we set $V_{sym} = 0.75V_{DD}$, $V''_T = 0.2V_{DD}$, $V_{po} = 0.8V_{DD}$, and $k_1 = 2k_2$. Alternatively, one could set $V_{sym} = 0.75V_{DD}$, $V''_T = 0.2V_{DD}$, $V_{po} = 0.55V_{DD}$, and $k_1 = k_2$.

C. Field-Dependent Mobility

In the above discussion of the gradual channel approximation, it was observed that the drain current, and hence the transconductance, was a function of the effective mobility of the electrons in the channel. In general, the effective mobility is reduced in these devices when compared to the mobility of the bulk material. The results of studies of the inversion layer mobility suggests that additional, surface-related, scattering centers dominate the mobility. These can arise from impurity scattering, surface roughness scattering, crystal defects, interface states, and increased field transverse to the channel (Ferry et al., 1981). Generally, the effective surface mobility is found to decrease with increasing inversion charge density, a result consistent with increased surface scattering in that the carriers are pushed closer to the surface–interface itself. In Fig. 3, the effective channel mobility is shown as a function of the effective surface (oxide) electric field, in which the latter is found by averaging the surface-normal field over the inversion layer, and (Sabnis and Clemens, 1979)

$$E_{eff} = \frac{e(N_a W + n_s/2)}{\epsilon} \tag{30}$$

where n_s is the inversion charge density per unit area. By taking such an average, it is found that such diverse effects as tailoring the channel by ion implantation and applying substrate bias seem to fit the same mobility dependence. The general falloff of mobility with carrier density as indicated in Fig. 3 would preclude interface charge scattering (Sah et al., 1972), as an opposite effect would be expected for such scattering. Rather, the mobility is most likely dominated by acoustic modes, both bulk and surface, and by surface roughness scattering (Ferry, 1976a; Ferry et al., 1981).

The falloff of mobility is often approximated by a replacement of μ_e by (Klaassen and de Groot, 1980)

$$\mu_e \rightarrow \mu_0[1 + \theta(V_G - V''_T)]^{-1} \tag{31}$$

although this is inexact within the MOSFET itself. While Eq. (31) is a good approximation for a MOS diode, the surface potential leading to E_{eff} in Eq.

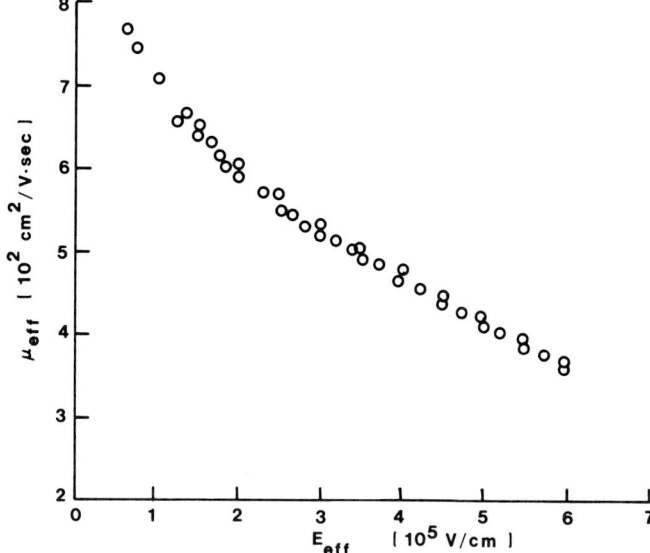

FIG. 3. The effective channel mobility as a function of the effective surface field. The data is taken from Sabnis and Clemens (1979), and is for several values of the substrate bias. The data is typical of modern high-quality MOSFETs.

(30) varies along the channel, even in the gradual channel approximation, so that $\mu = \mu(x)$ and varies along the channel.

The relevance of hot carriers to the characteristics of MOSFETs became clear when the typical size of the devices began to become small. For example, a device of 1.0-μm channel length operating at 5 V has an average field of 50 kV/cm in the channel, whereas a device of 0.1-μm channel length operating at 1 V has an average field of 100 kV cm^{-1}, and these fields can heat the electrons far above thermal equilibrium. The first evidence for the importance of hot electron effects for MOSFET characteristics was due to Hofstein and Warfield (1965), but hot electron effects had earlier been suggested as the mechanism of current saturation (Grosvalet et al., 1963). In comparison with bulk semiconductors, hot electron effects in the channel seem less in magnitude and occur at higher electric fields, although the resultant saturated velocity v_s is not considerably less. This is shown in Fig. 4. It is currently felt that the saturated velocity is dominated by bulk intervalley phonons, and that the weaker hot electron effects and lower value of the saturated velocity at intermediate fields are a result of the lower surface mobility (Ferry, 1976b), although the ultimate v_s (occurring at higher fields than in the bulk) is similar to the bulk value (Müller and Eisele, 1980). There is, however, some indication in the recent data of Cooper and

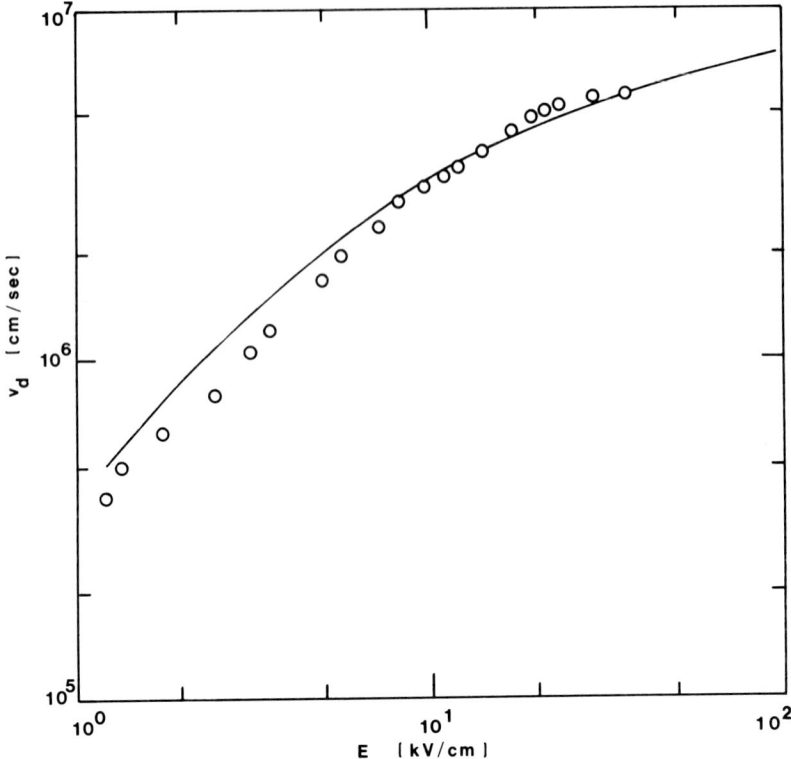

FIG. 4. The typical drift velocity for electrons in a Si(100) surface inversion layer. The solid curve is a theoretical fit to the data (circles) of Fang and Fowler (1970). The data is complicated due to saturation in the channel and more recent work by Cooper and Nelson (1980) suggests a value of v_s closer to the bulk value.

Nelson (1980) that the saturated velocity v_s in the surface channel is slightly lower than in the bulk, due probably to the presence of scattering by the oxide polar modes at the interface (Moore and Ferry, 1980).

The variation of the carrier mobility with the high channel field can be modeled by a formula of the form

$$\mu = \mu_0[1 + (\mu_0/v_s)(dV/dx)]^{-1} \tag{32}$$

where v_s is the saturated velocity (Trofimenkoff, 1965). In actual fact, however, μ_0 in Eq. (32) should be taken from Eq. (31). If, however, Eqs. (31) and (32) are introduced for μ_e in Eq. (16), the channel current can be written as (Klaassen and de Groot, 1980),

$$I_D = \frac{I_{D0}}{[1 + \theta(V_G - V_T'')][1 + \mu_0 V_D/v_s L]} \tag{33}$$

for $V_D \ll V_G - V_T''$, and I_{D0} is given by Eq. (19). In the saturation region, Eq. (33) is still a relatively good approximation near $\theta(V_G - V_T'') \simeq 1.5$, which fortuitously is close to the value in 1–2-μm devices today. However, for increased accuracy $\theta \rightarrow \theta'$, where θ' is actually measured in the device under consideration. The problem arises due to the fact that the integration of the modified Eq. (16) yields logarithmic functions, and Eq. (33) above results from an expansion of these functions. Equation (33) is a valid expansion in the linear region, but in the saturation region, an estimate for θ' in small devices is given by

$$a' = 2.5a - 2.25 \tag{34}$$

where $a = \theta(V_G - V_T')$ and $a' = \theta'(V_G - V_T')$. In this case, θ can be estimated from curves such as Fig. 3, and the estimate for θ' is good over the range $1.5 < \theta < 3.0$. In Eq. (33), we further note that μ_0 and v_s are the *bulk* semiconductor values. Thus for $V_D > v_s L/\mu_0$, the saturation currents are reduced by the saturated velocity effect of the carriers themselves. The variation of μ along the channel is apparently the cause of a variation of the observed v_s along the channel (Müller and Eisele, 1980). It should finally be remarked, however, that v_s is a size-dependent quantity and can vary as the transit time is reduced toward the carrier relaxation time. We return to this in a later section. Moreover there is some indication in recent data (Cooper and Nelson, 1980) that v_s is affected by polar modes in the SiO_2 (Moore and Ferry, 1980) as the carriers are pulled closer to the interface.

D. Scaling the MOSFET

Scaling down the size of the MOSFET in order to achieve increased packing densities implies a proportional reduction of device dimensions, pattern geometries, and the power supply voltage. For example, scaling down the size of a device by a factor of 4 can result in a 16 times greater integration density and even greater gain of performance, as indicated by the speed–power product. To achieve this factor, the channel length and width are each reduced by the factor of four. However, reducing the channel length is limited by the onset of punch-through, and to avoid this, the depletion widths are reduced by reducing the circuit supply voltage and increasing the substrate doping concentration. This creates a further complication, since as the substrate doping is increased, the field required for strong inversion is increased, necessitating a reduced-gate oxide thickness. All of these factors taken together lead to a general theory of scaling (Dennard *et al.*, 1974; Masuda *et al.*, 1979).

In general, the length and width of the channel are assumed to be reduced by a factor of α. In addition, the source voltage is also reduced by a factor of α and the sheet resistivity is reduced by α (N_a is increased by α). Thus, the drain depletion width is reduced by α. The surface electric field at turn-on is approximately eN_aW/ϵ; since ϕ_b scales only logarithmically with α, the surface field must increase approximately as $\alpha^{1/2}$, which also is a weak dependence. As V_G has gone down by α, the oxide thickness must also go down by at least α, and the oxide thickness is usually scaled by exactly this factor, since the threshold voltage will be reduced in the smaller device, as discussed below. Then, the gate capacitance scales as

$$C'_{OT} = \epsilon_0 \left(\frac{A/\alpha^2}{y_0/\alpha}\right) = \frac{C_{OT}}{\alpha} \qquad (35)$$

but the specific capacitance per unit area (which appears in all of the above equations) increases as

$$C_0' = C_0 \alpha \qquad (36)$$

The saturated drain current scales as

$$I'_{Dsat} = \frac{(Z/\alpha)\mu_e}{2L/\alpha} C_0 \alpha \left(\frac{V_G - V''_T}{\alpha}\right)^2 = \frac{I_{Dsat}}{\alpha} \qquad (37)$$

and the transconductance remains unchanged. Thus, the power is reduced by α^2. The transit time through the device is also reduced, assuming a saturated velocity situation, as is the switching speed. The latter scales as C_{OT}/g_m, which scales as $1/\alpha$ from Eq. (35). Thus the smaller devices are faster and the speed–power product, or equivalently the energy stored in the gate (or node) capacitance, goes as $1/\alpha^3$, although line capacitance can eventually hold this to $1/\alpha^2$ if it dominates the switching (we return to this in Section IV) and does not scale properly. It is possible to actually generalize these arguments to include the field patterns and current densities, but the field patterns and carrier velocities should be unchanged. Here is where second-order effects such as threshold voltage shift and mobility reduction, as in Eq. (33), become significant however.

But there are limitations to this scaling of device size. Short-channel effects lead to a reduction of the threshold voltage and to drain-induced subthreshold currents. These can significantly affect the usage of small devices in actual logic circuit functions. The significant subthreshold density, for example, is given by

$$\rho_{s,n} \simeq -\frac{k_B T}{2}\left[\frac{\epsilon N_a}{2|\phi_b|}\right]^{1/2} \exp\left[\frac{e(V_G - V_T')}{k_B T}\right] \qquad (38)$$

The prefactor scales as $\alpha^{1/2}$, since ϕ_b varies only weakly with α. The major problem lies in the argument of the exponential, however. The off-state resistance of the MOSFET varies as ($V_G = 0$)

$$R_{\text{off}} \sim 1/\rho_{s,n} \sim \exp(eV_T/k_B T) \tag{39}$$

as this resistance is dominated by the subthreshold effects. Normally, for logic applications, the threshold is set in the vicinity of $V_T = 0.2 V_{DD}$, where V_{DD} is the high state voltage and the drain supply voltage. For $V_{DD} = 5$ V, the argument of the exponential is about 200 and R is large. As V_{DD} is scaled down by α though, the off-state resistance goes as $R^{1/\alpha}$. Thus great care must be exercised to control subthreshold currents in order to maintain an adequate resistance in the off state for logic applications. The problem arises, of course, because the temperature does not scale. It is further complicated by short-channel effects which lead to threshold reductions. In attempts to hold the voltage at a higher level and control the subthreshold currents, nonlinear scaling factors have been suggested (Chatterjee et al., 1980).

A second limitation on scaling arises from attempts to maintain voltages higher than those required in strict scaling, as mentioned above. As fields increase in the device, hot electrons can be emitted from the silicon into the gate oxide. The hot electrons can originate either from the current in the surface channel or from the substrate. Subsequent trapping of the electrons in the oxide can lead to threshold instabilities due to oxide charging effects on the threshold voltage (Ning, 1978). Nonconstant field scaling laws have been suggested (Chatterjee et al., 1980). One such scenario allows the voltage to decrease slower than α and accepts the consequent increase in oxide field. The rationale for this lies in the increased drive currents available in the size reduced devices (in comparison to the constant field scaling above). The available drive current is a critical parameter in circuit performance, with the switching time being proportional to Q/I_{dm}, where Q is the necessary charge transfer.

A final limitation on the amount of size reduction that can be achieved is from the breakdown voltage of the oxide itself. The breakdown field is about 7×10^6 V cm^{-1} in SiO_2 and this sets a limit on how small devices can be made (Hoeneisen and Mead, 1972), due to the necessity of thinning the oxide.

E. Short-Channel Effects

It was assumed in the gradual channel approximation that the induced surface charge was a function only of the voltage differential between the gate and the substrate. This assumption is not valid near the source and drain depletion regions due to the additional fields and charge densities there,

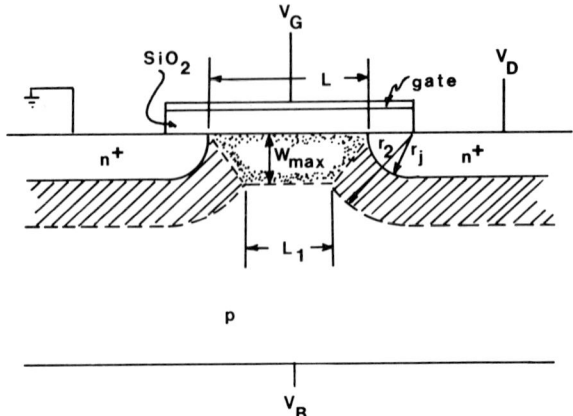

FIG. 5. A short-channel MOSFET, showing effective channel shortening and drain-induced variation of threshold due to reduced depletion charge.

thus requiring the full three-dimensional terms in Poisson's equation. If the channel is long, however, these edge regions do not affect the overall treatment of the device by these approximate equations. For short channels, however, these edge effects can no longer be ignored. Because of the interpenetrating depletion regions, a significantly smaller amount of charge (in the normal space-charge region) is actually linked to the gate potential than expected from the simple theory, the remainder being linked to the source and drain potentials. In Fig. 5 the geometry of this situation is illustrated. For devices that are not too short, a straightforward geometrical argument can be made from Fig. 5 in order to estimate the actual threshold voltage (Troutman, 1974, 1979; Taylor, 1978, 1979). The depletion charge induced by the gate voltage is taken to lie approximately within the trapezoidal area of height W and lengths L and L_1, where L_1 is at the substrate side of the gate depletion region. The size of this charge is then approximately (charge per unit area)

$$\rho_{\text{depl}} = eWN_a(L + L_1)/2 \tag{40}$$

This then is the depletion charge that must be induced by the gate at threshold, and if L is large, L_1 approaches L to give the long-channel results. For short channels, L_1 can be significantly smaller than L, and in fact $L_1 \to 0$ as punch-through occurs, so that punch-through currents can actually occur well below the surface (Troutman, 1979). The drain depletion width, in the corner region, is given by

$$r_2 - r_j = (2\epsilon V_{\text{DB}}/eN_a)^{1/2} \tag{41}$$

where $V_{DB} = V_D - V_B + V_{bi}$, V_{bi} is the built-in potential of the drain junction, r_2 is the drain depletion region boundary radius, and r_j is the drain metallurgical junction radius. A simple geometrical argument then gives

$$\rho_{depl}/\rho_L = 1 - (r_j/L)(g - 1) = f \quad (42)$$

where ρ_L is the long-channel value, and

$$g = \left\{1 + \frac{2}{r_j}\left(\frac{2\epsilon V_{DB}}{eN_a}\right)^{1/2} + \frac{1}{r_j^2}\left[\frac{2\epsilon V_{DB}}{eN_a} - \frac{2\epsilon(V_s - V_B)}{eN_a}\right]\right\}^{1/2} \quad (43)$$

where we have used Eqs. (41) and (11). Expression (43) usually appears under the simplifying assumptions of $V_D = 0$ and $r_2 = r_j + W$. The parameter f is a unique function of the device geometry and material characteristics. The effective threshold is then given by using Eq. (42) in Eq. (19), and

$$V_T = -2\phi_b k_B T/e + V_{FB} + (f/C_0)[2\epsilon N_a e(V_c - V_B - 2\phi_b k_B T/e)]^{1/2} \quad (44)$$

where $V_s = 0$ is usually taken ($V_D = 0$). The presence of a potential at the source and/or drain can lead to a reduction of the gate depletion charge and thus to a reduction of the threshold voltage. This effect also leads to a reduction of the threshold in short channels just due to the presence of the depletion regions. This effect is shown in Figs. 6 and 7. It is evident that

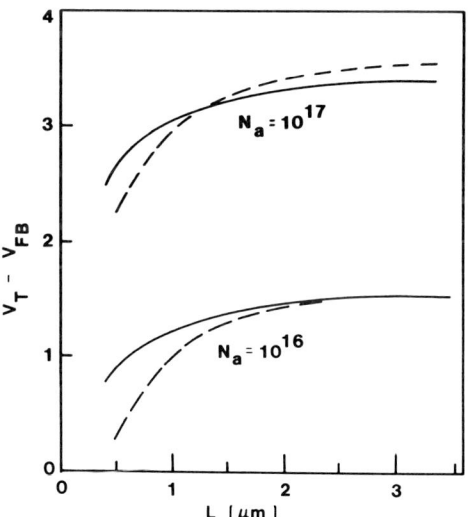

FIG. 6. Reduction in threshold voltage in short-channel devices according to (44). A substrate bias of -3 V is assumed. The solid curves are for $V_D = 0$ and the dashed curves are for $V_D = 3$ V. The region below 1.0 μm is shown expanded in Fig. 7.

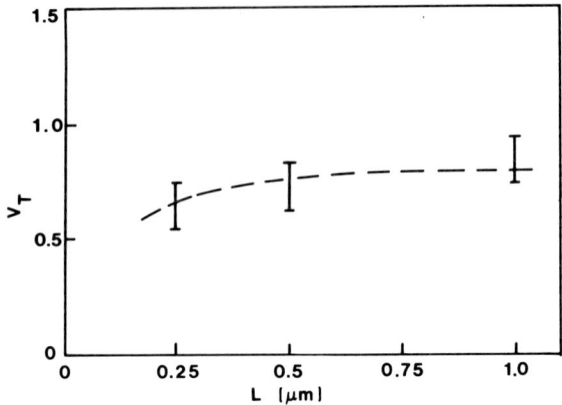

FIG. 7. The reduction in threshold voltage in short-channel devices according to Eq. (44). Here it is assumed that $N_a = 1 \times 10^{16}$ cm^{-3}, $r_j = 0.2$ μm, $W = 0.34$ μm, and $t_{ox} = 240$ Å. The error bars indicate the range of experimental data from Elliott et al. (1979), and $V_D = V_B = 0$.

threshold reduction in short-channel devices becomes a significant effect that must be considered. Further, this reduction in threshold is dependent upon the drain potential. If $V_{DB} \gg V_s$, then

$$g \simeq 1 + \frac{(2\epsilon V_{DB}/eN_a)^{1/2}}{r_j} \qquad (45)$$

$$f \simeq 1 - \frac{(2\epsilon V_{DB}/eN_a)^{1/2}}{L} \qquad (46)$$

In this case, the dominant threshold shift is drain-induced, or alternatively, substrate-bias-induced. Within the approximation, this corresponds to the case $r_2 = r_j + W$ as well. In general, it has been suggested that the threshold voltage in short channels can be found from the simple relation due to Troutman (1977)

$$V_T = V_{T,LC} - a - bV_D \qquad (47)$$

where a and b can be evaluated from two-dimensional computer simulations.

The threshold reduction in short-channel devices and its dependence upon the various potentials, such as drain and source, make it imperative to accurately control the device threshold. Thus, turn-on stability over a long period of operation is crucial, and accurate knowledge of interface states and substrate homogeneity is necessary in fabrication. Temperature also becomes important.

In the short-channel device, saturation is still assumed to occur when $\rho_{s,n}(L) = 0$. Extending Eq. (21) to the above case,

$$C_0(V_G - V_T' - V_{Dsat}) = f[2\epsilon eN_a(V_{Dsat} + V_{TB})]^{1/2} \qquad (48)$$

MATERIALS CONSIDERATIONS FOR SUBMICRON VLSI 331

where f is given by Eq. (42) with $V_{DB} = V_D - V_B + V_{bi} = V_{Dsat} - V_B + V_{bi}$ for the present case. If we assume that f is a slowly varying function of V_D, and take the approximate value at V_{Dsat}, then from Eq. (22)

$$V_{Dsat} = V_G - V_T' + f^2 K^2 \{1 - [1 + 2(V_G - V_T' + V_{TB})/K^2 f^2]^{1/2}\} \quad (49)$$

Thus we find that the actual saturation value of V_D is a function of V_D itself for short-channel devices. In Fig. 8, we shown the variation of V_{Dsat} on V_D for a typical short-channel device.

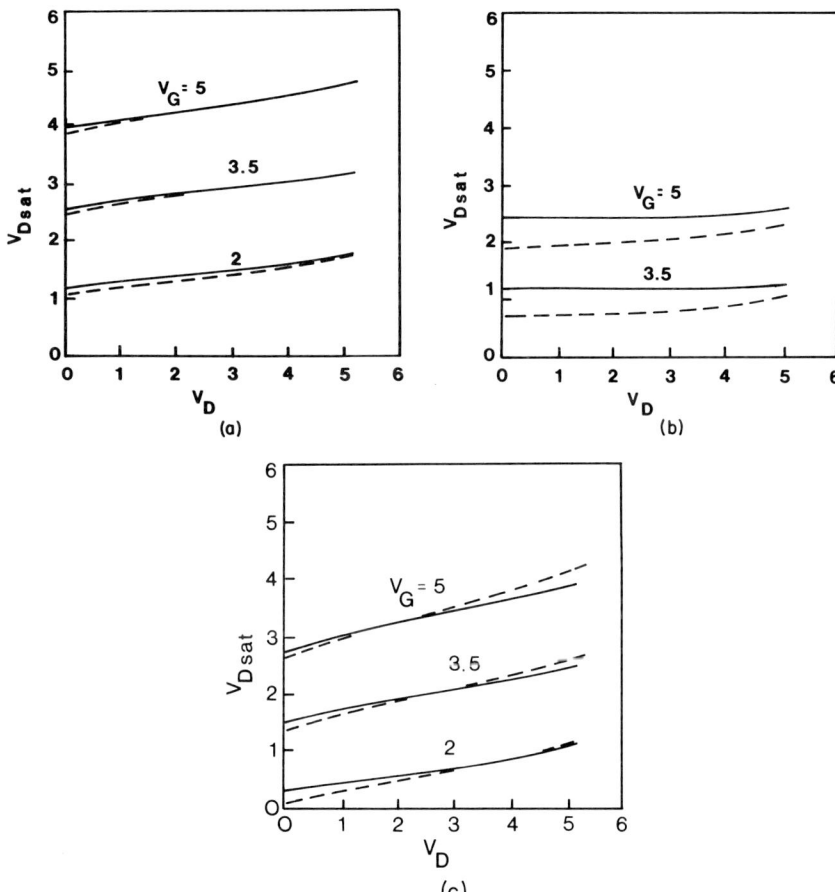

FIG. 8. Saturation voltage versus drain voltage V_D for a MOSFET with $t_{ox} = 250$ Å, $r_j = 0.3$ μm, and $V_{FB} = -0.6$ V, for $V_B = 0$ (solid curves) and $V_B = -3$ V (dashed curves). In (a) $N_a = 10^{16}$ cm^{-3} and $L = 1$ μm, whereas in (b) $N_a = 5 \times 10^{16}$ cm^{-3} and $L = 1$ μm. In (b) the curves are flat since the parameters are such that V_T is relatively insensitive to V_D. In (c) $N_a = 5 \times 10^{16}$ cm^{-3} and $L = 0.5$ μm.

The short-channel effects discussed here are for a standard, scaled n-channel MOS technology. Recent studies have however focused on improvements in small devices to minimize short-channel effects. Most of these center around lightly doped regions of the source and drain adjacent to the channel (Ohta et al., 1975, 1980; Ogura et al., 1980). The purpose of these regions is to induce a portion of the depletion width spreading to occur within the drain portion of the junction rather than entirely within the area under the gate.

III. Submicron MESFETs

The majority of semiconductor devices, even the MOSFETs of the previous section, are interconnected via metal–semiconductor contacts of one type or another. In fact, the earliest studies of space-charge effects at the semiconductor surface were in connection with such metal–semiconductor contacts (Braun, 1874; Bose, 1904; Schottky, 1938). The application of these contacts in integrated circuits is widespread, from "ohmic (tunneling)" contacts to diode logic arrays. These logic applications relay upon the fact that the devices are very fast and can be made with very small dimensions. In recent years, the depletion-width modulation control of such Schottky barriers has found application in the MESFET. This device is pictured in Fig. 9. Here, the current channel lies between the gate depletion region and the semiinsulating substrate. The MESFET is a special form of a JFET, differing primarily in the manner in which the gate is formed. As in the MOSFET, the MESFET can be either an enhancement or a depletion mode device. In the depletion-mode device, the channel is normally open and $V_G < 0$ is applied to the gate to increase the depletion width and reduce the

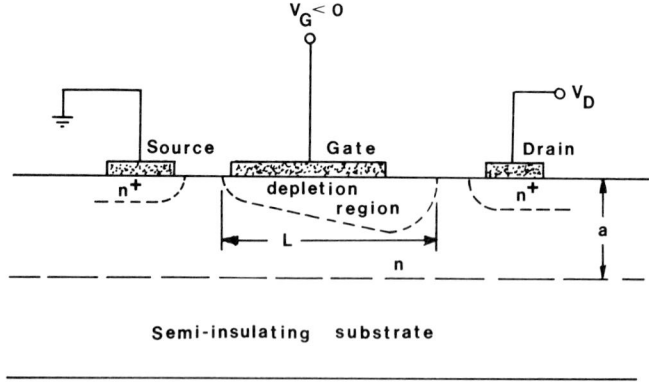

FIG. 9. The MESFET transistor.

current. In the enhancement-mode device, the built-in potential of the Schottky barrier is sufficient for the depletion region to punch entirely through to the substrate; i.e., the depletion region extends entirely across the active n-region. In this latter case, a positive V_G is required to turn on the device by opening the channel. Although the MESFET has been used for GaAs devices for some time (Mead, 1965), primarily for logic and microwave applications, it also can be used in Si circuits (Drangeid et al., 1972).

Because the electron mobility in GaAs is some four times greater than that of bulk Si, it is in principle possible to achieve much higher performance in GaAs, if velocity saturation is not important. However, GaAs does not possess sufficiently good native oxides or surface properties to permit operation in a MOSFET mode, so that it has primarily been used in JFET or MESFET devices. Because of the greater built-in potential barrier, $p-n$ junction FETs are more often used for enhancement-mode devices, but the principles are the same. We develop the theory below for the JFET, then turn to scaling and the inverter circuit.

A. The JFET

At the top of the device, a junction (either a $p-n$ junction or a Schottky barrier) has been generated. This junction is reverse biased so that the depletion region is fairly wide. The actual bias across the junction is a function of position, since the voltage on the channel side is not zero, but is a function of the position in the channel, as

$$V_n(x) = V_D(x/L) \tag{50}$$

For a negative voltage applied to the gate ($V_G < 0$), it is found that

$$V_a(x) = V_G - V_n(x) \simeq -(|V_G| + V_D x/L) \tag{51}$$

The width of the depletion region of the junction is

$$w(x) = [2\epsilon(\Psi_0 - V_a)/eN_d]^{1/2} \tag{52}$$

in the Schottky barrier or for $N_a \gg N_d$, where N_a is the doping of the gate region and N_d is the doping in the channel region, in the $p-n$ junction, and Ψ_0 is the built-in potential. The p-type (gate) regions in the JFET are heavily doped so that the junction width w is primarily in the channel region. It is in this regard that the $p-n$ junction FET and the Schottky barrier FET do not differ in detail. As the drain voltage is increased, the reverse bias across the junction is increased, and its width increases. At some critical value of V_D, the depletion region just touches the semiinsulating substrate. This value of V_D is termed the pinch-off voltage V_p. While the junction width is increasing, the cross-sectional area of the channel is decreasing and hence the channel resistance is increasing. When the depletion region punches through to the substrate, this area has gone to zero, and the resistance must be very large. The channel current, which has been increasing with the voltage V_D, saturates. The current does not go to zero, since if it did, the voltage drop at the channel would not be present. This would make the channel open again and current would flow. Thus some stable point for which the channel is almost but not quite pinched off must be reached, where the current flows through the portion of the channel that is open and is then *injected* into the gate-drain depletion region at the pinch-off point. The total device resistance is very high, and the current is at its saturated value and does not increase very much with further increases of V_D. By varying the gate voltage, one can vary the pinch-off voltage and hence the saturated current. Hence, since the reverse resistance of the *p–n* junction is very high, the input resistance to the gate is also very high, and one can control the output current by varying the gate voltage. Moreover, since the output current is saturated, the output impedance of the device is very high.

One can determine an expression which relates the drain current I_D to the drain voltage V_D and the controlling gate voltage V_G. From (52), one has, with $V_b = \Psi_0 - V_a$,

$$w = (2\epsilon V_b/eN_d)^{1/2} \tag{53}$$

When the voltage is sufficiently large, a pinch-off is achieved and $w = a$. This occurs at $x = L$ first, since this portion of the diode has the highest reverse bias. Here $V_b = V_{bp}$, where

$$V_{bp} = \Psi_0 - V_G + V_p = eN_d a^2/2\epsilon \tag{54}$$

$$V_a = V_G - V_p = \Psi_0 - eN_d a^2/2\epsilon \tag{55}$$

where we have introduced the pinch-off voltage V_p as the drain voltage such that reach-through of the gate depletion region just occurs. Thus

$$V_p = V_G - \Psi_0 + eN_d a^2/2\epsilon = V_{G0} + eN_d a^2/2\epsilon \tag{56}$$

where we have used $V_{G0} = V_G - \Psi_0$.

Consider now an increment of resistance dR along the x direction. This increment of resistance at point x contributes a voltage drop of

$$dV = I_D\, dR \tag{57}$$

For the resistance,

$$dR = (\rho/A)\, dx = dx/[\sigma_n Z(a - w)] \tag{58}$$

so that

$$[a - w(x)]\, dV = I_D\, dx/\sigma_n Z \tag{59}$$

MATERIALS CONSIDERATIONS FOR SUBMICRON VLSI   335

where Z is the lateral extent of the device (the gate width). From Eqs. (53) and (54), we can write the channel height as

$$a - w(x) = a - \{2\epsilon[V(x) - V_{G0}]/eN_d\}^{1/2} \quad (60)$$

In normal operation, $V_{G0} < 0$ for the depletion-mode device. Here $V(x)$ is the voltage at point x in the channel. Then Eq. (59) becomes

$$(I_D L)/(\sigma_n Z a) = \int_0^{V_D} \{1 - [(2\epsilon/eN_d a^2)(V - V_{G0})]^{1/2}\} \, dV \quad (61)$$

$$I_D = G_0\{V_D - \tfrac{2}{3}(2\epsilon/eN_d a^2)^{1/2}[(V_D - V_{G0})^{3/2} - (-V_{G0})^{3/2}]\} \quad (62)$$

where $G_0 = n_0 e \mu Z a/L$ ($n_0 = N_d$ if the impurities are fully ionized in the channel).

In the linear region, where $V_D \ll |V_{G0}|$ ($V_D \ll V_p$), we can expand the first term in the square brackets as

$$(V_D - V_{G0})^{3/2} \simeq (-V_{G0})^{3/2}[1 - 3V_D/2V_{G0} + \ldots] \quad (63)$$

$$I_D \simeq G_0 V_D[1 - (2\epsilon/eN_d a^2)^{1/2}(-V_{G0})^{1/2}] \quad (64)$$

At large drain bias, the current reaches a maximum and saturates. Using Eq. (56) for $V_{Dsat} = V_p$ gives

$$I_{Dsat} = G_0\left\{\frac{eN_d a^2}{6\epsilon} + V_{G0}\left[1 - \frac{2}{3}\left(\frac{2\epsilon}{eN_d a^2}\right)^{1/2}|V_{G0}|^{1/2}\right]\right\} \quad (65)$$

If we define the reduced turn-off voltage ($V_{T0} = V_T - \Psi_0$) as

$$V_{T0} = -eN_d a^2/2\epsilon \quad (66)$$

Eq. (65) can be written as

$$I_{Dsat} = G_0[(V_{G0} - V_{T0}) - 2\{(-V_{G0})^{3/2} - (-V_{T0})^{3/2}\}/3(-V_{T0})^{1/2}] \quad (67)$$

This form should be compared with Eq. (26). Although the functional forms are different in detail, they actually give very similar current levels for the same materials and an appropriate choice of parameters. This is illustrated below in Section IV,B. Thus, Eq. (65) is often approximated by Eq. (26) for convenience of comparison. Indeed, Zuleeg et al. (1978) use such a relation as Eq. (26) in enhancement-mode JFET modeling.

B. The Inverter Circuit

The logic gate design that is used for the fastest GaAs MESFET logic is the inverter that appears in Fig. 2, with the exception that both devices are depletion-mode devices. In this case, the input logical voltage swings negative to turn off the pull-down transistor. Since the output voltage is always positive, it is imperative that buffer output drivers–level shifters be used to give the negative-going output voltage. However, as pointed out in Section II, any circuit can be buffered, and we will not consider the buffer stage here.

There has been some effort to utilize normally-off enchancement-mode devices (p–n junction FETs primarily) for the pull-down transistor. However, these circuits appear to be inevitably slower than the depletion-mode devices because the limited voltage swing restricts the device to regions of operation in which the switching properties are far from optimum (Van Tuyl and Liechti, 1977; Liechti, 1977), although recent experiments seem to indicate that the difference in speed does not really exist (Zuleeg et al., 1978), and 1.0-μm gate length devices with delay times below 100 psec have been made by many groups (Lundgren et al., 1979; Suyama et al., 1980; Nuzillat et al., 1980). Devices of 0.5-μm gate length have speeds comparable to depletion-mode devices (Mizutani et al., 1980).

Because both pull-up and pull-down transistors are depletion-mode devices, the length–width ratio can be the same for both devices for equivalent current capabilities, although for fast logic, fine tuning to balance the pull-down current to twice that of the pull-up transistor can be made. Thus there is little difference between the two types of devices for the simple inverter circuit.

C. Field-Dependent Mobility

In the above discussion of the gradual channel approximation for the JFET, it was assumed that the drain current, and hence the transconductance, was a function of the effective mobility of the electrons in the channel. In the MOSFET, we found that surface scattering reduced the mobility. This does not occur in the JFET (or MESFET) as transport takes place in the bulk of the device. However, in the pinched-off channel current is constrained to flow in a very narrow channel, and this can lead to transverse quantization effects (Pepper, 1978) in submicron devices. This can lead to a reduction of the mobility just as in a MOSFET, but very little is known about these effects. Hot-electron effects can occur at much lower effective fields, especially in GaAs (or other material than silicon). The major competitors for high-speed logic are various III–V compounds such as GaAs, InGaAs, or InP. These semiconductors all have a number of common attributes.

The III–V compounds of interest are all direct band-gap semiconductors (see Section V) with relatively low values of the electron effective mass. In addition, higher-lying satellite valleys of the conduction band are situated so that intervalley transfer can occur under appropriate hot-electron conditions. Under such conditions, electrons are transferred from the light-mass central valley of the conduction band to satellite valleys (usually at the L-point in the Brillouin zone) that have a relatively higher effective mass. This leads to a differential negative conductance in the $v(E)$ characteristic. This is shown in Fig. 10. Such a $v(E)$ curve can lead to the Gunn effect, an unstable oscillatory behavior characterized by traveling high-field domains (Shaw et al., 1979). Such a $v(E)$ characteristic is important for several reasons.

The negative differential conductivity (NDC) illustrated in Fig. 10 for $E > E_p$ is important to the operation of FETs. In particular, it leads to two qualitatively different trapped domains and one traveling domain (Grubin et al., 1980). The first trapped domain forms within the conducting channel at the drain edge of the gate contact (Engelmann and Liechti, 1977). Its formation is a direct consequence of velocity saturation and can lead to current saturation at values of drain voltage significantly below that predicted by the gradual channel approximation (Turner and Wilson, 1968). In this regard, it does not differ from the velocity saturation phenomena of the previous section.

The second trapped domain is qualitatively different and forms in FETs capable of sustaining NDC. This domain forms near the drain contact at high drain bias (low gate bias) and is coincident with the cessation of traveling domains arising from the NDC (Grubin et al., 1980). This domain is qualitatively similar to the high anode field configuration in two-terminal devices (Shaw et al., 1979). This trapped domain mode and the traveling domain mode do not effectively influence the average $J(E)$ curve for the device and will not be considered further.

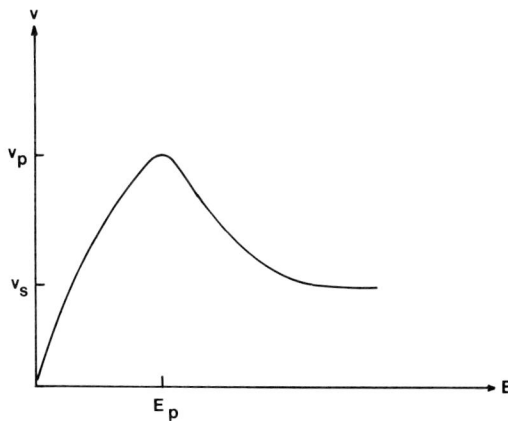

FIG. 10. The velocity–field relationship for generic III–V compounds similar in structure to GaAs or InP.

In general, an approximate relation can be developed to fit the $v(E)$ curve shown in Fig. 10. One model that is widely used is expressed as

$$v = \mu_0 E(1 + k_1 E)/(1 + k_2 E^2) \tag{68}$$

with k_2 and k_1 appropriately selected to fit the curve. In actual devices though, the peak velocity v_p is seldom seen. While a charge depletion–accumulation may form under the gate or at the drain contact, the general current–voltage or current–field behavior is dominated by the downstream velocity. In this general regard, the drain and source regions behave as control regions of the device (Kroemer, 1968; Shaw et al., 1969). The steady J–E curves of the FET do not show negative differential conductivity. Thus, the effective $v(E)$ curve is the same as that for silicon, i.e.,

$$v(E) = \mu_0 E [1 + \mu_0 E/v_s]^{-1} \tag{69}$$

However, in this case v_s is not a well-defined quantity, but

$$v_{\text{sat}} < v_s < v_p \tag{70}$$

and a value of $v_s = 1.4 \times 10^7$ cm sec^{-1} seems to fit a wide variety of devices (Fukui, 1979).

In long (or large) channel devices, $v_s \simeq v_{\text{sat}}$, where v_{sat} is the minimum velocity at high fields in Fig. 10, but this identity is less accurate in small devices and we will see in Section IV how v_s can approach or exceed v_p in very small devices. If we now introduce Eq. (69), the channel current can be written as

$$I_D = I_{D0}/[1 + \mu_0 V_D/v_s L] \tag{71}$$

where I_{D0} is given by Eqs. (62), (64), or (65). As in the MOSFET and as mentioned above, v_s is a size-dependent quantity and can vary as the transit time is reduced toward the carrier relaxation time.

D. Scaling the MESFET

Scaling down the size of the MESFET or JFET in order to achieve increased packing densities implies a proportional reduction of device dimensions, pattern geometries, and power supply voltages, just as in the MOSFET. In general, the length and width of the channel are assumed to be reduced by a factor of α. In addition, the drain voltage (and gate voltage) is also reduced by α and the sheet resistivity of the epitaxial n-layer is also reduced by α by increasing N_d by the factor α. Thus, the gate depletion region width is also reduced by α. Then, the saturated current scales as

$$I'_{D0} = I_{D0}/\alpha \tag{72}$$

from (65) as G_0 remains unchanged. Moreover, the factor $\mu_0 V_D/Lv_s$ also remains unchanged other than for the second-order effect of the increased N_d upon μ_0. Thus the transconductance remains unchanged.

The power dissipated in the device is reduced as $1/\alpha^2$. The transit time through the device is also reduced, assuming a saturated velocity situation, as is the switching speed. Line capacitance can also limit this latter quantity; we return to this in Section IV.

In the MOSFET, troubles inherent in scaling centered around the threshold voltage. Here, the problem rests with the built-in potential Ψ_0. In the JFET and MESFET, this potential is relatively independent of scaling, and actually increases in magnitude slightly. As Ψ_0 is almost 0.8 V in GaAs, it is not an insignificant quantity. Thus, if scaled voltages become of the order of 1 V or less, then design corrections must be taken. In this sense, scaling theory has failed. Nonconstant field scaling rules can alleviate this situation, but the major solution must lie with nonscaling approaches to design, barrier reduction techniques, or the use of different materials with reduced barriers.

E. Short-Channel Effects

In the resulting equation (65) for the source–drain current in saturation, it appears as if the current indeed is truly saturated. However, in the MOSFET this did not actually occur, nor does it here. In general, the slight potential barrier between the active channel epilayer (or implanted layer) and the semiinsulating substrate serves as a relatively effective channel stop. In addition, there is believed to be a significant density of interface states–traps at this interface which reduces the mobility slightly. The detailed behavior is likely to be a result of the combination of such mechanisms (Tanimoto et al., 1976; Rossel et al., 1978; Houng and Pearson, 1978). In such a situation, the electrons are strongly repelled by the interfacial electric field and do not normally penetrate to the substrate. Yet it is likely that substrate currents are the cause of nonsaturation of the source–drain current in the JFET–MESFET.

Eastman and Shur (1979) suggested that this additional current in GaAs could be a substrate current injected due to the high fields associated with a trapped domain situated between the gate and the drain. Such domains are found in power FETs (Engelmann and Liechti, 1977) and it is likely that they could occur in logic devices. Tsironis (1980) however has ruled out such injection from careful measurements of the dependence of the excess current upon V_D, as well as the presence of excess noise. These latter results suggest that the substrate current is more likely to be due to avalanching in the high fields arising from the domain being close to the interface. On the other hand,

Bonjour *et al.* (1980) have investigated this problem with two-dimensional computer simulations. Their results indicate a virtual widening of the channel in the neighborhood of the trapped domain. Moreover, the domain is found to be under the gate (at the drain end) and so the widening directly affects the pinch-off–saturation condition. This position for the domain agrees with the results of Grubin *et al.* (1980). Thus, substrate current is probably the cause of weak saturation, although its specific cause is still a topic of debate.

The presence of traps at the interface is likely a result of out-diffusion of Cr atoms from the semi-insulating substrate (Houng and Pearson, 1978). Random variations in the trapped charge density in these states can affect the channel properties, especially in the saturated region (Itoh and Yanai, 1980). Thus these traps represent a reliability–stability problem for the GaAs JFET–MESFET.

IV. Switching of High-Speed Logic

The decision whether or not to develop an alternative technology to conventional scaled *n*-channel MOSFETs depends largely upon the performance advantages that can be realized by such a technological move. Not surprisingly, considerable controversy surrounds this decision, particularly in regard to the manner in which performance is to be measured. Since the early days of logic integration, the speed–power (or delay–power) product has served as a figure-of-merit for digital integrated circuits (Josephs, 1965). Other concepts such as functional throughput, the product of the number of gates and the clock frequency, have also arisen. In any case, the pertinent parameters are typically calculated from static transfer characteristics and capacitances of the logic gate. The argument may be simply stated, as the power dissipation in a chip will be (Eden *et al.*, 1979)

$$P \simeq 2N_g f_c (P_D \tau_D) \qquad (73)$$

where N_g is the number of gates, f_c is the applied clocking frequency at the gate level, and $2P_D \tau_D$ is the gate energy required to switch *through a complete cycle*. By using the complete cycle definition, we average over variations between rise and fall transients. If we also have

$$P \simeq N_g V_{DD} I_{dm}/2 \qquad (74)$$

where $I_{dm} (= I_{Dsat})$ is the peak current through the pull-down transistor, then

$$P_D \tau_D \simeq V_{DD} I_{dm}/2 f_c \qquad (75)$$

Clearly, the minimum speed–power product will occur at the largest f_c, but this is limited by the cutoff frequency of the transistors, or

$$f_c \simeq g_m/C_G \tag{76}$$

but C_G should be replaced by the total node capacitance C_N (see below), and

$$P_D\tau_D \simeq V_{DD}I_{dm}C_N/2g_m \simeq V_{DD}^2 C_N^2/2g_m\tau_D \tag{77}$$

where we have approximated the charging current during switching as $I_{dm} \simeq C_N V_{DD}/\tau_D$. An alternative form is to write $I_{dm} = K(V_G - V_T'')^2 \simeq KV_{DD}^2$, $g_m \simeq 2KV_{DD}$, and

$$P_D\tau_D \simeq V_{DD}^2 C_N/4 \tag{78}$$

Although both Eqs. (77) and (78) are very crude approximations, they contain elements of validity and agree with each other if $f_c = 2/\tau_D$. We will examine these simple approximations further in the sections below.

The pertinent parameters are not independent of one another. For example, one could conclude that making τ_D smaller would hurt the speed–power product in Eq. (77). But to accomplish this reduction of delay time would require either a larger I_{dm} or a smaller C_N, which are countering influences. If τ_D is reduced by a reduction of C_N, both Eqs. (77) and (78) indicate that $P_D\tau_D$ is reduced. If I_{dm} is increased in order to reduce τ_D, g_m increases accordingly, so that little change in $P_D\tau_D$ is expected. Thus, all parameters must be considered together. Equation (78) also points out that if C_N is dominated by interconnection capacitances, then $P_D\tau_D$ is relatively independent of the technology used for the inverter devices, but τ_D itself need not be so limited.

In any case, we discover that if we are to increase the functional complexity of a system by increasing the number of gates, the power–speed product $P_D\tau_D$ and individual device size must be reduced. To meet these goals, node capacitance and operating voltages must be reduced. Moreover, to accomplish faster switching, the drive current capability I_{dm} must be increased. Thus, I_{dm}, τ_D, and $P_D\tau_D$ are all factors that are important to high-speed logic.

A. Transient Switching

If one is to compare different technological implementations for high-speed logic, it is necessary to find various figures of merit. One such is the static power dissipation P_s. Müller et al. (1976) have also suggested the mean value of the minimum energy per logic operation W_I, the intrinsic time period required to perform a logic operation τ_I, and the mean power delivered

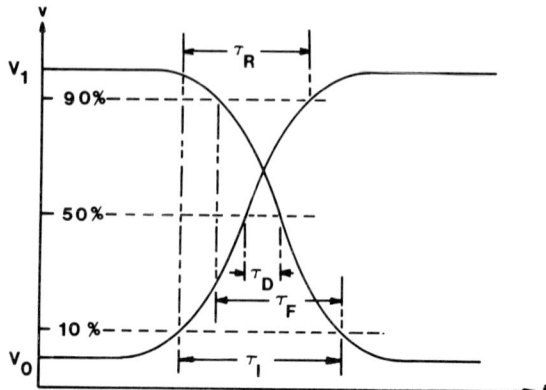

FIG. 11. Definition of the basic response times for the logic transitions. These definitions follow those of Müller et al. (1976).

during this period P_I as appropriate measures. These are illustrated along with τ_R, the rise time, τ_F, the fall time, and τ_D in Fig. 11. W_I, τ_I, and P_I are mean values considering the transient from $0 \to 1$ or from $1 \to 0$ and they are valid for an inverter with minimum interconnections within a chain of inverters (fan-out of one). In this sense, they represent the basic inherent performance of the inverter. The maximum frequency of operation is given by $f_{\max} \simeq (2\tau_I)^{-1}$.

We can estimate the power dissipated during the complete switching

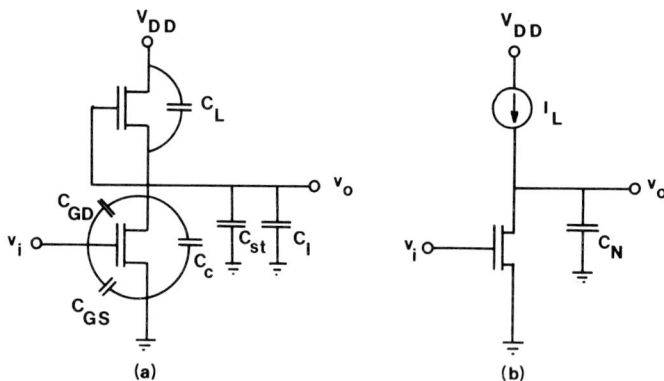

FIG. 12. (a) The basic inverter of Fig. 2 showing the various contributions to the node capacitance C_N. (b) The effective inverter. The capacitance definitions follow Lehovec and Zuleeg (1980).

cycle (one $0 \to 1$ and one $1 \to 0$ transition) for a repetition frequency f as (Müller et al., 1976)

$$P(f) = [2W_1 - P_{s1}(\tau_D + \tau_R) - P_{s0}(\tau_D + \tau_F)]f + P_S \qquad (79)$$

In most cases of high-speed logic, the pull-up and pull-down transistors have been matched to achieve equal drive current for both parts of the cycle. Then $\tau_F \simeq \tau_R$ and

$$\tau_D + (\tau_R + \tau_F)/2 = \tau_I \qquad (80)$$

Müller et al. (1976) have carried out simulations for a number of logic families and they found that $\tau_I \simeq 2.5\tau_D$ to be a very good approximation ($\tau_I \simeq 2.7\tau_D$ is good for n-MOS circuits, but this is a negligible difference). Using this result in Eq. (80) yields (with $\tau_F \simeq \tau_R$)

$$\tau_D = 2\tau_R/3 \qquad (81)$$

We can then rewrite Eq. (79) as

$$P(f) = 2[W_1 - P_S\tau_I]f + P_S = 2W_1 f - P_S \qquad (82)$$

or

$$W_1 = P_1\tau_I \simeq 5P_1\tau_D/2 \qquad (83)$$

in agreement with our earlier estimations. In fact, if the standby power P_S is small with respect to the switching power, then the speed–power product $P_D\tau_D$ is a good figure of merit to use in evaluating individual logic implementations.

The logic delay time τ_D can be split up into three components (Bosch, 1980). The first part is caused by the finite electron transit time. For a velocity-saturated flow in a silicon channel of 0.2 μm, this time is about 2 psec. The second part is due to parasitic reactances in the circuit and is usually of the same order as the previous part. Most of the total delay is caused by the interconnection loading of the output node in the inverter circuit.

Switching the inverter circuit from one level to another involves charging or discharging capacitance associated with the output node. In Fig. 12, we show these capacitances. The gate capacitance C_G is split into the gate–source C_{GS} and gate–drain C_{GD} portions. The channel capacitance C_C represents stored charge in the channel and includes particularly the above-mentioned transit-time delay factor. Including the transit-time delay in this fashion is valid so long as $C_C \ll C_N$. C_L is the load capacitance, C_I is the interconnection capacitance, and C_{st} represents the stray capacitances. Then (Lehovec and Zuleeg, 1980),

$$Q_N = (C_{GS} + C_I + C_{st})V_0 + C_{GD}(V_0 - V_i) - C_L(V_{DD} - V_0) \qquad (84)$$

The switching equation is just

$$\partial Q_N/\partial t = C_N(\partial V_0/\partial t) = I_L - I_D \tag{85}$$

where

$$C_N = C_{GS} + C_{GD}^* + C_i + C_{st} + C_L \tag{86}$$

is the lumped node capacitance. Here

$$C_{GD}^* = C_{GD}(1 - \partial V_i/\partial V_0) \simeq C_{GD} \tag{87}$$

is the effective gate-drain capacitance and differs little from C_{GD}, since the Miller effect is expected to be small.

The output voltage transient $v_0(t)$, after a change in v_i, is found from Eq. (85) to be

$$t = \int_{v_0(0)}^{v_0(t)} C_N(I_L - I_D)^{-1} \, dv_0 \tag{88}$$

If both pull-up and pull-down transistors are saturated during most of the switching cycle, then the currents are relatively independent of the output voltage. If the transistors are balanced so that $I_L - I_D = \pm I_{dm}/2$ for the rising and falling transients, the rise and fall times will be balanced as well. This requires $I_L = I_{dm}/2$, where I_{dm} is $\max[k_1(V_G - V_T'')^2]$ for the pull-down transistor. If we take V_m as the logic voltage swing ($=V_{DD}$ in most MOS logic), then the transition time is

$$\Delta t = 2C_N V_m/I_{dm} \tag{89}$$

and the rise time (or fall time) is

$$\tau_R = \tau_F = 1.6 C_N V_m/I_{dm} \tag{90}$$

Using Eq. (81), we find

$$\tau_D \simeq C_N V_m/I_{dm} \tag{91}$$

as supposed earlier. If we use $\tau_I \simeq 2.5\tau_D$, $\tau_R \simeq 0.4\tau_I$, and $\tau_I = 1/2 f_c$, we find $\tau_D \simeq 1/5 f_c$. The total energy required to switch is

$$W_I = C_N V_m^2/2 = P_I \tau_I \tag{92}$$

or

$$P_D \tau_D \simeq C_N V_m^2/5 \tag{93}$$

if $P_I \simeq P_D$, i.e., if the standby power dissipation is small. Here Eq. (93) should be compared with the earlier approximate result of Eq. (78).

In the previous section, we pointed out that the pertinent parameters are not independent of one another. The key equation for the power–delay product is Eq. (93), and although Eq. (77) appears more desirable, this latter one can be decieving. For example, one could conclude from Eq. (77) that making the delay time smaller would hurt the speed–power product, but references to Eq. (93) shows that this is not the case. To actually decrease the delay time, one must either decrease C_N or increase I_{dm}, as can be seen from Eq. (91). If C_N is decreased, the speed–power product is improved. On the other hand, if I_{dm} is increased, g_m is also increased, and the speed–power product is changed essentially not at all. Thus all parameters must be considered together.

If the interconnection capacitance C_I dominates the node capacitance C_N, then $P_D\tau_D$ is often thought to be independent of the inverter technology. However, τ_D need not be independent. If one can increase I_{dm} by appropriate transistor design or material choice, a smaller τ_D can be achieved without expense to $P_D\tau_D$, therefore giving faster circuits and a higher f_c. On the other hand, a technology giving a higher g_m (from higher I_{dm} at a given drive voltage V_m) can achieve a given drive current level at a lower value of V_m, hence giving a smaller τ_D and a considerably smaller $P_D\tau_D$, even if the node capacitance remains the same. It is therefore apparent that the drive current level I_{dm} is a critical parameter in choice of technology, and it is this drive current consideration which most suggests that alternatives to silicon NMOS can fruitfully be explored.

B. Drive Currents

In the preceeding section, it was concluded that one of the most important parameters is the available drive current of the pull-down transistor in the logic inverter. The current in the MOS device is given from Eqs. (19) and (33) as

$$I_d = \frac{k[V_G - V_T''(V_D)]^2}{[1 + \theta(V_G - V_T'')][1 + \mu_0 V_D/v_s L']} \tag{94}$$

in saturation, where $V_T''(V_D)$ is the drain voltage dependent threshold voltage for the channel and $k = Z\mu_0 C_0/2L'$. In very small devices, our concern here, the current is usually limited by the saturated velocity. The maximum drive current is then given approximately as ($V_D \simeq V_G - V_T''$)

$$I_{dm} \simeq k v_s L'/\mu_0 \theta \simeq Z v_s C_0/2\theta \tag{95}$$

In the MOSFET, current limitation arises from the onset of velocity saturation and surface scattering reduction of the effective mobility. However, we see that the drive current is directly proportional to the saturated velocity.

In a MESFET, the current is given from Eqs. (65) and (71) to be the resultant

$$I_d = \frac{G_0}{1 + \mu_0 V_D/v_s L} \left\{ \frac{V_p}{3} + V_{G0} \left[1 - \frac{2}{3}\left(-\frac{V_{G0}}{V_p}\right)^{1/2} \right] \right\} \quad (96)$$

in saturation, where $G_0 = N_d e \mu_0 Z a/L$ is the open channel conductance. Here, pinch-off of the channel is achieved at

$$-V_{G0} = V_p = e N_d a^2/2\epsilon \quad (97)$$

In Fig. 13, we plot Eqs. (65) and (19) for devices with $V_p = 4V_T''$ and k and G_0 adjusted to give the same current level at $G_0/3V_p = k$, or

$$2\epsilon_s/3a = \epsilon_{ox}/2t_{ox} \quad (98)$$

for otherwise equal dimensions and materials. In fact, Eq. (98) is often used to describe an equivalent MOSFET for the JFET–MESFET. It is readily apparent from Fig. 13 that there is little difference between the two relations. Rather, this figure actually overestimates the difference, since the current matching was taken at $V_G = 0(V_G = \Psi_0)$, a condition of very strong forward bias. Matching at a lower bias point would improve the comparison. Using

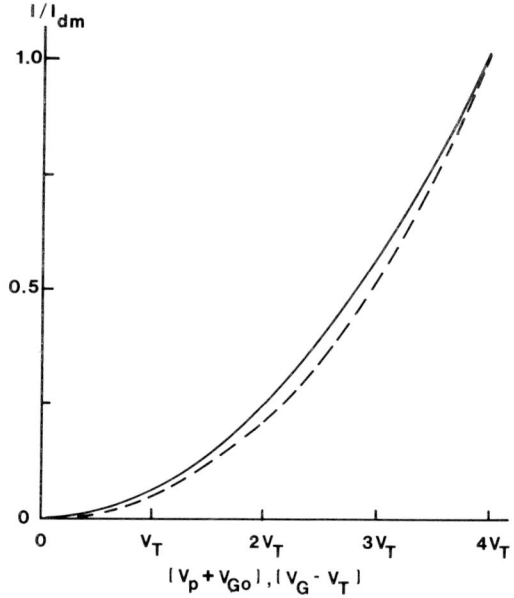

FIG. 13. Comparison of MESFET (dashed curve) and MOSFET (solid curve) current–voltage relations in saturation, for $V_p = 4V_T''$, and $G_0/3V_p = k$. Mobility reduction and velocity saturation have not been considered.

this value of I_D for I_{dm} however, is slightly more appropriate, since the maximum channel current is always greater than the zero-bias channel current (Fukui, 1979). While the maximum channel current is below $G_0 V_p/3$, there is a channel leakage contribution to the total device current. Thus $G_0 V_p/3$ is a reasonably good estimate, or with velocity saturation ($V_D \simeq V_p$)

$$I_{dm} = eN_d v_s Za/3 \tag{99}$$

If we now compare Eqs. (99) and (95) for the same material, using the matching equivalence of Eq. (98), we find that the maximum current drive from the MOSFET is reduced over the JFET–MESFET by the factor

$$I_{dm,MOS}/I_{dm,MES} = (\theta V_p)^{-1} = [\theta(V_G - V_T'')]^{-1} \tag{100}$$

However, *this is not a valid comparison between MOSFETs and MESFETs*, as it requires design parameters not often met in practice, when the two devices are individually optimized. It suffices to notice that both Eqs. (99) and (95) depend linearly upon the saturated velocity of the channel carriers. In submicron semiconductor devices, the average electron velocity cannot be easily represented as an explicit function of the electric field through a mobility. Rather, the important time processes such as transit time, energy and momentum relaxation times, and screening time are all comparable and the carrier velocity must be calculated as a function of space and time (Barker and Ferry, 1980; Ferry and Barker, 1980a; Grubin et al., 1979). In very small devices, transient velocity response effects can lead to enhanced effective velocities due to overshoot effects. While this overshoot effect plays a role in the delay time that is in fact small, due to the dominance of interconnection capacitances, the effect on available drive current can be significant.

C. Velocity Response

The Boltzmann transport equation (BTE) has long been the basis for semiclassical transport studies in semiconductors and other materials. Its utility also stems from the fact that it is readily transformable into a path-variable form which can be adapted to numerical solutions for complicated energy-dependent scattering processes. More importantly, it is useful as the basis for developing moment-balance equations which directly relate to the normal semiconductor equations. Thus, the dynamics of channel electrons may be obtained from moments of the BTE as (Ferry, 1979a),

$$\partial(np)/\partial t = neE - (np/\tau_m) \tag{101}$$

$$\frac{\partial}{\partial t}[np^2/2m + 3nk_B T_e/2] = nevE - nk_B T_e/\tau_e \tag{102}$$

where $p = m^*v$ is the average carrier momentum, v is the average (drift) velocity, T_e is the carrier temperature, and τ_m and τ_e are the momentum and energy relaxation times, respectively. For multivalley semiconductors, in which the sets of valleys are nonequivalent, such as GaAs or InP, equations such as Eqs. (101)–(102) must be written for each valley and supplemented with a particle continuity equation.

If a system of electrons is subjected to the combined influence of an electric field and scattering centers, the drift velocity of the particles asymptotically approaches the steady-state value

$$v = e\tau_m E/m^* \tag{103}$$

If we were to consider just the momentum balance equation (101), we could be satisfied that the particles would approach this value in a time approximately equal to $3\tau_m$, as illustrated in Fig. 14. But the momentum balance equation alone gives an incomplete picture of events. Energy balance tells us that the electron temperature increases with increasing electric field and departs significantly from room temperature in high electric fields. The effect of the increasing electron temperature is to decrease τ_m and hence to decrease the velocity given by Eq. (103). If the momentum and energy scattering times are similar in value, then both momentum and energy will follow changes in the electric field at approximately the same rate and the solid curve in Fig. 14 will result. On the other hand, if the energy relaxation time is significantly longer than the momentum relaxation time, the average velocity of the carriers will rise to its steady value and then decrease as the energy reaches its steady value, as the value of τ_m decreases as the energy

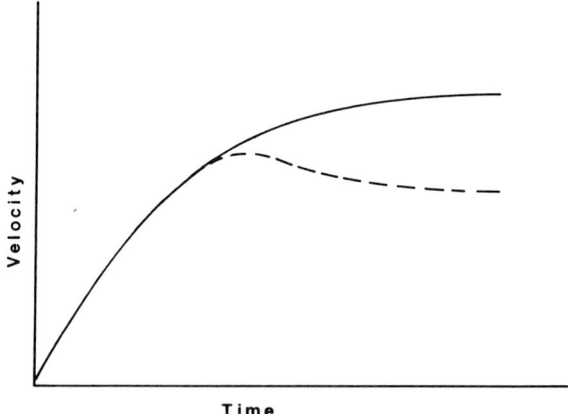

FIG. 14. The approach to the steady state, with and without transient relaxation effects in the energy which lead to overshoot effect in the velocity.

increases. This results in behavior such as the dashed curve in Fig. 14. It is clear that the velocity "overshoots" its actual steady-state value. Some authors have referred to the rising velocity region, as well as the overshoot region, as ballistic transport (Shur and Eastman, 1979), but this is clearly incorrect, since hundreds of collisions can occur during a single time span of τ_m (Barker *et al.*, 1980; Hess, 1981; Cook and Frey, 1981).

Most field-dependent velocities and values for the saturated velocity assume that steady-state conditions are reached. Clearly, this is not the case in very small devices. In Fig. 15, the average velocity as a function of distance is shown for electrons in silicon. Here it is assumed that the electrons see a homogeneous electric field of 50 kV cm^{-1}. Also shown is the small change induced by retardation in the velocity response (Ferry and Barker, 1980b), discussed below. It is evident however, that if the channel were only 200 Å long the velocity would never reach steady-state. While this is a very short distance, the effect is pronounced in the III–V materials and results in far different apparent velocity–field curves. In Fig. 16, a set of effective velocity–field curves is shown for GaAs. Again, the retarded solutions are actually used, and it is evident from this figure that there is a dramatic increase in the apparent saturated velocity in the short channel devices.

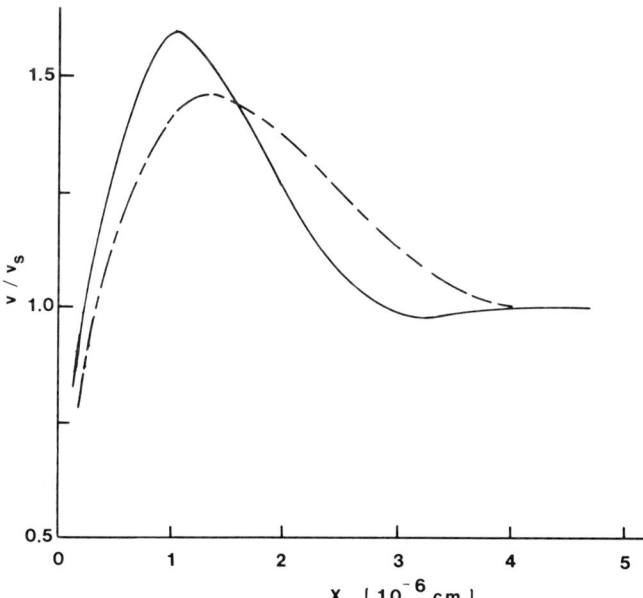

FIG. 15. The velocity response of electrons in Si at 300 K for an electric field of 50 kV cm^{-1}. The solid curve differs from the dashed curve by fully including the retardation effects.

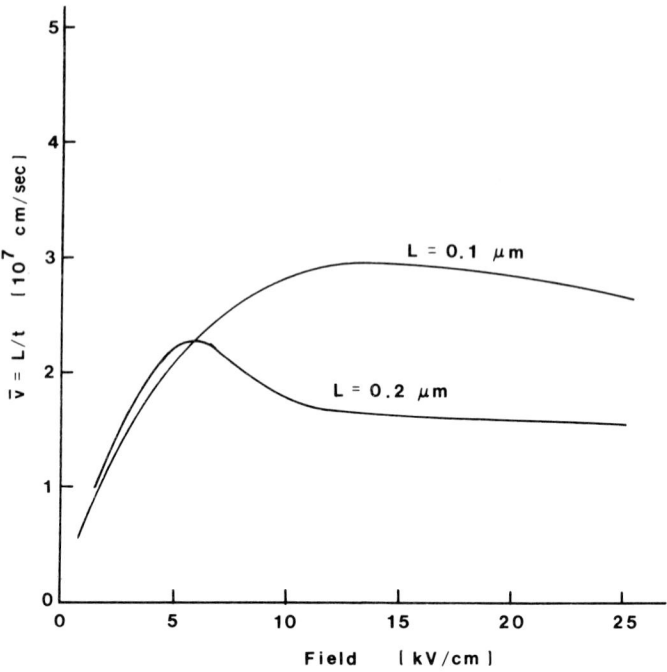

FIG. 16. The effective, or time-of-flight, velocity observed for very short transit lengths in GaAs at 300 K. These curves assume a steady homogeneous electric field with the carriers entering at $x = 0$ at $t = 0$ in thermal equilibrium with the lattice and no contact effects.

In Eqs. (101)–(102), it is apparent that the velocity response and the appropriate relaxation times are changing at very nearly the same rate. Why then do the equations not show retardation effects (in which the carrier velocity has some memory of the collision relaxation time at a previous time)? The answer is that the equations are wrong in that they are obtained from the BTE, which is itself not correct since it does not include these temporal retardation or memory effects.

The BTE is valid only in the weak-coupling limit under the assumptions that the electric field is weak and slowly varying at most, that the collisions are independent, and that the collisions occur instantaneously in space and time. Each of these approximations can be expected to be violated in future submicron-dimensioned semiconductor devices. We have previously shown that in such devices the time scales are such that collision durations are no longer negligible when compared to the relevant time scale upon which transport through the device occurs (Ferry and Barker, 1980a). In this situation, even for time-independent fields, a quantum kinetic equation which is nonlocal in time and momentum must be utilized. It may be recalled that the BTE can be rigourously derived from the quantum density matrix Liouville equation formulation in quantum transport (Kohn and Luttinger, 1957; Barker, 1973). In the BTE approach, the collision terms are derived under the assumption that the collisions occur instantaneously (which leads to the energy-conserving δ-function in the golden-rule transition rates for scattering), which is a reasonable approximation when the mean time between collisions is large. At high fields, such as will occur in very small devices, the collision duration is significant and correction terms must be generated (for the BTE) to account for the nonzero time duration of each collision. If the instantaneous collision approximation that leads to the BTE is relaxed, an additional field contribution appears as a differential superoperator term (see, e.g., the discussion in Barker and Ferry, 1980b) in the collision integrals evaluated in the momentum representation, resulting in an intracollisional field effect (Barker, 1973).

Moreover, on the short time scale of submicron devices, a truly causal theory introduces memory effects into the quantum kinetic equation which lead to convolution integrals in the transport coefficients (Green, 1952; Kubo, 1957; Mori, 1958; Zwanzig, 1961, 1964). That a memory effect should be included is evident from simple arguments based upon Langevin-type equations of which the balance equations are typical (nonstochastic) examples. If the damping terms in the balance equations are evolving in time with a characteristic time of the same order as the velocity, for example, then the product of the velocity and its damping must include a convolution which takes into account the damping at earlier times. Hence, the system has memory of these earlier times. We examine this in the following.

1. The Intracollisional Field Effect

The intracollisional field effect (ICFE) can be partially understood by the following simplified model. In the BTE, the collision occurs instantaneously, so that the carrier enters the collision sphere at one point and instantaneously exits at a second point, which we shall call P for reference. However, the collision does not occur instantaneously, but requires a nonzero collision duration τ_c. In this case, it can now be accelerated during the collision. Thus it exits not at P, but at P', some time $\Delta t = \tau_c$ later. The points P and P' differ by an amount corresponding to the acceleration of the electron during the collision, and this leads to a modification of the momentum conservation relations and the energy conservation relations that appear in the transition probability. When τ_c begins to become comparable to τ_m, the mean time between collisions, this ICFE will have a significant effect on the transport dynamics, particularly in the transient response region. Two major modifications of the scattering integral occur as a result of this intracollisional process.

First, the total energy-conserving δ-function is broadened by the presence of the field. Second, the threshold energy required for the emission of an optical phonon is modified, which causes a shift (in energy) of the δ-function. This latter process is easily understood in physical terms. The argument of the energy-conserving δ-function is just

$$E_f - E_i \pm \hbar\omega_0 = E(\mathbf{p}_f) - E(\mathbf{p}_i) \pm \hbar\omega_0 \tag{104}$$

but the initial and final momenta evolve during the collision as

$$\mathbf{p}(t') = \mathbf{p}(t) + \int_t^{t'} e\mathbf{E}(t'')\,dt'' \tag{105}$$

In the emission of an optical phonon, where the electron is scattered against the electric field, the field will absorb a portion of the electron's energy during the collision, and hence a reduction in energy loss to the lattice occurs. The opposite effect, and enhancement in energy loss to the lattice, occurs for emission along the electric field. These effects can be incorporated into the appropriate scattering integrals used in solutions to the BTE (Ferry 1979b).

Although the ICFE is exceedingly large in insulators such as SiO_2 because of the polar nature of the phonons there, it is also significant in the

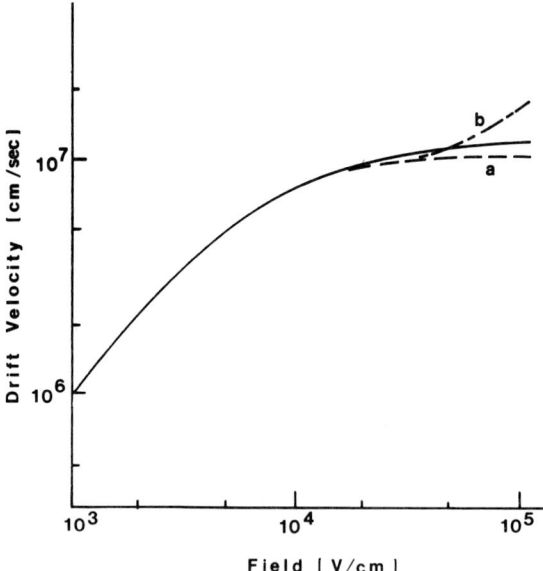

FIG. 17. The velocity–field curve for electrons in Si as calculated by a drifted Maxwellian distribution function. At high fields, the field weakens the collisions [dot–dashed curve (b)] causing an increase of velocity, although the collision retardation stops this effect (solid curve). The low-frequency curve normally encountered is the dashed curve (a).

case of silicon and the III–V compounds. In Fig. 17, the effect the collision duration has on the velocity–field curve is plotted. Also shown is the countering effect of collision retardation discussed below. The ICFE is especially noticable at high fields, where it essentially eliminates the scattering by the low-energy intervalley phonon (Long, 1960). This phonon is already weakly coupled since it scatters through a first-order interaction (Ferry, 1976b), but is normally an effective scatterer at high fields.

2. The Retarded Transport Equations

When the ICFE is included as a modification of the lowest-order kinetic equations, a high-field quantum kinetic equation, which also includes full memory retardation effects, arises to replace the BTE. This equation can be written as (Barker, 1973)

$$[\partial f(\mathbf{p}, t)/\partial t] + e\mathbf{F}(t) \cdot \nabla_p f(\mathbf{p}, t) = \int_0^t dt' \sum_{\mathbf{p}'} \{S(\mathbf{p}, \mathbf{p}'; t, t') f(\mathbf{p}', t')$$
$$- S(\mathbf{p}', \mathbf{p}; t, t') f(\mathbf{p}, t)\} \quad (106)$$

where the momenta $\mathbf{p}, \mathbf{p}'$ are explicit functions of the retarded time t' through the relationship of Eq. (105), and the transition terms S take the form, for inelastic phonon scattering,

$$S(\mathbf{p}, \mathbf{p}'; t, t') = (2\pi/\hbar) \operatorname{Re} \left\{ \int_0^t (1/\pi\hbar) \exp[-(t-t')/\tau_\Gamma] \right.$$
$$\left. \times [N_q + (1+\eta)/2] |V_q^2| \exp\left[-i \int_t^{t'} dt'' \, \beta(\mathbf{p}, \mathbf{p}'; t'')/\hbar\right] \right\} \quad (107)$$

where β is the argument of the normal δ-function and is given by Eq. (104). Not explicitly shown is the momentum-conserving δ-function. The two exponential factors in Eq. (107) are related to the joint spectral density function, which reduces to an energy-conserving δ-function in the instantaneous collision limit, $\hbar/\tau_\Gamma$ is the joint linewidth due to collisional broadening of the initial and final states, and η takes the values $+1, -1$ for phonon emission and absorption, respectively, in the in-scattering term. For the out-scattering term, the roles of $\mathbf{p}, \mathbf{p}'$ are reversed, although this does not upset detailed balance in the equilibrium sense. Now, Eq. (106) is the quantum kinetic equation and differs from the BTE only in the retardation of the collisional terms.

In small semiconductor devices, where the dimensional scale is of the order of 0.3 μm or less, the carrier concentration will in general be relatively high. In this case, we can use a drifted Maxwellian approach to developing a set of coupled balance equations, using Eq. (106) instead of the BTE. With this approach, a hierarchy of moment equations can be generated, from which the various parameters can be determined (Ferry and Barker, 1980b). These moment equations include first-order effects arising from the nonzero time duration of the collisions and the general retardation effects discussed above. Thus, the balance equations are modified in a straightforward fashion, although the details are much more complicated. This latter follows from the role of the intracollisional field effect, which both broadens and shifts the resonances, effectively lengthening the collision duration and weakening the effect of the collision itself. If Eq. (106) is Laplace-transformed, the moment equations can be developed by multiplying by an arbitrary function $\phi(\mathbf{p})$, integrating over the momentum, so that the moment equations are developed in the transform domain, and then retransforming. This yields (Barker and Ferry, 1980b),

$$\langle \partial \phi / \partial t \rangle - e\mathbf{F} \cdot \langle \nabla_p \phi \rangle = -(1/\tau_c) \int_0^t \exp(-t'/\tau_c) \langle \ddot{\phi}_c(t - t') \rangle \, dt' \quad (108)$$

where

$$\langle \ddot{\phi}_c \rangle = \int_0^t \phi(t - \tau) \{ d[\Gamma_\phi(\tau)]/d\tau \} \, d\tau \quad (109)$$

and Γ_ϕ is the relaxation rate for ϕ, so that $\dot{\phi}_c = \Gamma \phi$ is the time rate of change of $\phi(\mathbf{p})$ due to collisions. The result in Eq. (108) is particularly interesting, in that it allows $\langle \ddot{\phi}_c \rangle$ to be evaluated in the case of instantaneous scattering, and this result to be averaged into the proper convolution integrals with an effective collision duration τ_c. If, as is the case for low fields, $\langle \ddot{\phi}_c \rangle$ does not change much during a collision duration, the normal result ($t \gg \tau_c$) is obtained. However, in large fields, where the intra-collisional field effect is important, the variation of the relaxation during the collision becomes important. These effects will also be important in high-frequency transport where the collision duration becomes comparable to the relaxation times and the period of the wave (Das et al., 1981). From the form of Eq. (108), we note the right-hand side is such that the nonzero collision duration must be combined with the normal non-Markovian nature of the transport on these short time scales, so that the momentum relaxation time must be convolved as in Eq. (108) and the result reconvolved with the momentum. The resulting velocity–field curve for silicon is also shown in Fig. 17.

Calculations for the transient response have also been made for silicon. The details of the coupling constants and phonon parameters are those normally accepted. In Fig. 18, the transient response for a steady, homogeneous field of 20 kV cm^{-1}, applied at $t = 0$, is shown. The response for a retarded collisional interaction rises quicker and settles faster than that of the unretarded case. The quicker rise follows from the retarded momentum

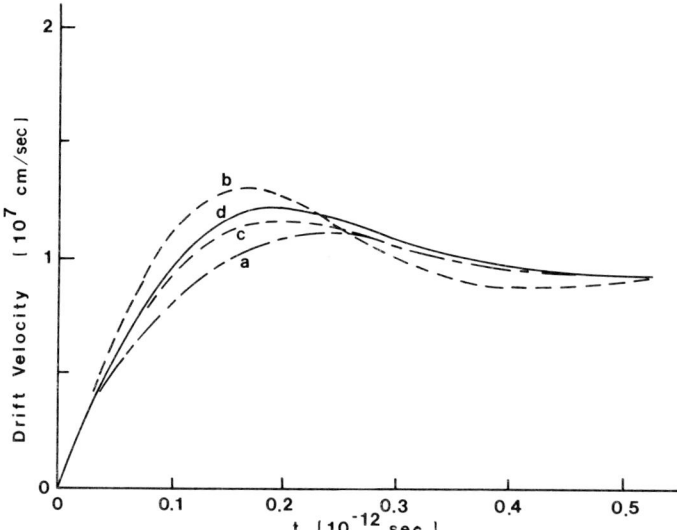

FIG. 18. Transient dynamic response of carriers in Si. It is assumed that the carriers see a homogeneous field of 20 kV/cm applied at $t = 0$. Curve a shows the response neglecting the short-time effects, but with the weakening of the collision in high fields. Curve b includes the effects of nonzero collision duration alone, curve c includes the effects of memory effects alone, and curve d includes all effects according to Eq. (108).

relaxation effects, while the faster settling occurs due to retarded energy relaxation effects which causes an overshoot to occur in the temperature as well.

The collisional retardation speeds up the transient process primarily due to the effect of slowing down changes in the effective momentum and energy via collisional relaxation as well. The small signal ac mobility is extremely sensitive to the energy distribution function, so it is extremely important that any simulation technique be very efficient in yielding this portion of the distribution function (Das *et al.*, 1981).

Most field-dependent velocities and values for the saturated velocity assume that steady-state conditions are reached, but this is not valid in short-channel devices due to the transient relaxation of the velocity. Since the retardation effects cause a change in the transient response of the carriers in the overshoot velocity region, they will strongly affect the effective time-of-flight velocity at saturation.

D. *Effective Saturation Velocity*

In the previous paragraphs, we pointed out how the effective saturation velocity in short-channel devices could differ markedly from the value at large dimensions. In Fig. 15, we showed the average velocity as a function of the distance, fully including the retardation effects important in this temporal region. It is evident that if the channel were only 200 Å long, the velocity would never reach steady-state. While this is a very short distance, the effect is pronounced in the III–V materials and results in far different apparent velocity–field curves, as can be seen from Fig. 16 for GaAs. What is evident here is the dramatic increase in the apparent saturated velocity in short-channel devices.

In Fig. 19, several semiconductors are compared to show how the effective saturated velocity increases in short channels. Here, the InP predictions were estimated from the Monte Carlo calculations of Maloney and Frey (1977), and are probably overoptimistic due to the lack of inclusion of the retardation effects. It becomes evident, however, from this figure that perhaps the most important aspect of short-channel overshoot effects is the indirect effect of increasing the effective saturated velocity and the concomitant increase in peak drive current I_{dm} that results.

The strong modification of the apparent velocity–field curve in GaAs (Fig. 16) and InP that leads to the results of Fig. 19 arises from the fact that the electrons effectively do not have sufficient time in which to transfer to the satellite valleys of the conduction band. Thus, in the very short channels

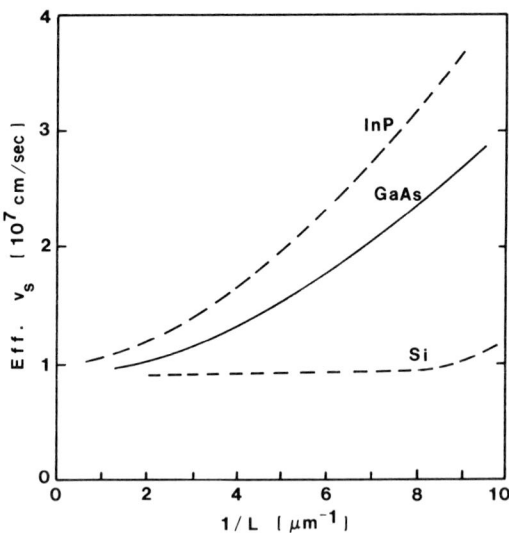

FIG. 19. The effective, or time-of-flight, saturated velocity as a function of the inverse channel length. The InP curve is estimated from the data of Maloney and Frey (1977) and may overestimate the increase in effective v_s.

($L < 0.2$ μm) that can exist, transport in the III–V compounds will principally be governed by the properties of the *central valley* of the conduction band. This will be true to a lesser extent in silicon, as the more rapid scattering will cause these effects to occur only on distance scales of 500–700 Å (Ferry and Barker, 1980b). We may then ask whether the curves depicted in Fig. 19 rise indefinitely as L is reduced or whether a limiting velocity within the central valley dominates and sets a limit on the increase of v_s.

In the central valley of the III–Vs (neglecting intervalley scattering), a single phonon, the LO polar mode, dominates the energy relaxation. In this case, a simple model can be used for estimating the saturated velocity within this valley. This can be expressed as (Wolff, 1954)

$$v_s^{(1)} = (8\hbar\omega_0/3\pi m^*)^{1/2} \tag{110}$$

Hilsum (1976) has suggested a further simple model. For the polar mode, Stratton (1958) has shown that

$$\mu = \frac{8q}{6\sqrt{\pi}}(am^*\omega_0)^{-1}\frac{\exp(z)-1}{\sqrt{z}}G(z) \tag{111}$$

where $z = \hbar\omega_0/k_B T$, $G(z)$ is a slowly varying function of z ($\simeq 0.72$ for $z = 1$), and a is the polaron coupling constant. Using this result, Hilsum (1976) estimates the saturated velocity in the central valley to be

$$v_s^{(2)} = (0.3 \times 10^7 \text{ cm sec}^{-1})\frac{\exp(z)-1}{(m^*/m_0)^{1/2}} \tag{112}$$

at 300 K. This simple relationship gives the velocity which would be attained by the carriers at the breakdown field in the central valley if the polar mobility were independent of field. In Table I, we list values for $v_s^{(1)}$ and $v_s^{(2)}$ for several semiconductors. The values of $v_s^{(2)}$ are expected to be more accurate for the polar semiconductors, whereas that for $v_s^{(1)}$ is expected to be more accurate for nonpolar silicon. However, the results of Table I generally support the findings of Fig. 19.

TABLE I

Material	$v_s^{(1)}$ (10^7 cm/sec)	$v_s^{(2)}$ (10^7 cm/sec)
Si	1.9	—
GaAs	2.9	3.7
InP	3.0	4.8
$In_{0.53}Ga_{0.47}As$	3.4	4.3

E. Interconnection Capacitance

Interconnections between devices in integrated circuits are usually accomplished by conducting metallizations (or polysilicon "metallizations"). In most respects these are rather crude implementations of microstrip transmission lines; crude in the sense that they are neither regularly nor neatly laid out and are thus difficult to analyze. However, we can get a good first impression of these interconnects, and their capacitance, by considering the case of microstrip on semiconductor substrates. The approximation as microstrip assumes that the semiconductor is semiinsulating so that the ground plane is the back contact. In doped material, the depletion under the metallization sets the line capacitance. However, the move to all implantation technology in both silicon and the III–Vs means that the approximation as microstrip will be quite good in future submicron ULSI.

Microstrip transmission lines for quasi-static TEM wave propagation in microwave circuitry was introduced as early as 1952 (Grieg and Engelmann, 1952). Although a difficult problem to solve directly, Wheeler (1964, 1965) developed a conformal transformation approach that allowed approximate solutions to be obtained. For the infinitely thin conductor, exact solutions are now available, as this case is a textbook example of the application of the Schwarz–Christoffel conformal transformation (Carrier et al., 1966). In Fig. 20, the microstrip line and its equivalent "double line" used for analysis are shown. A double Schwarz–Christoffel transformation is used to transform the double line into a uniform-field rectangle (equivalent fringeless parallel-plate capacitor). Then the complex potential (Carrier et al., 1966)

$$\Omega(x, y) = V(x, y) + iL(x, y) \tag{113}$$

is given as the solution of the transcendental equation-pair

$$\Omega = -(1/b)E(m, w) \tag{114a}$$

$$z = x + iy = (p/m^2)[E(m, w) + (1 - \lambda^2 m^2)F(m, w)] \equiv f(w) \tag{114b}$$

where $w = u + iv$ and E and F are incomplete elliptic integrals of the first and second kind, respectively. The parameters p, m, λ, b are given by the boundary conditions on $f(w)$ as

$$f(1) = f(1/m) = -i, \quad f(\lambda) = -i + a/2$$
$$b = E(m, 1) = K(m) \tag{115}$$

where $K(m)$ is the complete elliptic integral. The capacitance of the air-filled microstrip line is then

$$C_a = \epsilon_0 c/b \tag{116}$$

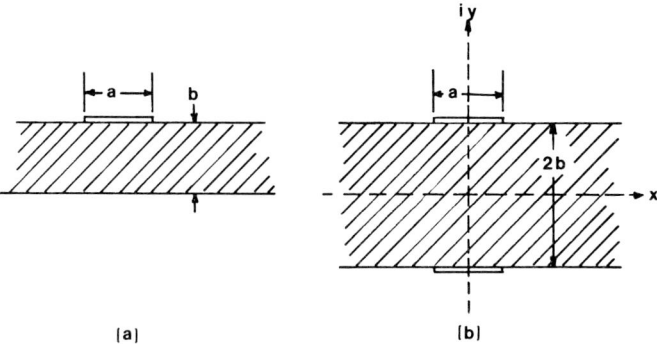

FIG. 20. The microstrip transmission line (a) and its double-line equivalent (b) used for analysis.

where
$$c = -iF(m, 1/m) - b \qquad (117)$$

For a dielectric-filled line, we define an effective dielectric constant as
$$\epsilon_{\text{eff}} = 1 + (c'/c)(\epsilon_r - 1) \qquad (118)$$
where
$$c' = -iF(m, \lambda) - b \qquad (119)$$

It is worth noting here that the quantity c'/c is identical to the effective filling fraction of Wheeler (1965). For a constant effective filling factor c'/c, the linear dependence on ϵ_r has been verified experimentally by Kaupp (1967). The effective dielectric then multiplies the capacitance of Eq. (116). In Fig. 21, the capacitance per unit length is shown for a range of width-to-height ratios that may be found in integrated circuits. A dielectric constant of 12 is used, as this is quite near the values for Si, GaAs, InP, and similar materials. The low degree of curvature for increasing a/b ratios arises from dominance of the total capacitance by the fringing capacitance.

The dominance of the capacitance in Fig. 21 by the fringing capacitance is indicative of a problem neglected in the above analysis, namely the finite thickness of the conductors. Such thick conductors cannot be easily handled by the conformal transformation used above, so that numerical solutions are still pursued. Such approaches have included variational techniques (Yamashita and Mittra, 1968) and method of moments (Farrar and Adams, 1970). However, an adequate approximation to use for $a/b < 1$ is to consider the capacitance of a round wire over a ground plane. For an air-filled line, this gives approximately ($a/b \ll 1$)
$$C_a = 2\pi\epsilon/\ln[1 + 16(0.5 + b/a)^2] \qquad (120)$$

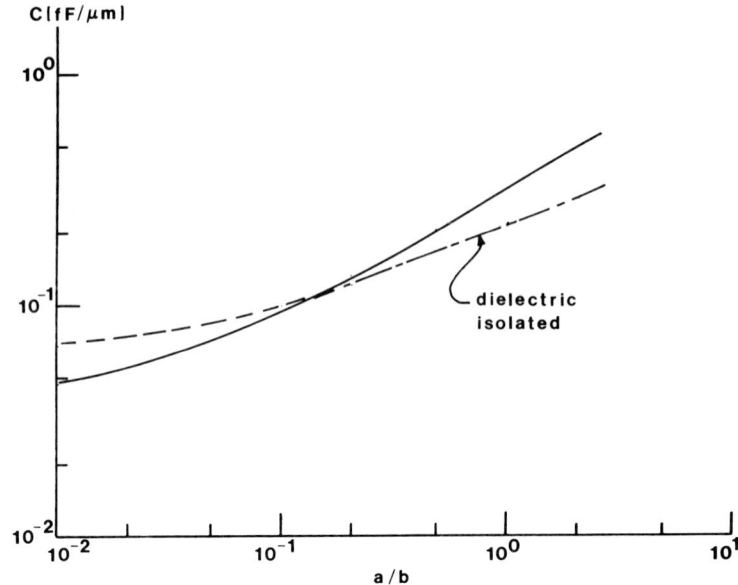

FIG. 21. Capacitance of a microstrip transmission line on a semiconductor substrate ($\epsilon_r = 12$). The dashed curve is the parallel-wire approximation. Also shown is the effect of dielectric isolation, in which a thin oxide ($b_{ox} = a/5$) is placed under the microstrip (dot–dashed curve).

and the effective dielectric constant can also be used here. This limit is also shown in Fig. 21.

In many applications, the interconnections are laid over an oxide for isolation purposes. The problem of the multiple dielectric is a much more complicated system (Farrar and Adams, 1974). For the quasi-static TEM mode, however, a good approximation is to replace the dielectric constant of the semiconductor by (Hasegawa et al., 1971)

$$\epsilon_{\text{eff}} = [(1/b)(b_1/\epsilon_{0x} + b_2/\epsilon_r)]^{-1} \tag{121}$$

where $b = b_1 + b_2$ and ϵ_{0x} is the dielectric constant of the oxide. In Fig. 21, the capacitance of a microstrip laid over an isolation oxide is also illustrated for comparison. Here, the case of $b_1 = a/5$ is taken. The use of such an oxide for isolation is useful for relatively wide strips and improves for larger b_1, but its effectiveness is very limited at narrow strip widths.

Also of interest is line-to-line capacitance, as this represents parasitic coupling between devices. This is essentially the problem of coupled pairs of microstrip lines. Such lines can have even or odd modes of transmission (Bryant and Weiss, 1968), depending upon the relative sign of the voltage on the two strips. The odd mode is the primary one for our interest, as it has

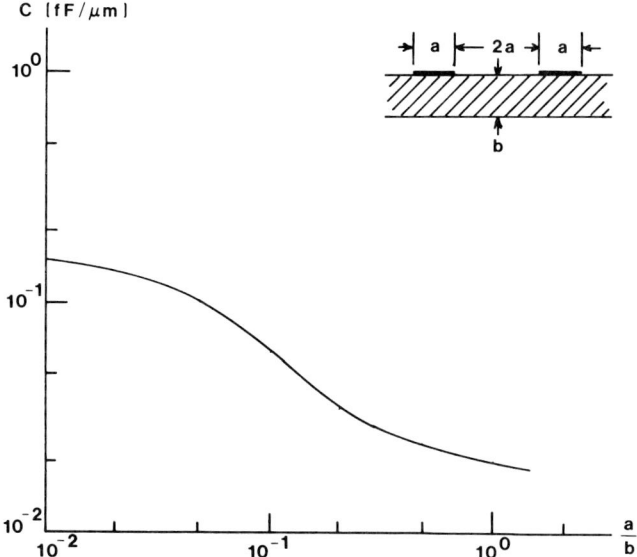

FIG. 22. The line-to-line parasitic capacitance between two microstrip lines on a semiconductor substrate ($\epsilon_r = 12$).

not only the largest coupling between lines, but also represents the strong coupling mode. In Fig. 22, the line-to-line capacitance (parasitic capacitance in our parlance) is shown for the case of a separation of $2a$. Comparison of these results to those of Fig. 21 shows that the line-to-line parasitic capacitance dominates the direct line capacitance for $a/b < 0.075$, at least for a separation of $2a$. For larger spacing, this parasitic capacitance is of course reduced.

V. Band Structure Considerations

If we are to properly make a material choice that will result in improved performance of submicron semiconductor devices, it is necessary to understand the requisite properties and their dependence upon the band structure of the material. Some of these properties are the energy gap, effective mass, satellite valley separations, and the behavior of these properties upon alloying if the material is to be a "random" alloy such as a ternary or a quaternary. Fortunately, silicon and the relevant III–V compounds are basically very similar. They are tetrahedrally coordinated with a common lattice structure. Silicon possesses the diamond structure and the III–V compounds possess the zincblende structure, which differs from the former structure only in having two dissimilar atoms per unit cell instead of two similar silicon atoms. As a consequence, these materials are characterized more by their similarities than by their differences. Phillips (1968, 1973) and Van Vechten (1969a, 1969b) have utilized fully these similarities among tetrahedrally coordinated compounds to develop a general quantum dielectric theory of electronegativity in covalent systems which describes not only the relative ionicity and dielectric constant of these materials, but also gives an excellent description for the ionization potential and interband transition energies. While others, notably Harrison (1980), have also developed general theories for a description of the variations in these materials, the dielectric method remains the most extensively developed, and it will be utilized here. The reason for this is that the dielectric method offers an excellent interpolative method without having to resort to extensive computer calculations. Thus, the theoretical basis for these treatments will be the dielectric method, as it is conceptually the easiest to understand, not requiring detailed appreciation of the subleties of the tight-binding (LCAO) or pseudopotential band calculations. In this section the general band structure variation for silicon and the relevant III–V compounds will be reviewed. The theory of random alloys and the band parameter variations, both for ternary and for quaternary alloys, will also be discussed.

The dielectric method is based upon a universal semiconductor model—a dielectric two-band model. In this model, an average energy gap E_g is deduced from the real, static, electronic dielectric constant $\epsilon_1(0)$. Here E_g represents an average energy between bonding (valence) and antibonding (conduction) hybridized sp^3 orbitals. Moreover, this average gap can be decomposed into homopolar (covalent) and heteropolar (ionic) contributions: E_h and C, respectively. These represent, in turn, the symmetric and antisymmetric parts of the atomic potentials within a unit cell. These may be combined as

$$E_h^2 + C^2 = E_g^2 \qquad (122)$$

In a homopolar (diamond structure) material such as Si or Ge, $C = 0$, and $E_g = E_h$ is fixed at the Si (or Ge) value for the group IV materials by the dielectric constant through (Penn, 1962)

$$\epsilon_1(0) = 1 + (\hbar\omega_p)^2 DA/E_g^2 \qquad (123)$$

where

$$A = 1 - B + B^2/3 \qquad (124)$$

$$B = E_g/4E_F \qquad (125)$$

E_F and ω_p are the valence electron Fermi energy and plasma frequency, and D is a correction factor to account for core d-level interactions with the valence electrons (Van Vechten, 1969a). The heteropolar potential C is related to the valence difference and is given by

$$C = b[(Z_a/r_a) - (Z_b/r_b)] \exp[-\tfrac{1}{2}k_s(r_a + r_b)] \qquad (126)$$

where k_s is the linearized Fermi–Thomas screening wave vector, and r_a, r_b are the covalent radii of elements a, b. The factor b also contains a correction factor of 1.5 (in addition to $e^2/4\pi\epsilon_0$ in MKS units) to account for the fact that the Fermi–Thomas screening is generally stronger than actually found (Phillips, 1973). In Table II, the values of $\epsilon_1(0)$, E_F, $\hbar\omega_p$, E_h, C, and D are listed for the major compounds of interest.

The strength of the dielectric method lies in its ability to also give good fits to the variations among elements of these compounds. The results of the dielectric method can be expressed in seven postulates (Van Vechten, 1969a, 1969b):

(1) Any direct energy gap E_i, in the absence of d-state perturbations is given in analogy to Eq. (122) as

$$E_i = [E_{i,h}^2 + C^2]^{1/2} = E_{i,h}[1 + (C/E_{i,h})^2]^{1/2} \qquad (127)$$

where $E_{i,h}$ is the value of the gap for the corresponding homopolar crystal (throughout, we use the subscript h for the homopolar value).

(2) The $E_{i,h}$ values, and all other homopolar variables, are assumed to be simple power-law functions of the nearest neighbor distance d, relative to the silicon value, as

$$E_{i,h} = E_{i,h,\mathrm{Si}}(d/d_{\mathrm{Si}})^{s_i} \qquad (128)$$

where the various s_i are parameters.

TABLE II

Material	$\epsilon_1(0)$	E_F (eV)	$\hbar\omega$ (eV)	E_h (eV)	C (eV)	D
Si	12.0	12.5	16.6	4.8	0	1.0
GaAs	10.9	11.5	15.6	4.3	2.9	1.24
GaP	9.1	12.4	16.5	4.7	3.3	1.15
GaSb	14.4	9.85	13.9	3.5	2.1	1.31
InAs	12.3	10.1	14.2	3.7	2.74	1.35
InP	9.6	10.7	14.8	3.9	3.34	1.27
InSb	15.7	8.8	12.7	3.1	2.1	1.42

TABLE III

Parameter	Si value	s_i
I_h	5.17	−1.31
X_4	8.63	−0.32
$E_{0,h}$	4.1	−2.75
$E_{1,h}$	3.6	−2.22
$E_{2,h}$	4.5	−2.38

$(X_3-X_1)/C = 0.14$

(3) The ionization potential, i.e., the energy difference between the top of the valence band at Γ and the vacuum level, is given by

$$I = I_h[1 + (C/I_h)^2]^{1/2} \qquad (129)$$

where I_h is scaled as in Eq. (128).

(4) The energy at the top of the valence band at the symmetry point X (zone edge along any of the [100] directions), the X_4 state in silicon and X_5 state in the zincblende structure, relative to the vacuum level is independent of C, and is given as a function of d only, as per Eq. (128).

(5) The energy at the top of the valence band at the symmetry point L, the L_3 state, is midway between the values of the energies at Γ and X (for the valence band):

$$E_{L_3} = (I + E_X)/2 \qquad (130)$$

(6) The splitting of the conduction-band X levels, X_1 and X_3, is proportional to C as

$$E_{X_3} - E_{X_1} = \text{const} \cdot C \qquad (131)$$

(7) Finally, the perturbative effect of the d-band on the s-like levels of greatest interest, Γ_1^c and L_1^c, is expressed by decreasing the $E_0(\Gamma_{15}^v - \Gamma_1^c)$ and $E_1(L_3^v - L_1^c)$ direct optical transitions from the values given by (129) to

$$E_i = [E_{i,h} - (D - 1) \Delta E_i][1 + (C/E_{i,h})^2]^{1/2} \qquad (132)$$

where here $i = 0, 1$ and ΔE_i is a parameter which is a function of d alone, as in (128). We tabulate and discuss the various parameters below.

A. Silicon

The valence and conduction band spectra for silicon are shown in Fig. 23, which is the result of a tight-binding calculation such as that of Pandey and Phillips (1967). The lowest energy gap is the indirect $\Delta_1^c - \Gamma_{25'}^v$ transition (the Δ_1^c minima occur at a point approximately 85% of the distance to X_1^c).

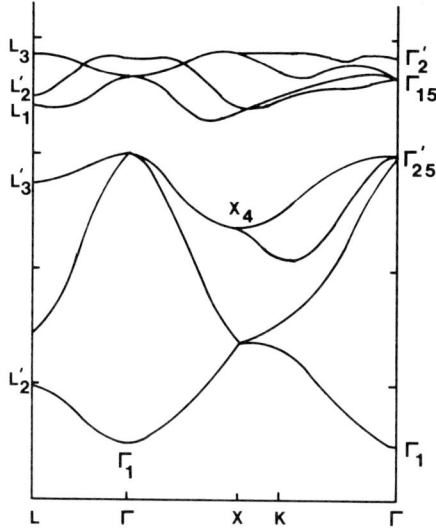

FIG. 23. The energy-band structure for Si.

The lowest direct transition at the zone center is the $\Gamma^c_{15}-\Gamma^v_{25'}$, but the usually quoted direct gap is the $\Gamma^c_{2'}-\Gamma^v_{25'} = 4.1$ eV (at 300 K), as it is the $\Gamma^c_{2'}$ state which converts to the usual Γ^c_1 lowest conduction band in the direct gap zincblende structured materials. The E_0, E_1, E_2 optical transitions occur at the Γ, L, and X points, respectively, and these play an important role in estimating the positions of the upper conduction-band minima in the ternary and quaternary compounds.

The six equivalent minima of the conduction band are ellipsoids of revolution, characterized by a longitudinal mass $m_L = 0.91 m_0$ and a transverse mass $m_T = 0.19 m_0$. These give rise to a density-of-states effective mass

$$m_d = 6^{2/3}(m_T^2 m_L)^{1/3} = 1.06 m_0 \tag{133}$$

and a conductivity mass

$$m_c = 3(1/m_L + 2/m_T)^{-1} = 0.26 m_0 \tag{134}$$

In Table III, the various band-gap and symmetry point energies are shown for silicon along with the index s_i. Two differences appear in this table from the original values of Van Vechten (1969b). First, the value of $(X_3-X_1)/C$ is twice the original value quoted, due to what appears to be a systematic loss of a factor of two throughout much of this latter work. Secondly, the index s_i for E_{X_4} has been reduced substantially from its earlier value, and this gives a much better representation of the various conduction-band minima for the binaries of the next section.

B. The III–V Compounds

The basic band structure of the III–Vs differs little from silicon's other than in the ordering of the various conduction band minima. The band structure of GaAs is shown in Fig. 24, but this could also be the structure of InAs, InP, InSb, or GaSb. Each of these is a direct gap semiconductor with Γ, L, X ordering of the conduction band minima. Only GaP differs, and it is essentially Si-like with X, L, Γ ordering. In Table IV, the principle energy levels and transitions are listed for these compounds. Also shown are the effective masses for the spherically symmetric Γ minimum of the conduction band, and the Γ–X and Γ–L separations of the conduction band. Many of these parameters depend upon the absolute X_5 level in the material. Using the s_i value given in Table III gives good results except for the Sb compounds, and these have been adjusted from the Ge-based extrapolation rather than the Si-based extrapolation. These values then give good $\Delta E_{\Gamma L}$ and $\Delta E_{\Gamma X}$ separations that are in good agreement with experimental results, except for GaP, but this material is not of great interest here.

The lowest satellite valleys of the direct-gap III–Vs are generally the L-point valleys. These are Ge-like, and thus can nearly all be characterized by four equivalent ellipsoids of revolution with $m_L = 1.64 m_0$ and $m_T = 0.082 m_0$, so that the density-of-states effective mass in this satellite valley is

$$m_d = 4^{2/3}(m_L m_T^2)^{1/3} = 0.56 m_0 \tag{135}$$

and the conductivity effective mass is

$$m_c = 0.12 m_0 \tag{136}$$

The last column in Table IV, for $P^2/2m_0$, arises from $\mathbf{k} \cdot \mathbf{p}$ theory as it is

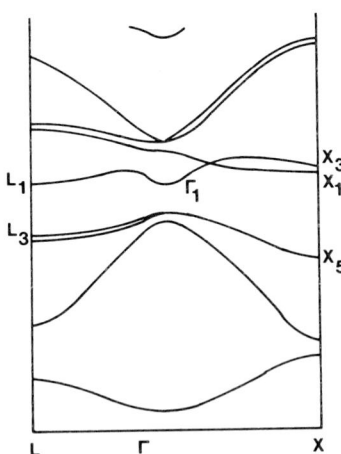

FIG. 24. The band structure of GaAs.

used in band structure calculations. In this approach, for a spherical band (Dresselhaus et al., 1955)

$$m/m_0 = 1 + (P^2/2m_0 E_0) \qquad (137)$$

As one can see from the values of $P^2/2m_0$ across the various materials, P^2 is a relatively constant parameter for these tetrahedrally coordinated compounds. The function P^2 is the square of the momentum matrix element in this formulation and shows weak chemical trends (Phillips, 1973). The slight variation of $P^2/2m_0$ observed can be attributed to a slight variation in wave functions. We have used this quantity in estimating the conduction-band effective mass in the alloys discussed below.

C. Pseudobinary Alloys

The zincblende structure is composed of two interpenetrating fcc lattices. Ternary alloys such as InGaAs have been formulated by a smooth mixing between the two constituents. In such $A_x B_{1-x} C$ alloys, all of the points of one fcc sublattice are occupied by type C atoms, but the points of the second sublattice are shared by atoms of type A and type B such that (Harrison and Hauser, 1976)

$$N_A + N_B = N_C = N \qquad (138)$$

$$x = N_A/N_C = C_A \qquad (139)$$

$$1 - x = N_B/N_C = C_B \qquad (140)$$

TABLE IV

Compound	$E_{0,h}$	ΔE_0	E_0	l_h	l	X_5	$E_{2,h}$	E_2
GaP	4.06	12.67	2.75	5.15	6.12	8.62	4.46	5.55
GaAs	3.67	10.9	1.42	4.91	5.7	8.42	4.09	5.01
GaSb	2.69	7.81	0.7	4.42	4.9	8.0	3.39	3.99
InP	3.31	8.75	1.35	4.67	5.74	8.42	3.74	5.01
InAs	3.07	7.92	0.35	4.5	5.27	8.34	3.5	4.44
InSb	2.52	5.72	0.2	4.1	4.7	7.94	2.96	3.74

Compound	L_3	ΔE_1	$E_{1,h}$	E_1	$\Delta E_{\Gamma X}$	$\Delta E_{\Gamma L}$	m_c/m_0	$P^2/2m_0$
GaP	7.37	4.69	3.57	3.89	−0.16	−0.11	—	—
GaAs	7.11	4.07	3.29	3.13	0.5	0.3	0.063	21.1
GaSb	6.45	2.75	2.76	2.41	0.19	0.16	0.045	14.9
InP	7.08	3.33	3.03	3.17	0.98	0.5	0.072	17.4
InAs	6.81	2.74	2.85	2.61	1.02	0.72	0.022	15.6
InSb	6.32	2.07	2.43	2.16	0.32	0.36	0.013	15.2

In this arrangement, a type C atom may have all type A neighbors or all type B neighbors, but on the average will have x type A neighbors and $1 - x$ type B neighbors. In effect, the structure is a fcc structure of mixed A–C and B–C molecules, complete with interpenetrating molecular bonding. This structure composes a "pseudobinary" alloy with the properties determined by the relative concentrations of A and B atoms.

In recent years, quaternary alloys have also appeared as $A_xB_{1-x}C_yD_{1-y}$. Here, C and D atoms share the first sublattice. This can also be considered as a pseudobinary alloy of $A_xB_{1-x}C$- and $A_xB_{1-x}D$-type complex molecules. Then a general theory of pseudobinary alloys can be applied equally as well to quaternaries as well as to ternaries. If these compounds are truly smooth mixtures, then the alloy theory holds, but if ordering occurs, we can expect changes. For example, $In_{0.5}Ga_{0.5}As$ may be a smooth alloy composed of a random mixture of InAs and GaAs. However, if perfect ordering were to occur, the crystal structure would become a chalcopyrite—a superlattice on the zinc blende structure. In this latter case, we would expect changes to occur in the band structure due to Brillouin-zone folding about $(0, 0, \pi/2c)$. For the present, we will assume no ordering.

Consider a pseudobinary disordered lattice of A–C and B–C molecules on a Bravais lattice. We sum the crystal potentials over the A and B atoms as

$$U(\mathbf{r}) = \sum_A U_A(\mathbf{r} - \mathbf{\tau}) + \sum_B U_B(\mathbf{r} - \mathbf{\tau}) \tag{141}$$

where $\mathbf{\tau}$ defines the lattice sites of the A–B sublattice. We can now decompose $U(\mathbf{r})$ into a virtual crystal (symmetric) part

$$U_1(\mathbf{r}) = \sum_\tau [C_A U_A(\mathbf{r} - \mathbf{\tau}) + C_B U_B(\mathbf{r} - \mathbf{\tau})] \tag{142}$$

and an antisymmetric (random) potential

$$U_2(\mathbf{r}) = \sum_\tau [U_A(\mathbf{r} - \mathbf{\tau}) - U_B(\mathbf{r} - \mathbf{\tau})] C_\tau, \tag{143}$$

where $C_\tau = 1 - C_A$ for an A atom at τ and $-C_A$ for a B atom at τ (Flinn, 1956). The virtual crystal potential [Eq. (142)] is the basis for smoothly varying the properties from the A–C crystal to the B–C crystal.

The function C_τ was introduced by Flinn (1956) primarily to discuss local ordering in binary alloys. It has the properties

$$\sum_\tau C_\tau = 0, \quad \sum_{\tau'} C_{\tau'} C_{\tau+\tau'} = NC_A(1 - C_A)\alpha_\tau \tag{144}$$

where α_τ is the Warren–Cowley order parameter. The matrix element of U_2 is

$$|\langle \mathbf{k}'|U_2|\mathbf{k}\rangle|^2 \simeq \sum_\tau \alpha_\tau I(\mathbf{k}', \mathbf{k}, \tau) \tag{145}$$

where I is an overlap integral. For truly random solutions, all $\alpha_\tau = 0$ for $\tau \neq 0$, so that

$$|\langle \mathbf{k}'|U_2|\mathbf{k}\rangle|^2 \simeq C_A(1 - C_A)\alpha_0 I(\mathbf{k}', \mathbf{k}, 0) \tag{146}$$

represents a perturbation on the virtual crystal. This perturbation is the so-called "alloy scattering" effect.

Hall (1959), however, has shown that even short-range order can invalidate Eq. (146). If Eq. (145) is used in an energy calculation, it yields only pair interactions with α_τ giving the number of pairs separated by τ. Here the A–A, A–B, B–B pairs have the weighting specified by C_τ. In this case, α_0 corresponds to a self-energy correction, rather than to a scattering term. This self-energy term can in fact work to stabilize an ordering trend in the alloy, a point to which we return below. With these limiting thoughts, we now turn to the virtual crystal approximation for the alloy.

Van Vechten and Bergstresser (1970) have shown that the dielectric method discussed above can be applied to the alloy problem. In the spirit of the virtual crystal approximation, the various parameters (C, D, $E_{0,h}$, $E_{i,h}$, etc.) are assumed to vary linearly with composition just as the crystal potential in Eq. (142) varies linearly with composition. In Fig. 25, we illustrate this for $\text{In}_{1-y}\text{Ga}_y\text{As}$ by plotting the direct and indirect gaps for this material. We also use Eq. (137) to estimate the effective mass in the alloy.

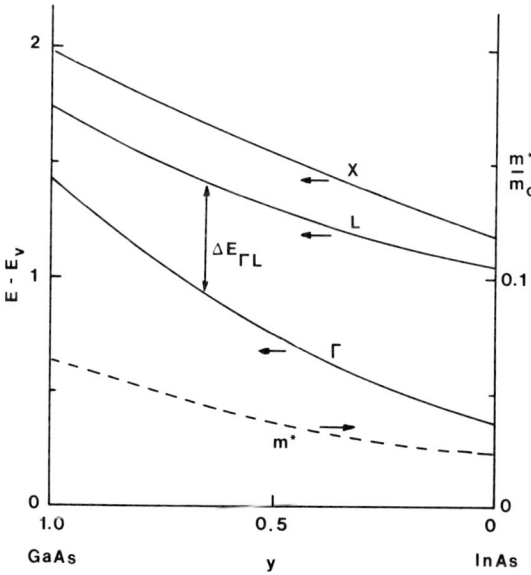

FIG. 25. Variation of the principal direct gaps and effective mass in the alloy InGaAs.

The pure compound data are taken from Table IV. The triangle at $y = 0.45$ is data from Sankarar et al. (1976). The accepted gap variation in this compound is (Woolley et al., 1961)

$$E_0 = 0.35 + 1.07y - 0.28y(1 - y) \tag{147}$$

although some (Wu and Pearson, 1972) give a bowing parameter of 0.46 eV rather than 0.28 eV. The difference, however, is small and Eq. (147) agrees with Fig. 25 to within a few percent over the entire alloy range.

The L-valley separation from the Γ minimum is only about 0.72 eV on the InAs side ($y = 0$). This is considerably lower than some had earlier estimated (Littlejohn et al., 1977), but this is in general agreement with the experimental results of Kwan and Woolley (1968) and El-Sabbahy and Adams (1979) and is in keeping with the generally lower values of $\Delta E_{\Gamma L}$ found, e.g., in GaAs than earlier expected.

In the quaternary alloys, it is necessary to extrapolate the band-gap and lattice-constant data from that of the ternaries. There are many possible quaternary materials. However, to grow these properly it is necessary to lattice match the alloy to a suitable binary substrate, typically InP or GaAs. In Fig. 26, the various binaries are plotted on a band-gap versus lattice-constant plane. Now, InGaAsP can be lattice matched to either GaAs or InP, whereas InGaAsSb can be lattice matched to InP. Both of these quaternary compounds have been suggested for high-speed (or high-frequency) devices. For reasons to be discussed below, only $In_{1-y}Ga_yAs_xP_{1-x}$ will be considered further, and this is only considered for the case of lattice-matching to InP.

Vegard's law provides for a linear extrapolation of lattice constant in the alloy. Moon et al. (1974) have applied this to $Ga_yIn_{1-y}As_xP_{1-x}$, with the result

$$a_0 = 5.87 + 0.18x - 0.42y + 0.02xy \text{ Å} \tag{148}$$

For lattice matching to InP, we require

$$0.18x - 0.42y + 0.02xy = 0 \tag{149}$$

or

$$y = 0.43x/(1 - 0.048x) \tag{150}$$

while $0 < x < 1$ (reference to Fig. 26 shows that the entire range of x can be covered while only a portion of the range of y can be covered). While Eq. (150) indicates that InGaAs matched to InP has $y = 0.45$, more recent indications are that $y = 0.47$ is more accurate, but we will not dwell further on this. The dielectric theory can be equally applied to the quaternary, and

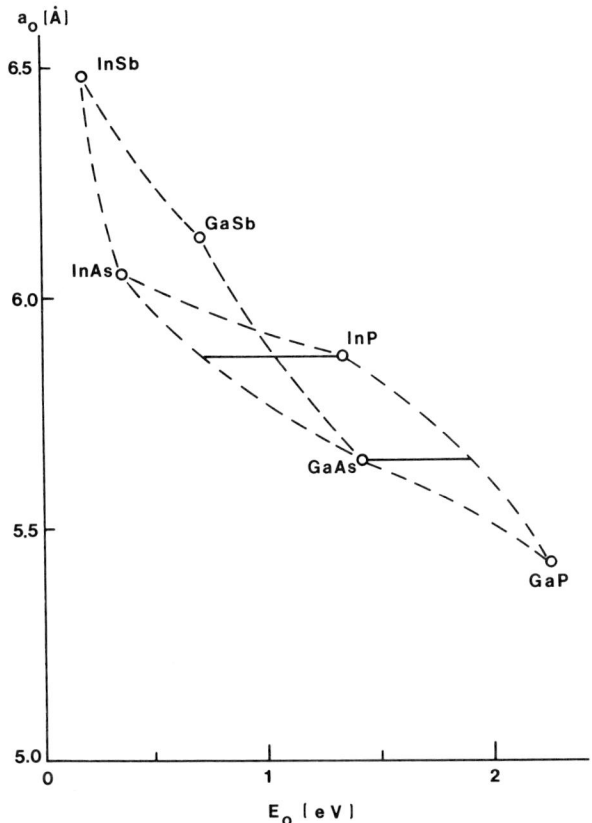

FIG. 26. The values of band-gap versus lattice-constant planes for the various binaries are plotted. The solid lines represent the quaternary lattices matched to InP and GaAs.

the direct and indirect gaps are shown in Fig. 27 along with the effective mass. Moon et al. (1974) give the direct gap as

$$E_0 = 1.35 + (0.101x - 1.101)x + (0.758 - 0.28x)y(1 - y)$$
$$- (0.101 + 0.109y)x(1 - x) \qquad (151)$$

and this agrees with Fig. 27 to within 3% over the range of x.

For comparison, the energy gap data of Grinyaev et al. (1980) are shown in Fig. 27 as they represent nonlocal pseudopotential calculations for reference. The agreement is good, not only for the direct gap but also for the indirect gap to the X-point. Many experimental measurements have been made of the direct gap and the spread in these measurements is more than the

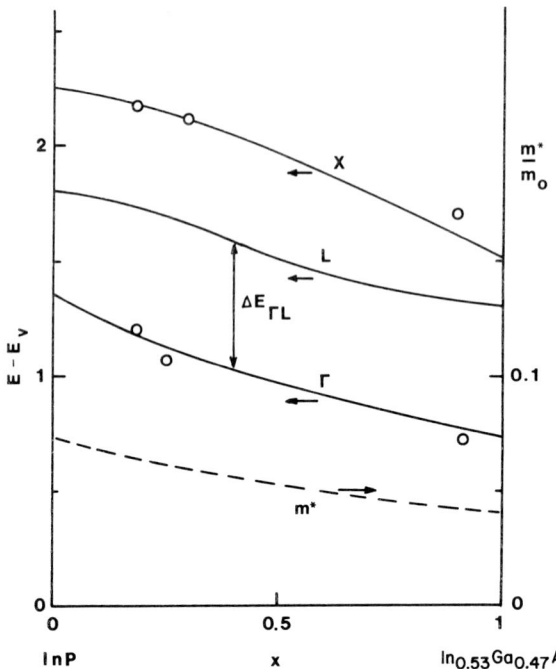

FIG. 27. Variation of the principal direct gaps and effective mass of the quaternary alloy InGaAsP lattice matched to InP.

error listed above in the theory, so all told the fit is exceedingly good. Further, the $E_1(L_3{}^v-L_1{}^c)$ transitions agree with the predictions of the dielectric method to within 5% for the three values calculated by Grinyaev et al. This suggests that the overall estimate of the dielectric method is within 10% for the indirect gaps. This is significant for the following reason. Littlejohn et al. (1977) have suggested that $Ga_{0.27}In_{0.73}P_{0.4}As_{0.6}$ would be a better material than GaAs or InP due to a much higher peak velocity (as a function of electric field). However, in their calculations, they estimated $\Delta E_{\Gamma L}$ to be 0.82 eV in this alloy composition. From Fig. 27, however, $\Delta E_{\Gamma L} = 0.55 \pm 0.05$ eV over the entire composition range. This lower value of $\Delta E_{\Gamma L}$ would result in a peak velocity lower than that expected, as intervalley transfer would occur much sooner. Now, lattice matching to GaAs would not improve this as the higher GaP concentration would lower the satellite valleys (GaP is indirect). Similarly, the choice of InGaAsSb is no better due to the lower satellite valley separation in GaSb. Consequently, from considerations only of the separation of the L-valley from the central valley, no quaternary alloy is expected to yield great improvements, while from considerations of the separation and the effective mass, it appears that only the ternary end of the quaternary alloy, $In_{0.53}Ga_{0.47}As$, has any advantage over InP. However, reference to Table I shows that the limiting velocity in the central valley is lower in the alloy than in the pure InP, a consequence of the much lower phonon energy involved in scattering.

D. Alloy Ordering

The basis of ordering in otherwise random alloys lies in the fact that the ordered lattice, whether it has short-range or long-range order, may be in a lower energy state. In a random alloy $A_xB_{1-x}C$, the cohesive energy will change by

$$E_{coh} = E_{coh}^B + x(E_{coh}^A - E_{coh}^B) \tag{152}$$

but the total cohesive energy does not change; i.e., while the A–C compound is losing energy, the B–C compound is gaining energy. Thus the cohesive energy of the random alloy is a simple average of its constituents. For example, in $In_{0.5}Ga_{0.5}As$, the cohesive energy is the average of those of InAs and GaAs, but the gain of energy by InAs in the alloy is exactly offset by the loss by GaAs, at least within the virtual crystal approximation which leads to the linear extrapolation of Eq. (152).

If at least short-range order persists, however, this argument does not hold any longer. Rather, the ordered GaAs regions undergo a loss of energy as their bonds are stretched in the alloy, while the ordered InAs regions gain energy as the bonds are compressed (here we mean gaining energy in the sense that the crystal is compressed and the equilibrium state has a lower energy). Since the cohesive energy varies as d^{-2} (Harrison, 1980), a net increase in cohesive energy can result for the ordered alloy. Now, the concept of cohesive energy in semiconductor crystals is a very simplified one, but the dielectric method can be used to calculate an "average" energy of the valence electrons (Van Vechten, 1968). This is given by

$$E_{av} - E_b = E_F[3/5 + 3B^2(1 + \ln B/2) - 4B^3] \tag{153}$$

where B is given in Eq. (125) and E_b is the absolute level of the bottom of the valence band, i.e., the Γ_1^v level in Figs. 22 and 23. Thus, we can use (153) if we can find the total valence bandwidth $\Gamma_{25'}^v - \Gamma_1^v$.

The valence bandwidth is just adequate to account for the $4n_0$ electrons, where n_0 is the atomic density, and thus must be related to the Fermi energy E_F. In practice, $E_{val} = \Gamma_{25'}^v - \Gamma_1^v = E_F$ for silicon and diamond, but is increased in Ge due to mixing of the wave functions with d-core states. Thus, we can use a form similar to Eq. (132) if we identify $E_{val,h}$ with E_F, so that

$$E_{val} = [E_F + (D-1)\Delta E_{\Gamma_1}]/[1 + (C/E_{val,h})^2]^{1/2} \tag{154}$$

TABLE V

Material	$\Gamma^v_{25'}-\Gamma^v_1$ (eV)	$E_{av}-\Gamma^v_1$ (eV)	E_{coh} (eV)	Material	$\Gamma^v_{25'}-\Gamma^v_1$ (eV)	$E_{av}-\Gamma^v_1$ (eV)	E_{coh} (eV)
Si	12.5	6.75	2.32	GaSb	11.2	5.18	1.48
Ge	12.7	5.59	1.94	InP	11.4	5.66	1.74
AlAs	11.9	6.09	1.89	InAs	11.3	5.2	1.55
GaP	12.4	6.37	1.78	InSb	10.6	4.6	1.4
GaAs	12.3	6.0	1.63				

Note that here the heteropolar correction is in the denominator due to the tendency of ionic bonding effects to narrow the valence band (just as they increase the energy gaps). Using Eq. (154), the calculated values of E_{val} may be found, along with the cohesive energies from Harrison (1980). These values are shown in Table V for several of the diamond and zincblende semiconductors of interest.

In Table VI, we list the changes in average energy for the ordered alloy as compared with the random alloy for $x = 0.5$ (equal concentrations). Recall that a decrease of the value of E_{av} or an increase of E_{coh} indicates that ordering in the alloy is energetically favored. The data on GaAlAs is mixed, but the energy change is so small that ordering would occur only at low temperatures (below 25 K), if at all. On the other hand, if ordering were incipient in this alloy, the self-energy correction discussed above could work to stabilize the ordered structure at much higher temperatures. In this case, only total energy calculations can shed much light on the question, and these are nonexistent at this time.

In the case of InGaAs, GaInSb, and InAsP, however, ordering is favorable, although in the latter case the transition temperature may be well below room temperature, depending upon which data are the most accurate. In the other ternaries, ordering does not appear to be energetically favorable.

If ordering does occur, one axis in the Brillouin zone will have a new zone-edge boundary (only for long-range ordering though), and this will

TABLE VI

Alloy ($x = 0.5$)	ΔE_{av} (meV)	ΔE_{coh} (meV)	Alloy ($x = 0.5$)	ΔE_{av} (meV)	ΔE_{coh} (meV)
GaAlAs	−1.0	−2.1	InGaP	652	−110
InGaAs	−89	23.4	GaAsP	101	−18.5
InAsP	−48	8.2	InAsSb	122	−17.8
GaInSb	−137	29.7	GeSi	4	−32.9

occur at $\pi/2c$, as opposed to π/c ($a = b = c$ in the zincblende structure if there is no tetragonal distortion in the ordered structure, although this latter occurs in chalcopyrites such that $c \simeq 0.9 - 0.98a$). The consequence of this is likely to be negligible on the transport properties as the compounds of interest are direct-gap materials and the zone folding is unlikely to change the satellite valley separations, band gaps, or effective masses to any noticable degree. However, the lattice dynamics will be modified and can be probed by Raman scattering. While the data on InGaAs is mixed, it appears to be a single LO-mode system and thus is probably a smooth alloy (Pearsall, 1981). However, InAsP shows two strong LO modes and is more likely ordered, at least on the short-range, local atomic level (Nicholas et al., 1979) and this behavior carries into the quaternaries (Portal et al., 1979; Pinczuk et al., 1978). This tendency to order may be the basis of the suggestion that a region of solid immiscibility of the alloy exists in the region $0.7 < x < 0.9$ (Marsh et al., 1981). Support for this conjecture is found in the work of Pinczuk et al. (1978) as smooth phonon structure is not observed; rather, a broad structure is found in the Raman scattering for this alloy range.

VI. Comparisons and Limitations of Logic for Ultra-Large-Scale Integration

The general study of scaling of MOS silicon devices, in order to gain improvements in performance (speed) and packing density, has historically centered on the *n*-channel device. However, integrated circuits operating at high data rates, in the gigabit-per-second region, are of interest for many applications. It is this latter driving force that has led to the interest in GaAs circuits, yet silicon is not out of the picture in this region. It is the main purpose of this last section to compare the prospects of high-speed logic, using silicon and the various III–V compounds, in a meaningful manner such that some trends can be drawn. The central result of the previous chapters lies in the necessary requirements of any logic system to be able to switch fast. Obviously, the devices must be of reduced size if they are to achieve high speeds. To this end, we will adopt an effective channel length of 0.12 μm and a gate width of 0.5 μm as standards in our comparisons. While of very small dimensions, devices of this size are no more than a factor of two beyond present laboratory devices.

It is well worth asking as to the expected validity of the device models developed in Sections II and III. These models are essentially long-channel models with appropriate corrections included for the *known* short-channel effects. And yet, the MOS models give quite good qualitative descriptions of devices in the 0.2–0.5-μm channel length regime (Elliott et al., 1979; Watts *et al.*, 1980; Lepselter, 1980; Fichtner *et al.*, 1980). While agreement is not as good with a 0.1-μm channel length device (Hunter *et al.*, 1981), this latter device is believed to be dominated by source and drain parasitic resistances which restrict the current. Similarly, the JFET–MESFET models agree well with devices in the 0.5–1.0-μm gate length region (Suyama *et al.*, 1980; Nuzillat *et al.*, 1980), and extrapolation to small gate lengths is expected to be more valid here than in the MOS devices. One may therefore conclude that these models give a good qualitative description of devices on the somewhat smaller scale, and it is in this spirit that we proceed. It should be remembered, however, that source–drain parasitic resistances have not been included, so that the models are for the intrinsic device and expectations of agreement beyond the qualitative level are not justified. Yet, the qualitative nature does allow a valid quantitative comparison between different devices to be made.

In the following, we examine the expected performance of our standard 0.12×0.5-μm gate devices of various technologies. Then we examine limitations on speed and packing density which arise from thermal considerations. Finally, we spend considerable time discussing a new problem which arises from the parasitic coupling between interconnection lines; a problem which could have profound impact on the prospects of ULSI as we know it.

A. Device Parameters

1. The MOS Device

The general argument in favor of silicon generally rests on the basis of the higher thermal conductivity of silicon and the well-established technology in this material. Yet, for channel lengths of 0.12 μm, the devices are below the effective maximum drive current capability (Chatterjee *et al.*, 1980). It is in this regime that the higher drive current capability of a material such as InP can lead to enhanced performance capability. Moreover, the carrier velocity is at its saturated value over much of the region of operation of these devices, and the higher effective saturated velocity of the III–Vs offers some advantage. In this regard, we consider only Si and InP for the MOS devices. For this device, as well as those that follow, quaternaries and ternaries are not expected to offer any advantages over InP, as discussed in Section V. Therefore, they will not receive extensive discussion.

The effective 0.12-μm-channel-length MOSFET is characterized as follows: a substrate doping of 7×10^{16} cm^{-3} is assumed, the source and drain regions are 10^{18} cm^{-3} with a depth of 0.15 μm. Drain currents for $V_G = V_D$ are calculated from the equations of Section II. From, Section IV, it is

TABLE VII

MOS DEVICES

Parameter	Si		InP	
V_{DD}	1.5	1.5	1.0	0.75
V_B	0	−1.0	−1.0	−1.5
V_T ($V_D = V_{DD}$)	0.27	0.16	0.33	0.2
I_{dm} (μA)	14.6	52.4	20.9	16,7
τ_D (psec)	82.0	23.0	38.0	36.0
$P_D\tau_D$ (fJ)	0.36	0.36	0.16	0.09
f_m (GHz)	2.4	8.7	5.2	5.6

ascertained that the node capacitance is largely dominated by the interconnection capacitance. This latter is taken as 0.6 fF, a reasonable value as scaled from the switching behavior of larger devices, and the total node capacitance for the MOS device is 0.8 fF. (The interconnection capacitance is kept at the above value for all of the devices considered in order to provide a meaningful comparison.) The effective saturated velocity is taken from Fig. 19. The calculated parameters are given in Table VII and Fig. 28. It is clear from the falloff of the performance of the InP device when the drain voltage is reduced from 1.5 V to 1.0 V that the higher threshold is holding back this

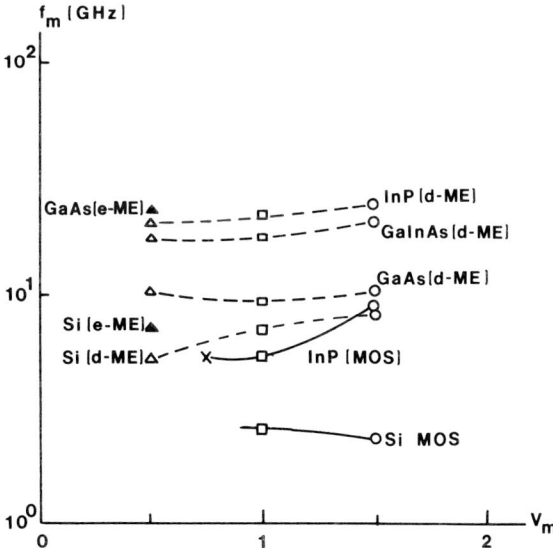

FIG. 28. Maximum operating frequency, as a function of operating logic voltage swing, for the various technologies considered here.

device and better performance might be obtained at higher values of the substrate bias.

2. *The d-MESFET Device*

The depletion MESFET in GaAs is currently the leading candidate to supplement silicon MOS circuits in VLSI. As in the MOSFET, the carrier velocity is at its saturated value for these devices. We consider here d-MESFETs fabricated in Si, GaAs, InP, and $In_{0.53}Ga_{0.47}As$. Although the discussion of Section V suggested that the ternaries and quaternaries will not produce behavior comparable to InP, the ternary is included here to further illustrate this point.

The devices are considered to be fabricated on an 0.08-μm epitaxial (or implanted) layer, except in the 0.5-V device where a 0.06-μm epitaxial layer is assumed. This layer is doped to 4×10^{17} cm^{-3}. The equations of Section III are used to calculate the drain characteristics, and again the saturated velocities are taken from Fig. 19. The calculated parameters are shown in Table VIII and Fig. 28. Gate metallizations are taken to be Au on Si and Al on the remainder.

It is clear from these results that the operating region is not as favorable to the GaAs device as it might otherwise be, as the offset voltages are assumed to be the same for each device. The fact that V_{bi} in the GaAs device is a much larger fraction of the V_p value works against this device. In fact, in

TABLE VIII

d-MESFET Devices

V_m	C_N (fF)	$P_D\tau_D$ (fJ)	Parameter	Si	GaAs	InP	InGaAs
			V_{bi}	0.55	0.8	0.45	0.4
1.5	0.8	0.36	V_p	1.98	1.85	1.87	1.71
			I_{dm} (μA)	48.0	63.0	154.0	125.0
			τ_D (psec)	25.0	19.0	7.8	9.6
			f_m (GHz)	8	10.5	26.0	21.0
1.0	0.78	0.16	V_p	1.98	1.85	1.87	1.71
			I_{dm} (μA)	28.0	24.0	88.0	70.0
			τ_D (psec)	28.0	33.0	8.9	11.0
			f_m (GHz)	7.0	6.0	23.0	18.0
0.5	0.75	0.038	V_p	1.11	1.04	1.05	0.96
			I_{dm} (μA)	10.0	12.0	40.0	35.0
			τ_D (psec)	38.0	33.0	9.4	11.0
			f_m (GHz)	5.3	6.0	21.0	18.7

this size range (of device), the smaller values of built-in potential are a decided advantage. Moreover, it should be noted that the d-MESFETs offer approximately a factor of 2–2.5 improvement in speed over the MOSFET, primarily due to the greater value of I_{dm}. We examine this further below.

3. The e-MESFET Device

The enhancement-mode MESFET devices are constrained to operate at a small V_m due to the necessity of using V_{bi} to deplete through to the substrate. While circuit design can be improved to partially alleviate this problem, such effects would take the device away from its optimum operating region. Because of the low voltage, these devices inherently possess a low speed–power product. However, the low built-in potential of the In compounds essentially works against these materials, although junction FETs could be used. Therefore, only Si and GaAs are included in the tabulated data, although an InP device is illustrated later. Here, we assume an epitaxial layer doped to 10^{17} cm^{-3} and 0.08 μm thick. Operation at 0.5 V is assumed and the results are given in Table IX and Fig. 28. A total node capacitance of 0.8 fF results and yields a speed–power product of 0.035 fJ.

A surprising result of these calculations is that the enhancement MESFET, on this dimensional scale, is as fast or faster than the depletion-mode device. In larger devices, the opposite is true (Liechti, 1977). This latter follows primarily from the larger voltage swing and hence larger drive currents available in the d-MESFET. At the smaller dimensions, however, the enhancement-mode devices actually utilize a larger portion of the epitaxial layer and are thus more efficient. This result could have been predicted from available data on 0.5–0.6-μm devices, as Mizutani et al. (1980) achieved a gate delay of 30 psec in a 0.6-μm-gate-length enhancement-mode device, while Greiling (1979) achieved a gate delay of 34 psec in a depletion-mode device with 0.5-μm gate length, both devices being fabricated in GaAs. While it is difficult to directly compare devices prepared at different sites, these results certainly serve as a precursor to the results of Tables VIII and IX. Below gate lengths of 0.5 μm, the enhancement-mode device is expected to be faster than the depletion-mode device of the same or comparable logic voltage swing.

4. Power-Delay Considerations

In Fig. 29, the results of Fig. 28 and Tables VII–IX are replotted on a power versus delay time plot. One clear point from these results is the dominance of the interconnection capacitance. This results in an essentially constant node capacitance throughout the various technologies and the speed–power product is thereby governed by the operating voltage. However, the

TABLE IX

e-MESFET Devices

Parameter	Si	GaAs
V_{bi}	0.7	0.8
V_p	0.49	0.46
I_{dm} (μA)	13.0	41.0
τ_D (psec)	27.0	8.6
f_m (GHz)	7.4	23.0

speed (delay) itself is not so limited, and the technologies which can achieve considerably enhanced drive currents can be exceedingly fast, with delay times in the 10-psec range. These times are clearly fast enough to compete favorably even with Josephson technology (Gheewala, 1980).

B. Thermal Considerations

It is often pointed out that high-speed circuits are usually constructed so that an appreciable fraction of the circuits are running at top speed concurrently. As a consequence, the power into the chip can be rather large, so

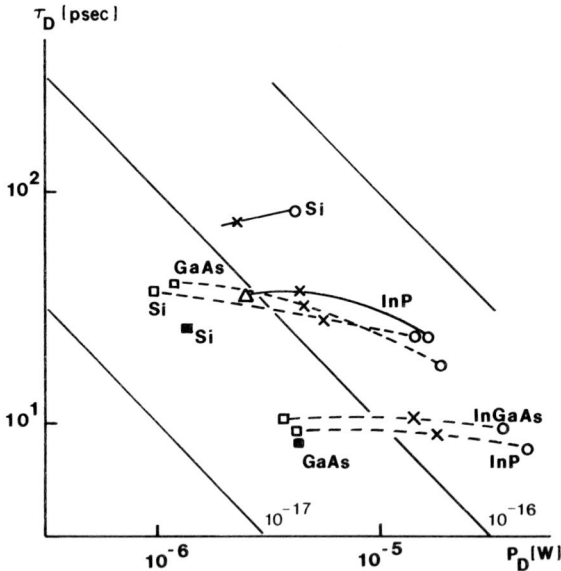

FIG. 29. Comparison of the various logic technologies in terms of speed and power: (○) 0.5 V, (×) 1.0 V, (△) 0.75 V, (■ □) 0.5 V. (—) MOS, (---) d-MES, (■) e-MES.

that the top speed is limited by power dissipation rather than intrinsic device speed. Where this is true, it may be expected that the superior thermal conductivity of silicon, with respect to the III–V compounds, will allow higher clock rates for the silicon devices if the circuit is packaged with proper thermal design considerations. This is generally true if the logic circuits are constrained to the same operating voltages. However, if one can increase I_{dm} by appropriate technology or material choice, a smaller τ_D can be achieved without expense to $P_D\tau_D$, therefore giving a higher f_c. On the other hand, a technology giving a higher g_m (from higher I_{dm} at a given drive voltage V_m) can achieve a given drive current level at a lower value of V_m, hence giving a smaller τ_D and a considerably smaller $P_D\tau_D$, even if the node capacitance remains the same. It is apparent that these trade-offs might result in performance from the III–Vs that is improved *for comparable packing densities*, even though the thermal conductivity may be inferior to that of silicon.

The power dissipation in a logic chip, operating at a frequency f_{op}, is found from (73) to be

$$P = 2N_g f_{op}(P_D\tau_D) \tag{155}$$

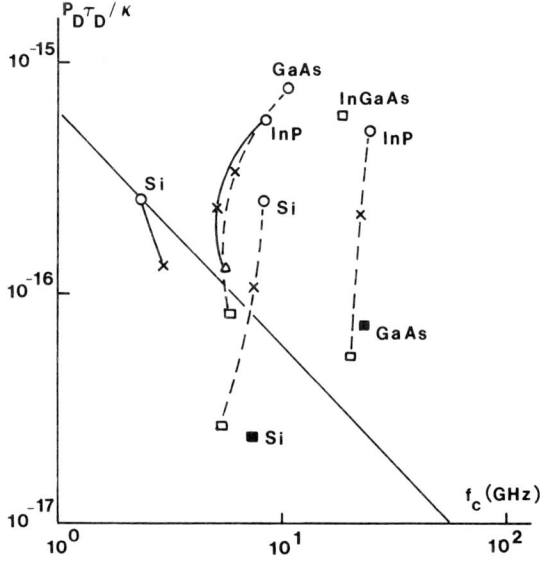

FIG. 30. The performance of the various technologies on a power-delay versus maximum frequency plot, with normalization on the thermal conductivity. According to Eq. (156) this should give a constant (solid curve) for a given power per gate. (○) 1.5 V, (×) 1.0 V, (△) 0.75 V, (■ □) 0.5 V. (—) MOS, (---) d-MES, (■) e-MES.

If we assume that an equal packing density N_g of gates cm^{-2} is used, the power per gate is

$$P/N_g = 2f_{op}(P_D\tau_D) \tag{156}$$

In Fig. 30, the power–delay product is plotted against f_c, the maximum frequency allowed by τ_D ($1/5\tau_D$). However, the power–delay product is normalized to a thermal conductivity of 1 W K^{-1} cm^{-1} in order to compare various materials. The individual thermal conductivities are taken from the review of Holland (1966) and are listed in Table X. The solid curve is a curve of constant $f_{op}(P_D\tau_D)/\kappa$ at the value corresponding to the silicon MOSFET with $V_{DD} = 1.5$ V. Only the silicon MESFETs with $V_m = 0.5$ V and GaAs d-MESFET with $V_m = 0.5$ V fall below this line. These latter technologies offer improvements over the silicon MOSFET, on an absolute basis, even when operating at their maximum frequencies, a result of the dramatically lower $P_D\tau_D$ achieved when operating at $V_m = 0.5$ V.

The results of Fig. 30, however, are for operation at the maximum possible frequency, a limit set by τ_D. Is it possible to reduce f_{op} slightly and thus achieve a performance superior to silicon? In Fig. 31, we plot f_{op} versus the operating logic voltage V_m. Here, the individual technologies are assumed to have the same packing density N_g and to be limited to a power level equal to κ/κ_{Si} of the equivalent silicon MOSFET device operating at $V_{DD} = 1.5$ V. The dashed portions of the curves are where f_{op} is limited by f_c, the maximum frequency for a particular technology. Clearly, InP offers superior performance, at the same packing density, for operation with $V_m < 1.0$ V, regardless of the technology implemented. Similarly, GaAs MESFETs are also suitable for $V_m < 0.8$ V, as is the silicon MESFET. The ternary is observed to be totally unsuitable for any operating conditions.

The superior thermal conductivity of silicon is therefore found to be a good, but perhaps misleading indicator. The improved speed–power product achieved by lowering the voltage to 0.5–0.67 of the level in silicon n-MOS allows the superior performance, in terms of operating speed, to be achieved while maintaining the same packing density of the VLSI system.

TABLE X

Material	κ (W K^{-1} cm^{-1})
Si	1.5
GaAs	0.46
InP	0.68
In$_{0.53}$Ga$_{0.47}$As	0.06

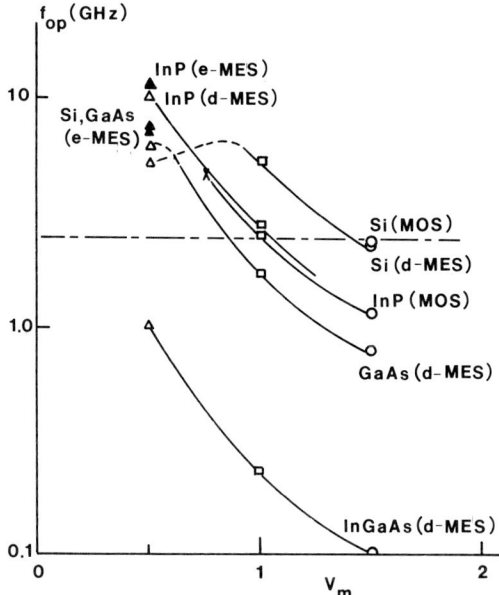

FIG. 31. Peak operating frequencies achieved in various technologies when designed with the same packing densities of the equivalent silicon circuit.

This packing density is quite high. Indeed, the dashed line in Fig. 31 represents, for only a 1 K cm^{-1} rise through the wafer, a packing density of 8.2×10^5 gates cm^{-2} or 5.3×10^6 gates in.$^{-2}$. If a temperature rise of 10°C is allowable across a 10-mil-thick wafer mounted with proper thermal design, then this level represents a packing density of more than two billion gates in.$^{-2}$, which is in principle achievable, assuming of course that a reasonable yield is also achievable.

C. Further Considerations

1. Cooled Logic

Schottky-gate FETs operated at a temperature of 77 K have been suggested as having power–delay products which make them worth investigating for low–power, high–speed logic applications (Rees et al., 1976). At this temperature, reduced voltage operation results in higher mobility and higher saturated velocity, so that f_c can be expected to improve. In the FETs discussed above, the small gate length and relatively high source–drain potential is already sufficient to raise the electron velocity to its saturated value. In cold FETs, however, velocity saturation occurs at smaller values of the electric field because of the generally higher mobility and enhanced carrier heating. In silicon devices, the saturated velocity of electrons is approximately the same at 77 K as at room temperature, a result expected from Eq. (110). This is not the case in the III–Vs however, as can be seen from Eq. (111). Rather, v_s increases by some 20–30%. Moreover, the retardation effects of Section IV,C occur over a larger distance so that the effective saturation velocity can rise much more rapidly than Fig. 19 indicates. As a consequence, performance advantages of the III–V compounds are magnified if cooling to 77 K is adopted.

2. Device–Device Interactions

In Section IV,E, the problem of line-to-line parasitic capacitance was introduced. For device sizes in the 0.12-μm or less region, the line-to-line parasitic capacitance begins to dominate the direct-line capacitance in setting the node capacitance. This parasitic capacitance leads to a *direct device–device interaction outside of the normal circuit or architectural design*. The constraints imposed by device–device interactions will have to be included in future architectural design of compact VLSI systems, and this will most easily be accomplished if these constraints are reflected in the system theory description of the architecture itself.

In conventional descriptions of LSI circuits, each device is assumed to behave in the same manner within the total system as it does when it is isolated. The full function of the system is determined once the interconnection matrix is specified to join the individual devices together. A different function can only be assigned to the system by redesigning the interconnecting metallizations—a practical impossibility for most systems, but in practice accomplished in PLA modules. The conventional clean separation of device design from system design thus depends ultimately on being able to isolate each individual device from the environment of the other devices except for planned effects occurring through the interconnection matrix. This simplification is likely to be seriously in error for submicron-configured VLSI systems, where the isolation of one device from another (and by generalization, from the surrounding environment of metallizations) will be difficult to achieve.

The possible device–device coupling mechanisms are numerous and include such effects as capacitive coupling, of which the line-to-line parasitic capacitance is one example, and wave-function penetration (tunneling and charge spillover) from one device to another. Formally however, one may describe these effects on system and individual device behavior by assuming the simplest form of interdevice coupling.

In the following, the basic principles of a system restructuring that can occur are illustrated by a simple circuit theoretic analogy utilizing component-connection type approaches. First, a special case of isolated devices is developed. Then a connection function (such as the network of desired metallizations for interconnecting logic gates) is introduced to describe the system in terms of the devices and to show how the properties of the connection function can alter the system's dynamics. As the individual device dynamics and connections will be nonlinear, it may be expected that although the equations used here are linear, the general result will admit of synergetic responses for the system.

First, the state equations are examined for an isolated integrator with input–output conditioning. The applicability of this example is examined later. The state equations for the ith device are then

$$\dot{u}_i = a_i u_i + b_i y_i, \qquad z_i = c_i u_i \tag{157}$$

where u_i is the state variable and y_i and z_i are the input and output variables for the single device, respectively. For an ensemble of N devices, these become

$$\dot{\mathbf{U}} = \mathbf{A}\mathbf{U} + \mathbf{B}\mathbf{Y}, \qquad \mathbf{Z} = \mathbf{C}\mathbf{U} \tag{158}$$

where $\mathbf{A}$, $\mathbf{B}$, and $\mathbf{C}$ are square diagonal matrices and $\mathbf{U}$, $\mathbf{Y}$, and $\mathbf{Z}$ are column matrices. Solving for the transfer function of the single devices gives (in the Laplace transform domain with relaxed initial state)

$$\mathbf{Z} = \mathbf{C}(s\mathbf{I} - \mathbf{A})^{-1}\mathbf{B}\mathbf{Y} \tag{159}$$

So far, it has been considered that each device was isolated from the others. If we describe the connection matrix $\mathbf{F}$ through the ansatz (Ransom and Saeks, 1975)

$$\mathbf{Y} = \mathbf{F}\mathbf{Z} + \mathbf{L}\mathbf{G}, \qquad \mathbf{H} = \mathbf{M}\mathbf{Z} \tag{160}$$

where $\mathbf{G}$ and $\mathbf{H}$ are the total system input and output matrices and $\mathbf{L}$ and $\mathbf{M}$ are conditioning matrices, the connection matrix $\mathbf{F}$ generally describes how the input of a particular device is related to the outputs of other devices, and is thus the system representation of the metallization interconnect chosen for a particular VLSI chip. Equations (159) and (160) can then be combined to yield the system transfer function as

$$\mathbf{H} = \mathbf{M}\mathbf{C}\mathbf{B}(s\mathbf{I} - \mathbf{A} - \mathbf{F}\mathbf{C}\mathbf{B})^{-1}\mathbf{L}\mathbf{G} \tag{161}$$

where we have used the fact that $\mathbf{A}$, and hence $(s\mathbf{I} - \mathbf{A})^{-1}$, is diagonal.

Equation (161) for the system transfer function is a special case of the connection function theory of systems applicable to the integrator. Although we have used this special case, the approach is far more general and is applicable to arbitrary circuits. Further, even though we have assumed an analog signal approach by employing the Laplace transformation for the time variation, the technique is currently extendable to a class of digital circuits, linear sequential circuits, through the description of the system dynamics in the abstract mathematical concept of an extension field (Sangani et al., 1976). However, this simple case is adequate to illustrate the major points discussed above.

The quantity $(s\mathbf{I} - \mathbf{A} - \mathbf{FCB})^{-1} = \mathbf{S}^{-1}$ plays the conceptual role of a resolvent for the system and the zero of $\det\mathbf{S}$ define the various modes of operation. Since $\mathbf{B}$ and $\mathbf{C}$ are diagonal, any deviation of the system response from that defined by $\mathbf{A}$ must arise through the structure of $\mathbf{F}$. For example, if we consider that the system is logically connected, i.e., y_i is connected only to z_j if $j < i$, then $\mathbf{F}$ has elements only in the lower triangle below the main diagonal. Since $\mathbf{A}$ is diagonal, $\mathbf{F}$ does not modify the modes determined by $\mathbf{A}$, i.e., $\det\mathbf{S} = \det(s\mathbf{I} - \mathbf{A})$. Only when $\mathbf{F}$ has entries across the main diagonal does this change. For example, if $y_i = z_{i-1}$, then $\mathbf{F}$ has entries along the diagonal just below the main diagonal. If now the last stage is fed back to the first stage, an entry appears in the upper right corner of $\mathbf{F}$ and one new mode is generated, the collective ring-oscillator mode. This last follows as the last entry altered the topology of the directed graph of $\mathbf{S}$, allowing a new, strongly connected graph to arise.

In general, the connection function $\mathbf{F}$ can be divided into two parts, $\mathbf{F}_1$ and $\mathbf{F}_2$, where $\mathbf{F}_1$ is the portion of $\mathbf{F}$ that represents the desired metallizations, i.e., the designed architectural circuit yielding

$$\mathbf{S}_1 = s\mathbf{I} - \mathbf{A} - \mathbf{F}_1\mathbf{CB} \tag{162}$$

Then $\mathbf{F}_2$ represents the parasitic interactions—the parasitic device–device couplings which arise from, e.g., the line-to-line coupling capacitance. Thus, a new resolvant $\mathbf{S}_2$,

$$\mathbf{S}_2 = \mathbf{S}_1 - \mathbf{F}_2\mathbf{CB} \tag{163}$$

arises with a new set of eigenmodes given by $\det\mathbf{S}_2$. Thus, the structure of the system is altered in the presence of $\mathbf{F}_2$. As $\mathbf{F}_2$ depends upon the states of $\mathbf{U}$ (voltages, for example), as well as the inputs $\mathbf{G}$, it is entirely conceivable that the system is now strongly nonlinear. In large-scale systems, where sizes are more than a micron in scale, $\mathbf{F}_2$ may reasonably be assumed to be negligible. In future VLSI and ULSI systems of submicron dimensions, this is no longer the case, and the presence of $\mathbf{F}_2$ will have to be accounted for in design.

D. Summary

While the rhetorical debate over various technologies is not expected to abate in the near future, a number of conclusions can be drawn from the material presented in the above review and which will have impact on these debates. The major conclusions can be summarized as follows:

(1) Devices fabricated from materials such as GaAs and InP offer definitive speed advantages over silicon circuitry.

(2) For comparable packing densities, this speed advantage can be maintained if operation at logic voltages slightly less than that of the reference silicon n-MOS is acceptable. Indeed, with this mode of operation the superior thermal conductivity of silicon is not a dominating factor, and comparable packing densities can be achieved in the faster III–V-based systems.

(3) The ternaries and quaternaries offer no advantage over InP in comparable devices and are severely restricted in packing density by their exceedingly poor value of thermal conductivity. However, this conclusion is based upon operation in which v_{sat} is achieved. There may be other modes of operation in which performance is dominated by the mobility, and the ternaries can be fruitfully employed in these circuits.

(4) Performance advantages of the III–Vs with respect to silicon will be enhanced if operation at reduced temperatures are acceptable. However, even at room temperature, high-speed semiconductor logic in the submicron region will compete favorably in speed with Josephson junction technology.

(5) The dominance of parasitic line-to-line capacitance in future VLSI–ULSI systems will have important implications for architectural circuit design and may well prove to be the limiting factor in the move to next-generation ULSI.

REFERENCES

Barker, J. R. (1973). *J. Phys. C* **6**, 2663.
Barker, J. R., and Ferry, D. K. (1980). *Solid-State Electron.* **23**, 519.
Barker, J. R., Ferry, D. K., and Grubin, H. L. (1980). *IEEE Electron Dev. Lett.* **edl-1**, 209.
Bonjour, P., Castagne, R., Pone, J.-F., Courat, J.-P., Bert, G., Nuzillat, G., and Peltier, M. (1980). *IEEE Trans. Electron Dev.* **ed-27**, 1019.
Bosch, B. G. (1980). *IEE Proc. I* **127**, 254.
Bose, J. C. (1904). U.S. Patent 775,840.
Braun, F. (1874). *Ann. Phys. Pogg.* **153**, 556.
Bryant, T. G., and Weiss, J. A. (1968). *IEEE Trans. Microwave Theory Tech.* **mtt-16**, 1021.
Carrier, G. F., Krook, M., and Pearson, C. E. (1966). "Functions of a Complex Variable," p. 147. McGraw-Hill, New York.
Chatterjee, P. K., Hunter, W. R., Holloway, T. C., and Lin, Y. T. (1980). *IEEE Electron Dev. Lett.* **edl-1**, 220.
Cook, R. K., and Frey, J. (1981). *IEEE Trans. Electron Dev.* **ed-28**, 951.
Cooper, J. A., Jr., and Nelson, D. F. (1980). *IEEE Trans. Electron Dev.* **ed-27**, 2179.
Das, P., Ferry, D. K., and Grubin, H. L. (1981). *Solid State Commun.* **38**, 537.
Dennard, R. H., Gaensslen, F. H., Yu, H. N., Rideout, V. L., Bassous, E., and Leblanc, A. R. (1974). *IEEE J. Solid-State Circuits* **sc-9**, 256.

Drangeid, K. E., Broom, R. F., Jutzi, W., Mohr, T. O., Mosei, A., and Sasso, G. (1972). *IEEE J. Solid-State Circuits* **sc-7**, 277.
Dresselhaus, G., Kip, A. F., and Kittel, C. (1955). *Phys. Rev.* **98**, 368.
Eastman, L. F., and Shur, M. (1979). *IEEE Trans. Electron Dev.* **ed-26**, 1359.
Eden, R. C., Welch, B. M., Zucca, R., and Long, S. I. (1979). *IEEE Trans. Electron Dev.* **ed-26**, 299.
Elliott, M. T., Splinter, M. R., Jones, A. B., and Reekstin, J. P. (1979). *IEEE Trans. Electron Dev.* **ed-26**, 469.
El-Sabbahy, A. N., and Adams, A. R. (1979). *Conf. Ser.–Inst. Phys.* **43**, 589.
Engelmann, R. W. H., and Liechti, C. A. (1977). *IEEE Trans. Electron Dev.* **ed-24**, 1288.
Fang, F., and Fowler, A. B. (1970). *J. Appl. Phys.* **41**, 1825.
Farrar, A., and Adams, A. T. (1970). *IEEE Trans. Microwave Theory Tech.* **mtt-18**, 65.
Farrar, A., and Adams, A. T. (1974). *IEEE Trans. Microwave Theory Tech.* **mtt-22**, 889.
Ferry, D. K. (1976a). *Phys. Rev. B* **14**, 5364.
Ferry, D. K. (1976b). *Phys. Rev. B* **14**, 1605.
Ferry, D. K. (1979a). In "Physics of Nonlinear Transport in Semiconductors" (D. K. Ferry, J. R. Barker, and C. Jacoboni, eds.), p. 134. Plenum, New York.
Ferry, D. K. (1979b). *J. Appl. Phys.* **50**, 1422.
Ferry, D. K., and Barker, J. R. (1980a). *Solid-State Electron.* **23**, 545.
Ferry, D. K., and Barker, J. R. (1980b). *J. Phys. Chem. Solids* **41**, 1083.
Ferry, D. K., Hess, K., and Vogl, P. (1981). In "Microstructure Science and Technology/VLSI" (N. Einspruch, ed.). Vol. 2, p. 68. Academic Press, New York.
Fichtner, W., Fuls, E. N., Johnston, R. L., Sheng, T. T., and Watts, R. K. (1980). *Proc. Int. Electron Device Meet., 1980* p. 24.
Flinn, P. A. (1956). *Phys. Rev.* **104**, 350.
Frankl, D. A. (1961). *Solid-State Electron.* **2**, 71.
Fukui, H. (1979). *Solid-State Electron.* **22**, 507.
Gheewala, T. R. (1980). *IEEE Trans. Electron Dev.* **ed-27**, 1857.
Green, M. S. (1952). *J. Chem. Phys.* **21**, 1281.
Greiling, P. T. (1979). *Proc. Int. Electron Device Meet., 1979* p. 670.
Grieg, D. D., and Engelmann, H. F. (1952). *Proc. IRE* **40**, 1644.
Grinyaev, S. N., Il'in, M. A., Lukomskii, A. I., Chalyshev, V. A., and Chupakhina, V. M. (1980). *Sov. Phys.–Semicond. (Engl. Transl.)* **14**, 446.
Grosvalet, J., Motsch, C., and Tribes, R. (1963). *Solid-State Electron.* **6**, 65.
Grubin, H. L., Ferry, D. K., and Barker, J. R. (1979). *Proc. Int. Electron Device Meet., 1979* p. 394.
Grubin, H. L., Ferry, D. K., and Gleason, K. R. (1980). *Solid-State Electron.* **23**, 157.
Hall, G. L. (1959). *Phys. Rev.* **116**, 604.
Harrison, J. W., and Hauser, J. R. (1976). *Phys. Rev. B* **13**, 5347.
Harrison, W. A. (1980). "Electronic Structure and the Properties of Solids." Benjamin, New York.
Hasegawa, H., Furukawa, M., and Yanai, H. (1971). *IEEE Trans. Microwave Theory and Tech.* **mtt-19**, 869.
Heil, O. (1935). British patent 439,475.
Hess, K. (1981). *IEEE Trans. Electron Dev.* **ed-28**, 937.
Hilsum, C. (1976). *J. Phys. C* **9**, L629.
Hoeneisen, B., and Mead, C. A. (1972). *Solid-State Electron.* **15**, 819.
Hofstein, S. R., and Warfield, G. (1965). *IEEE Trans. Electron Dev.* **ed-12**, 129.
Holland, M. G. (1966). In "Semiconductors and Semimetals" (R. K. Willardson and A. C. Beer, eds.), Vol. 2, p. 3. Academic Press, New York.
Houng, Y. M., and Pearson, G. L. (1978). *J. Appl. Phys.* **49**, 3348.

Hunter, W. R., Holloway, T. C., Chatterjee, P. K., and Tasch, A. F., Jr. (1981). *IEEE Electron Dev. Lett.* **edl-2**, 4.
Hyltin, T. M. (1965). *IEEE Trans. Microwave Theory Tech.* **mtt-13**, 777.
Itoh, T., and Yanai, H. (1980). *IEEE Trans. Electron Dev.* **ed-27**, 1037.
Josephs, H. C. (1965). *Microelectron. Reliab.* **4**, 345.
Khang, D., and Atalla, M. M. (1960). Unpublished.
Kaupp, H. R. (1967). *IEEE Trans. Electron. Comput.* **ec-16**, 185.
Klaassen, F. M., and DeGroot, W. C. J. (1980). *Solid-State Electron.* **23**, 237.
Kohn, W., and Luttinger, J. M. (1957). *Phys. Rev.* **108**, 590.
Kroemer, H. (1968). *IEEE Trans. Electron Dev.* **ed-15**, 889.
Kubo, R. (1957). *J. Phys. Soc. Jpn.* **12**, 570.
Kwan, C. C. Y., and Woolley, J. C. (1968). *Can. J. Phys.* **46**, 1669.
Lehovec, K., and Zuleeg, R. (1980). *IEEE Trans. Electron Dev.* **ed-27**, 1074.
Lehovec, K., Slobodsky, A., and Sprague, J. L. (1963). *Phys. Status Solidi* **3**, 447.
Lepselter, M. (1980). *Proc. Int. Electron Device Meet, 1980* p. 42.
Lilienfeld, J. E. (1930). U.S. patent 1,745,175.
Liechti, C. A. (1977). *Conf. Ser.-Inst. Phys.* **33a**, p. 227.
Lindner, T. (1962). *Bell Syst. Tech. J.* **41**, 803.
Littlejohn, M. A., Hauser, J. R., and Glisson, T. H. (1977). *Appl. Phys. Lett.* **30**, 242.
Long, D. L. (1960). *Phys. Rev.* **120**, 2024.
Lundgren, R. E., Krumm, C. F., and Pierson, R. L. (1979). *IEEE Trans. Electron Dev.* **ed-26**, 1827.
Maloney, T. J., and Frey, J. (1977). *J. Appl. Phys.* **48**, 781.
Marsh, J. H., Houston, P. A., and Robson, P. N. (1981). *Conf. Ser.-Inst. Phys.* **56**, 621.
Masuda, H., Nakai, M., and Kubo, M. (1979). *IEEE Trans. Electron Dev.* **ed-26**, 980.
Mead, C. A. (1965). *Proc. IEEE* **54**, 307.
Mizutani, T., Kato, N., Ishida, S., Osafune, K., and Ohmani, M. (1980). *Electron. Lett.* **16**, 315.
Moll, J. L. (1948). *In* "IEEE WESCON Conv. Rec.," Part 3, p. 32. IEEE Press, New York.
Moon, R. L., Antypas, G. A., and James, L. W. (1974). *J. Electron. Mater.* **3**, 635.
Moore, B. T., and Ferry, D. K. (1980). *J. Appl. Phys.* **51**, 2603.
Mori, H. (1958). *Phys. Rev.* **112**, 1829.
Müller, R., Pfleiderer, H.-J., and Stein, K.-U. (1976). *IEEE J. Solid-State Circuits* **sc-11**, 1677.
Müller, W., and Eisele, I. (1980). *Solid State Commun.* **34**, 447.
Nicholas, R. J., Stradling, R. A., Portal, J. C., and Askenazy, S. (1979). *J. Phys. C* **12**, 1653.
Ning, T. H. (1978). *Solid-State Electron.* **21**, 273.
Ning, T. H., Cook, P. W., Dennard, R. H., Osburn, C. M., Schuster, S. E., and Yu, H. N. (1979). *IEEE Trans. Electron Dev.* **ed-26**, 346.
Nuzillat, G., Bert, G., Ngu, T. P., and Gloanec, M. (1980). *IEEE Trans. Electron Dev.* **ed-27**, 1102.
Ogura, S., Tsang, P. J., Walker, W. W., Critchlow, D. L., and Shepard, J. F. (1980). *IEEE Trans. Electron Dev.* **ed-27**, 1359.
Ohta, K., Morimoto, M., Saitoh, M., Fukuda, T., Morino, A., Shimizu, K., Hayashi, Y., and Tarui, Y. (1975). *IEEE J. Solid-State Circuits* **sc-10**, 314.
Ohta, K., Yamada, K., Saitoh, K., and Tarui, Y. (1980). *IEEE Trans. Electron Dev.* **ed-27**, 1352.
Pandey, K. C., and Phillips, J. C. (1976). *Phys. Rev. B* **13**, 750.
Pearsall T. (1981), private communication.
Penn, D. (1962). *Phys. Rev.* **128**, 2093.
Pepper, M. (1978). *Philos. Mag., Part B* **38**, 515.
Pfann, W. G., and Garrett, G. C. B. (1959). *Proc. IRE* **47**, 2011.
Phillips, J. C. (1968). *Phys. Rev. Lett.* **20**, 550.
Phillips, J. C. (1973). "Bonds and Bands in Semiconductors." Academic Press, New York.

Pinczuk, A., Worlock, J. M., Nahory, R. E., and Pollack, M. A. (1978). *Appl. Phys. Lett.* **33,** 461.
Portal, J. C., Perrier, P., Renucci, M. A., Askenazy, S., Nicholas, R. J., and Pearsall, T. (1979). *Conf. Ser.–Inst. Phys.* **43,** 829.
Ransom, M. N., and Saeks, R. (1975). *Circuit Theory Appl.* **3,** 5.
Rees, H. D., Sanghera, G. S., and Warriner, R. A. (1976). *Electron. Lett.* **33,** 461.
Reiser, M. (1973). *IEEE Trans. Electron Dev.* **ed-20,** 35.
Rossel, P., Nuzillat, G., Tranduc, H., Bert, G., Graffeuil, I., and Azizi, C. (1978). Unpublished.
Russo, P. M. (1980). *IEEE Trans. Electron Dev.* **ed-27,** 1332.
Sabnis, A. G., and Clemens, J. T. (1979). *Proc. Int. Electron Device Meet., 1979* p. 18.
Sah, C. T., Ning, T. H., and Tschopp, L. L. (1972). *Surf. Sci.* **32,** 561.
Sangani, S. H., Saeks, R., and Liberty, S. R. (1976). *J. Franklin Inst.* **302,** 239.
Sankarar, R., Moon, R. L., and Antypas, G. A. (1976). *J. Cryst. Growth* **33,** 271.
Schottky, W. (1938). *Naturwissenschaften* **26,** 843.
Shaw, M. P., Solomon, P. R., and Grubin, H. L. (1969). *IBM J. Res. Dev.* **13,** 587.
Shaw, M. P., Grubin, H. L., and Solomon, P. R. (1979). "The Gunn–Hilsum Effect." Academic Press, New York.
Shockley, W., and Pearson, G. L. (1948). *Phys. Rev.* **74,** 232.
Shur, M., and Eastman, L. F. (1979). *IEEE Trans. Electron Dev.* **ed-26,** 1677.
Stratton, R. (1958). *Proc. R. Soc., Ser. A* **246,** 406.
Suyama, K., Kusakawa, H., and Fukuta, M. (1980). *IEEE Trans. Electron Dev.* **ed-27,** 1092.
Tanimoto, M., Suzuki, K., Itom, T., Ikoma, T., Yanai, H., Kaufmann, L., Nievenctick, W., and Heime, K. (1976). Unpublished.
Taylor, G. W. (1978). *IEEE Trans. Electron Dev.* **ed-25,** 337.
Taylor, G. W. (1979). *Solid-State Electron.* **22,** 701.
Terman, L. M. (1962). *Solid-State Electron.* **5,** 285.
Trofimenkoff, F. N. (1965). *Proc. IEEE* **53,** 1765.
Troutman, R. (1974). *IEEE J. Solid-State Circuits* **sc-9,** 55.
Troutman, R. (1979). *IEEE Trans. Electron Dev.* **ed-26,** 461.
Troutman, R., and Fortino, A. G. (1977). *IEEE Trans. Electron Dev.* **ed-24,** 1266.
Tsironis, C. (1980). *IEEE Trans. Electron Dev.* **ed-27,** 2160.
Turner, J. A., and Wilson, B. L. H. (1968). *Conf. Ser.–Inst. Phys.* **7,** 195.
Van Tuyl, R., and Liechti, C. (1977). *Spectrum*, March, p. 41.
Van Vechten, J. A. (1968). *Phys. Rev.* **170,** 773.
Van Vechten, J. A. (1969a). *Phys. Rev.* **182,** 891.
Van Vechten, J. A. (1969b). *Phys. Rev.* **187,** 1007.
Van Vechten, J. A., and Bergstresser, T. K. (1970). *Phys. Rev. B* **1,** 3351.
Watts, R. K., Fichtner, W., Fuls, E. N., Thibault, L. R., and Johnston, R. L. (1980). *Proc. Int. Electron Device Meet., 1980* p. 772.
Wheeler, H. A. (1964). *IEEE Trans. Microwave Theory Tech.* **mtt-12,** 280.
Wheeler, H. A. (1965). *IEEE Trans. Microwave Theory Tech.* **mtt-13,** 172.
Wolff, P. A. (1954). *Phys. Rev.* **95,** 1415.
Woolley, J. C., Gillet, C. M., and Evans, J. A. (1961). *Proc. Phys. Soc.* **77,** 700.
Wu, T. Y., and Pearson, G. L. (1972). *J. Phys. Chem. Solids* **33,** 409.
Yamashita, E., and Mittra, R. (1968). *IEEE Trans. Microwave Theory Tech.* **mtt-16,** 251.
Zuleeg, R., Notthoff, J. K., and Lehovec, K. (1978). *IEEE Trans. Electron Dev.* **ed-25,** 628.
Zwanzig, R. (1961). *Phys. Rev.* **124,** 983.
Zwanzig, R. (1964). *J. Chem. Phys.* **40,** 2527.

Author Index

Numbers in parentheses are reference numbers and indicate that an author's work is referred to although his name is not mentioned in the text. Numbers in italics indicate the pages on which the complete references are given.

A

Abagyan, S. A., 112, 116, 117, 127, 129, *135*
Abakumov, V. N., 83, *135*
Abele, M., 272(15), *310*
Aberth, W., 147, *189*
Adams, A. C., 237(73), 238(73), *268*
Adams, A. R., 370, *388*
Adams, A. T., 359, 360, *388*
Ahiezer, A., 272(15), *310*
Aitken, J. M., 12(54, 56, 59), 67(108), *78, 79*
Akamats, T., 208(86), *268*
Akasaka, Y., 237, *268*
Albritton, D. L., 177, 182, *187*
Allègre, J., *135*
Allen, J. W., 112, 114, 115, 119, *135, 138, 140*
Allen, S. D., 231(102), *268*
Almassy, R. J., 104, 108, *140*
Almeleh, N., 107, *136*
Altarelli, M., 92, *135*
Anderson, C. L., 225(94), *268*
Andersen, H. H., 203, 204(30), *267*
Anderson, L. W., 173, *188*
Anderson, W. W., 117, *138*
Ando, R., 243(104), *268*
Andre, B., 12(48), *77*
Andrianov, D. G., 115, *135*
Ankudinov, V. A., 179, *188*
Annis, B. K., 180, 186, *187, 188*
Antoniadis, D. A., 219(61), 220(61), 221(92), *267, 268*
Antypas, G. A., 370, 371, *389, 390*
Aoki, K., 263(130), *269*
Appels, J. A., 241(78, 79), *268*
Arnold, G. W., Jr., 6, *77*
Asbeck, P. M., 81, 119, *141*
Ashikawa, M., 241(83), *268*
Askenazy, S., 375, *389, 390*
Atalla, M. M., 314, *389*
Aubuchon, K. C., 53(99), *79*
Aubuchon, K. G., 71(114), *79*, 192, *266*
Ausman, G. A., Jr., 12(49), *77*

Aven, M., *135*
Averuns, M., *135*
Azizi, C., 339, *390*

B

Bailey, T. L., 147, 166, 176, 178, 180, 183, 184, 185, 186, *187, 189*
Balkanski, M., 124, 126, *137, 139*
Baranowski, J. M., 112, 115, *135*
Barbee, T. W., 273(19), *310*
Bardeen, J., 4, *77*
Bardsley, J. N., 161, 162, *187*
Barker, J. R., 347, 349, 350, 351, 353, 354, 357, *387, 388*
Barnett, C. F., 179, *189*
Barry, A. L., 9, 10(43), *77*
Barthruff, D., 103, *137*
Bass, S. J., 126, 127, *141*
Bassani, F., *135*
Bassous, E., 315, 325, *387*
Bates, D. R., 161, 162, *187*
Baxter, R. D., 12(63), 53(63), 67(63), 71(63), 72, *78*, 103, *140*
Bay, H. L., 203(30), 204(30), *267*
Bayazitov, R. M., 192(11), *266*
Bayerl, P., 232(69), 248(134), *268, 269*
Bebb, H. B., 102, 103, *141*
Bednar, J. A., 179, *188*
Beer, A. C., 81, *141*
Behle, A. F., 257(122), 259(122), *269*
Bell, G., 237(74), *268*
Bell, K. L., 161, 162, *187*
Bell, R. L., 134, *135*
Bennett, R. A., 147, 176, 183, 185, 186, *187*
Benz, K. W., 103, *137*
Bergh, A. A., 81, *135*
Bergstresser, T. K., 369, *390*
Berkeyheiser, J. E., 134, *139*
Bernholc, J., 97, *140*
Bernstein, R. B., 151, *187*

Bert, G., 120, 124, *141*, 336, 339, 340, 376, *387*, *389*, *390*
Bhargava, R. N., 105, *136*
Biersack, J. P., 195, 196(26), 197(25), 220(90), *266*, *268*
Bimberg, D., 88, 89, 109, 112, 114, *135*, *141*
Bishop, S. G., 100, 113, *135* *138*
Blätte, M., 112, *141*
Blair, R. R., 6(32), *77*
Blakemore, J. S., 114, 116, 124, *135*
Blanc, J., 103, 107, 108, *135*
Blenkinsop, I. D., 126, 127, *141*
Bloch, F., 2(2), *76*
Blum, S. E., 92, 105, *139*
Boccon-Gibod, D., 120, 124, *139*
Boesch, H. E., Jr., 12(49, 50, 51, 52), 26, *77*
Bois, D., 124, 125, *135*
Bolotovski, B. M., 272, *310*
Bonjour, P., 340, *387*
Boris, D., 83, *139*
Bosch, B. G., 343, *387*
Bosch, J., 246(109), *269*
Bose, J. C., 314, 332, *387*
Bourgeois, J. M., 120, 124, *139*
Bourgoin, J. C., 101, *139*, 250(114), 263(114), 264(114), *269*
Bowen, D. J., 166, 177, *188*
Bower, R. W., 192, 252(6), *266*
Brattain, W. H., 4, *77*
Brauman, J. I., 144, *188*
Braun, F., 314, 332, *387*
Bristol, S. P., 237, *268*
Brodsky, M. H., 195(24), *266*
Broom, R. F., 333, *388*
Brown, D. R., 184, *188*
Brown, M. R., 134, *135*
Brown, W. J., 114, 116, *135*
Brown, W. L., 6(32), *77*
Browne, J. C., 173, *189*
Brozel, M. R., 119, *135*
Brucker, G. L., 12(66), *78*
Brunwin, R. F., 127, *135*
Bryant, T. G., 360, *387*
Bube, R. H., 84, 103, 107, 108, 119, *135*, *138*
Buck, T. M., 232(64), *267*
Buckley, R. R., 237(72), *268*
Bury, P., 122, *135*
Busby, K. O., 273(22, 23), *310*
Butler, J., 119, *135*
Butler, J. K., 81, *138*

Butlin, R. S., 124, *135*, *136*
Buxo, J., 12, *77*
Bydin, Yu. F., 147, 165, 166, 167, 184, *187*

C

Cardwell, M. J., 124, *136*
Carrier, G. F., 358, *387*
Carter, A. C., 89, *135*
Carter, G., 205(34), *267*
Caruso, R., 103, *137*
Casey, H. C., 81, 134, *135*
Cass, T. C., 205(32), 206(32), *267*
Castagne, R., 340, *387*
Cavenett, B. C., 94, 126, *136*, *138*
Celler, G. K., 192 (13), 225(94), *266*, *268*
Čerenkov, P. A., 271, *310*
Chaing, S. Y., 99, *135*
Challis, L. J., 122, *135*
Chalyshev, V. A., 371, *388*
Champion, R. L., 147, 148, 151, 152, 153, 165, 166, 167, 168, 169, 170, 171, 172, 177, 178, 180, 182, 183, 184, 186, *187*, *188*, *189*
Chang, L. L., 99, *135*
Chaplygin, Yu. A., 68(109), *79*
Chapman, R. A., 116, *135*
Chatterjee, P. K., 100, *135*, 259(126), 260(126), *269*, 313, 327, 376, *387*, *389*
Chen, C. K., 273(18), *310*
Chen, J. C. Y., 164, *187*, *189*
Chen, J. W., 133, *139*
Cheung, J. T., 147, 183, 184, *187*
Chiu, K. Y., 15(76), 20(76), 71(76), *78*
Chrenko, R. M., 115, *140*
Chu, A. N., 273(19), *310*
Chu, W. K., 206(40), 211(47), 213(47), *267*
Chupakhina, V. M., 371, *388*
Churchill, J. N., 12(70, 71, 72, 73, 74), 22(71), 26(74), 28(74, 85), 29(74), 32(72), 34(90), 38(90), 40(91), 42(90, 93, 94), 46(93), 57 (74), 61(74, 94), 62(73), 64, 67(93), 69(85), 71(85, 93), 72, *78*, *79*
Clampitt, R., 244(108), *269*
Clemens, J. T., 252(116), 253(116), *269*, 322, 323, *390*
Clerjaud, B., 119, 124, 126, 127, 128, *136*, *137*, *139*
Cobié, B., 176, *189*
Cohen, J. S., 161, 162, *187*
Colby, J. W., 220(63), *267*

Coleman, P., 272, *310*
Collins, T. W., 12(69, 70, 71, 72, 73, 74), 22 (71), 26(74), 28(74, 85), 29(74), 32(72), 34(90), 38(90), 40(91), 42(90, 93, 94), 46 (93), 57(74), 58(69), 61(74, 94), 62(73), 64(93), 67(93), 69(85), 71(85, 93), 72 (85, 93, 94), *78*, *79*
Comer, J. J., 215(53), *267*
Compton, W. D., 6, *77*
Cook, R. K., 349, *387*
Cooke, P. W., *389*
Cooke, R. A., 89, *136*
Cooper, J. A., Jr., 323, 324, 325, *387*
Corbett, J. W., 3(9), *76*, 250(114), 263(114), 264(114), *269*
Corderman, R. R., 144, *187*
Costello, W. R., 215(56), *267*
Courat, J.-P., 340, *387*
Cox, A. F. J., 134, *135*
Cramer, A., 252(117), 255(117), *269*
Crawford, J. H., 6, *77*
Crawford, O. H., 171, *188*
Critchlow, D. L., 332, *389*
Cronin, G. R., 109, 119, 130, *137*
Crossley, I., 124, *135*, *136*
Crowder, B. L., 195(24), 209(45), 232(65), 248 (132), 252(117), 255(117), *266*, *267*, *269*
Csanky, G., 247(112), *269*
Csepregi, L., 211, 213(47, 48), 214, 247(111), *267*, *269*
Cullis, A. G., 205(33), 218(58), 232(66), *267*
Curie, E., 271, *310*
Curtis, O. L., Jr., 15(76), 20(76), 71(76), *78*
Cuthbert, J. D., 94, *137*

D

Danos, M., 272, *310*
Das, P., 354, 355, *387*
Datz, S., 147, 180, 183, 184, 186, *187*, *188*
Davies, J. A., 191(2), 193(2), *266*
Davis, R. E., 5(22), *77*
Deacon, D. A., 307(25), *310*
Deal, B. E., 7(37), 57(105), *77*, *79*, 220(91), *268*
Dean, P. J., 114, *137*, *140*, *141*
Dearnaley, G., 191, 192, 193, *266*
De Groot, W. C. J., 322, 324, *389*
de Heers, F. J., 158, 159, *189*
Dehlinger, U., 2, *76*
De Keersmaecker, R. F., 12(57), *78*

Delos, J. B., 147, 148, 151, 152, 153, 155, 157, *188*, *189*
Demkov, Yu. N., 153, *188*
Dennard, R. H., 193(14), *266*, 315, 325, *387*, *389*
de Pooij, N. F., 49(98), 68(98), *79*
Derbenwick, G. F., 11, 28, 53(84), 71(46), 72, *77*, *78*
Devdarianni, A. Z., 153, *188*
Deveaud, B., 127, *136*, *139*
de Vreugd, C., 153, 168, 169, 170, 171, 172, *188*, *189*
Dewangan, J. P., 158, *188*
De Wit, M., 112, *137*
Dexter, W. H., 248(133), 264(133), *269*
Dhuicq, D., 158, 159, 160, *188*
Dill, H. G., 192(6), 252(6), *266*
DiMaria, D. J., 12(54, 55, 57, 58), 67(58), *78*
Dmitriev, I. S., 162, *188*
Doklan, R. H., 252(116), 253(116), *269*
Doverspike, L. D., 147, 148, 151, 152, 153, 165, 166, 167, 174, 175, 177, 178, 180, 182, 183, 184, 186, *187*, *188*, *189*
Dow, J. D., 83, *137*
Drangeid, K. E., 333, *388*
Dresselhaus, G., 367, *388*
Drum, C. M., 232(67), *267*
Dümke, R., 200(29), 221(29), 225(29), *267*
Duffy, M. T., 134, *139*
Dukel'skii, V. M., 144, 147, 165, 166, 167, 184, *187*, *188*
Dunkin, D. B., 177, *189*
Dunlap, W. C., 5, 58(19), *77*
Dunning, T. H., 175, 176, *188*
Dunse, J. U., 99, *138*
Durschlag, M. S., 100, *135*
Dutton, R. W., 219(61), 220(61), 221(92), *267*, *268*
Dyson, F., 3(10), *77*

E

Eastman, L. F., 339, 349, *388*, *390*
Eaves, L., *136*
Eden, R. C., 81, 119, *141*, 257(121), *269*, 312, 313, 340, *388*
Edwards, A. K., 160, *188*, *189*
Edwards, W. R., 147, 165, 166, 167, 177, *189*
Eernisse, E. P., 12, 69(62), *78*
Eichinger, P., 247(111), 250(137), *269*

Eisele, I., 323, 325, *389*
Elias, L. R., 307(25), *310*
Elliott, C. T., 99, *141*
Elliott, M. T., 330, 375, *388*
El-Sabbahy, A. N., 370, *388*
Enderby, C., 272(11), *310*
Enea, G., 12(48), *77*
Engelmann, H. F., 358, *388*
Engelmann, R. W. H., 337, 339, *388*
Engemann, D., 119, 127, *136*
Englade, J., 261(128), 262(128), *269*
Englert, T., *136*
Ennen, H., 105, 102, 113, 114, 115, 117, 127, 128, 130, 131, 132, 133, *136*, *137*, *138*, *140*
Enomoto, T., 243, *268*
Ephrath, L. M., 252(117), 255(117), *269*
Erikson, L., 191(2), 193(2), *266*
Esaki, L., 99, *135*
Esaulov, V., 148, 158, 159, 160, 162, 163, *188*
Esaulov, V. A., 174, 175, *188*
Esteve, D., 12(48), *77*
Estle, T. L., 112, *137*
Evans, E. W., 177, *189*
Evans, J. A., 370, *390*
Evans, S. A., 261, 262(128), *269*
Everhart, T. E., 46, *79*
Evwaraye, A. O., 116, 117, *136*

F

Fabre, E., 105, *136*
Fair, R. B., 219(60), 220(62), 221(60), *267*
Fairfield, J. M., 232(65), *267*
Faist, M. B., 171, *188*
Fang, F., 324, *388*
Fang, F. F., 46(96), *79*
Farges, J. P., *139*
Farrar, A., 359, 360, *388*
Faulkner, R. A., 88, 89, 91, *136*
Fehsenfeld, F. C., 177, *189*
Felch, K. L., 273(22, 23), *310*
Ferguson, E. E., *189*
Fern, A. M., 215(51), *267*
Ferris, S. D., 225(95), *268*
Ferry, D. K., 20(80), 69(80), *78*, 322, 323, 324, 325, 337, 340, 347, 349, 350, 352, 353, 354, 355, 357, *387*, *388*, *389*
Fichtner, W., 375, 376, *388*, *390*
Firsov, O. B., 158, *188*
Fischer, G., 205, *267*

Fistul, V. I., 115, *135*
Fite, W. L., 147, 161, 162, *188*
Fitzgerald, D. J., 11(44), 46(44), 74(44), *77*
Fleurov, V. N., 110, *136*
Flinn, P. A., 368, *388*
Fogel', La. M., 179, *188*
Forbes, L., 84, *140*
Fortino, A. G., *390*
Forward, J. E., 215(54), *267*
Fowler, A. B., 324, *388*
Fowler, R. H., 55(100), *79*
Frank, I. M., 271, *310*
Frankl, D. A., 314, *388*
Franklin, J. L., 144, *188*
Frankovsky, F., 57, *79*
Freeman, J. H., 191(3), 192(3), 193(3), *266*
Freeman, R., 28, 73(87), *78*
Freemann, J. H., 241(81), 248(81), *268*
Frenkel, J., 3, *76*
Frey, J., 349, 356, *387*, *389*
Friebertshauser, P. E., 257(122), 259(122), *269*
Fritsche, C. R., 241(82), *268*
Frosch, C. J., 94, *136*
Fukuda, T., 332, *389*
Fukui, H., 338, 347, *388*
Fukuta, M., 124, *141*, 336, 376, *390*
Fuller, C. S., 99, *140*
Fuls, E. N., 375, 376, *388*, *390*
Fung, S., 127, *136*
Furukawa, A., 263(130), *269*
Furukawa, M., 360, *388*
Furukawa, S., 200, 206(39, 41), *267*
Furuya, T., 206(43), 207(43), 208(43, 86), *267*, *268*

G

Gaensslen, F. H., 193(14), *266*, 315, 325, *387*
Gal, M., 94, *136*
Gallagher, T. J., 214(49), *267*
Galland, D., 127, *139*
Galyatudinov, M. F., 192(11), *266*
Gard, G. A., 241(81), 248(81), *268*
Gardner, K. R., 215(56), *267*
Gardner, M. A., 172, 174, *188*
Garrett, G. C. B., 314, *389*
Garrett, W. R., 171, *188*
Gat, A., 229(100), *268*
Gauyacq, J. P., 148, 153, 154, 155, 158, 160, 174, 175, *188*

AUTHOR INDEX

Geballe, R., 147, 153, 158, 160, 178, 179, *188, 189*
Geddes, J., 147, 161, 162, *188*
Gendron, F., 127, *136*
Gershenzon, M., 93, *137*
Gdula, R. A., 12(60), *78*
Gheewala, T. R., 380, *388*
Gibb, R. M., 112, 113, *136*
Gibbons, J. F., 194(20), 195, 197, 198, 227(99), 229, 230(23), 242(23), *266, 268*
Gilbody, H. B., 147, 161, 162, *188*
Gillespie, G. H., 162, *188*
Gillet, C. M., 370, *390*
Ginzburg, V. L., 272, *310*
Gippius, A. A., 134, *140*
Giroux, R. R., 6, *77*
Gleason, K. R., 337, 340, *388*
Glisson, T. H., 370, 372, *389*
Gloanec, M., 336, 376, *389*
Gloriozova, R. I., 127, *136*
Goddard, W. A., 153, *189*
Godlewski, M., 114, *136*
Gölzlich, J., 237, *268*
Goetzberger, A., 205(37), *267*
Goffe, T. V., 147, 161, 162, *188*
Goldstein, B., 107, *136*
Golja, B., 232(68), 236, 237(70), *267, 268*
Goltzene, A., 124, *139*
Gonzalez, A. G., 219(61), 220(61), *267*
Goodman, A. M., 55(100, 101), 56(102), *79*
Goodridge, I. H., 124, *136*
Goodyear, C. C., 147, 176, 183, 184, 185, *189*
Gorey, E. F., 250(115), *269*
Goswami, N. K., 121, 129, *136*
Graffeuil, I., 339, *390*
Green, D., 7, *77*
Green, M. S., 351, *388*
Gregory, B. L., 11, 12, 28, 53(84), 71(46), 72, *77, 78*
Greiling, P. T., 379, *388*
Grieg, D. D., 358, *388*
Griffing, K. M., 172, *188*
Grinshtein, P. M., 115, *135*
Grinyaev, S. N., 371, *388*
Grimmeiss, H. G., 84, 105, *136*
Grobman, W., 252(117), 255(117), *269*
Groombridge, I., *79*
Gross, E. F., 112, *136*
Grosvalet, J., 323, *388*
Grove, A. S., 7(37), 11(44), 46(44), 74(44), *77*

Grubin, H. L., 337, 338, 340, 347, 349, 354, 355, *387, 388, 390*
Guillot, G., 127, *136, 138*
Gwyn, C. W., 12, 53(68), *78*

H

Haberger, K., 194(19), 195(26), 196(26), 200(29), 221(29), 225(29), 244(110), 246(109), *266, 267, 269*
Hagston, W. E., 126, *138*
Haisty, R. W., 109, 119, 130, *137*
Hall, G. L., 369, *388*
Hall, R. N., 5, 58(19, 20), *77*
Hallais, J. P., *139*
Ham, F. S., 115, *137, 140*
Hamilton, B., 112, 113, 127, 132, 133, *135, 136, 137*
Handy, R. M., 5(27), *77*
Hansen, S. E., 221(92), *268*
Hanson, G. R., 244(107), *269*
Harland, P. W., 144, *188*
Harrison, J. W., 367, *388*
Harrison, W. A., 362, 373, 374, *388*
Hartnagel, H., 215(54), *267*
Hasegawa, H., 360, *388*
Hasegawa, S., 215(54), *267*
Hasted, J. B., 144, 147, 167, 176, 184, 185, *188*
Haszko, S. E., 237(72), *268*
Hauser, J. R., 367, 370, 372, *388, 389*
Hawrylo, F. Z., 99, *138*
Hay, P. J., 175, 176, *188*
Hayashi, Y., 332, *389*
Hayes, W., 113, 114, *137, 140*
Haynes, J. R., 92, *137*
Haywood, S. E., 166, 177, 186, *188*
Heaviside, O., 271, *310*
Heil, O., 314, *388*
Heiman, F. B., 5, *77*
Heime, K., 339, *390*
Heinemeier, J., 178, *188*
Heinz, O., 147, *189*
Hemstreet, L. A., 110, *137*
Henkelmann, R., 194(19), 195(26), 196(26), *266*
Henkelmann, R. A., 220(90), *268*
Hennel, A. M., 114, 119, 124, 126, 127, *136, 137, 139*
Henry, C. H., 91, 94, 95, 106, *136, 137*
Henry, H., 94, *138*

Henry, M. O., 126, *138*
Henry, R. L., 127, *140*
Herbert, D. C., 85, 104, 106, *136*
Herbst, E., 184, 186, *188*
Herzenberg, A., 157, 160, 161, 163, *187*, *188*
Hess, K., 56(103), *79*, 322, 349, *388*
Hijya, S., 206(43), 207(43), 208(43), *267*
Hill, J., 147, 161, 162, *188*
Hilsum, C., 357, *388*
Hirao, T., 205, 207(36), *267*
Hiskes, J. R., 172, 174, *188*
Hjalmarson, H. P., 83, *137*
Ho, C. P., 220(91), *268*
Hodgkinson, J., 127, *135*
Hoeneisen, B., 193(15), *266*, 313, 327, *388*
Hoepfner, J., 237, *268*
Hoffmann, K., 194(19), 200(29), 221(29), 225(29), *266*, *267*
Hofker, W. K., 194(18), *266*
Hofstein, S. R., 5, 7, *77*, 323, *388*
Holland, M. G., 382, *388*
Holloway, T. C., 313, 327, 376, *387*, *389*
Holm, C., 105, *136*
Holmes-Siedle, A. G., 28, 71(112), 73(87, 112), 74(113), *78*, *79*
Holmstrom, F. E., 12(71, 72, 73, 74), 22(71), 26(74), 28(74, 85), 29(74), 32(72), 34(90), 40(91), 42(90, 93, 94), 46(93), 57(74), 61(74, 94), 62(73), 64(93), 67(93), 69(85), 71(85, 93), 72(85, 93, 94), *78*, *79*
Holton, W. C., 112, *137*
Hong, J. D., 227(99), *268*
Hopfield, J. J., 93, 95, *137*
Horie, K., 237(75), *268*
Hornung, T., 119, 127, *136*
Hotop, H., *188*
Hoult, R. A., 89, *136*
Houng, Y. M., 120, 124, *137*, 339, 340, *388*
Houston, P. A., 375, *389*
Hsieh, C. M., 232, *267*
Hu, K. L., 257(122), 259(122), *269*
Huber, S. A. M., 119, *137*
Hughes, G. W., 11, 12, 28, 44(47), *77*, *78*
Hughes, H. L., 6, 12, 53(63), 67(63), 71(63), 72, *77*, *78*
Hughes, R. C., 12, 20(79), 34(79), 69, 71, *78*, *79*
Hummer, D. G., 147, 161, 162, *188*
Hunter, W. R., 12(55, 58), 67(58), *78*, 252(117), 255(117), *269*, 313, 327, 376, *387*, *389*
Huq, M. S., 148, *188*
Hurle, D. T. J., 103, *137*, *138*

Hutchinson, W. G., 116, *135*
Hvelplund, P., 178, *188*
Hwang, C. J., 99, 100, *137*
Hyltin, T. M., 313, *389*

I

Iberl, F., 247(111), *269*
Igelmund, A., 99, *137*
Il'in, M. A., 371, *388*
Inouh, K., 205(36), 207(36), *267*
Instone, T., *136*
Ippolitova, G. K., 115, 124, 125, *135*, *137*, *140*
Irene, E. A., 12(57), *78*
Iseler, G. W., 127, *137*
Ishida, S., 336, 379, *389*
Ishikawa, H., 124, *141*
Ishiwara, H., 206(39, 41), *267*
Ito, T., 206, 207(43), 208(43), *267*
Itoh, T., 340, *389*
Itom, T., 339, *390*
Ivanov, G. A., 112, 116, 117, 127, 129, *135*
Iwamatsu, S., 259(125), *269*

J

Jacob, G., 124, *139*
Jahnel, F., 195(26), 196(26), *266*
Jakowetz, W., 103, *137*
James, L. W., 370, 371, *389*
Janousek, B. K., 144, *188*
Jansen, J. A. J., 124, 125, 126, *139*
Janzen, E., 105, *136*
Jaros, M., 83, 102, 105, *137*, *138*
Jefferies, D. K., 244(108), *269*
Jelley, J. V., 272, 273(7), *310*
Johnson, N. L., 194(21), *266*
Johnson, W. E., 5(22), *77*
Johnson, W. S., 195(23), 197(23), 198(23), 230(23), 242(23), *266*
Johnston, R. L., 375, 376, *388*, *390*
Jones, A. B., 330, 375, *388*
Jordan, A. S., 103, *137*
Jordan, K. D., 170, 171, 172, *188*
Josephs, H. C., 340, *389*
Jutzi, W., 333, *388*

K

Kaitna, R., 247(113), *269*

Kameyama, S., 262(129), 264(129), 269
Kanzaki, K., 263(130), 269
Kapilow, D., 273(22, 23), 310
Karo, A. M., 172, 174, 188
Kasatkin, V. A., 134, 137
Kato, N., 336, 379, 389
Katz, L. E., 220(63), 267
Kaufmann, L., 339, 390
Kaufmann, U., 88, 89, 97, 98, 104, 105, 106, 108, 109, 112, 113, 114, 115, 117, 121, 122, 123, 127, 128, 129, 130, 131, 132, 133, 135, 136, 137, 138, 140
Kaupp, H. R., 359, 389
Kawazu, S., 237(75), 268
Keck, J. C., 177, 189
Kennedy, E. F., 213(48), 214(48, 49, 50), 267
Kennedy, T. A., 97, 98, 105, 106, 108, 130, 131, 137, 138
Kenney, J., 172, 188
Kerkdijk, C. D., 158, 159, 189
Kern, W., 237(71), 268
Kesamanly, F. P., 134, 137
Khaibullin, I. B., 192(11), 266
Khang, D., 314, 389
Kiblik, V. Y., 14(75), 78
Kikoin, K. A., 110, 136
Killoran, N., 126, 138
Kim, C. K., 103, 137
Kimerling, L. C., 84, 101, 138, 139
Kimerling, L. S., 192(13), 266
Kimura, S., 91, 136
King, E. E., 12(64), 78
King, P. J., 122, 135
Kingston, A. E., 161, 162, 187
Kip, A. F., 367, 388
Kirel, B., 177, 189
Kirillov, V. I., 115, 128, 138
Kirkman, R. F., 89, 136
Kittel, C., 367, 388
Klein, P. B., 100, 117, 118, 138
Klaassen, F. M., 322, 324, 389
Kleinfelder, W. J., 248(132), 269
Köhl, F., 130, 138
Kohn, W., 83, 88, 89, 138, 351, 389
Kolesnik, L. I., 127, 136
Kolos, W., 162, 188
Kooi, E., 6, 77, 241(78), 268
Kopylov, A. A., 90, 138
Koroleva, G. A., 116, 117, 135
Koroleva, K. A., 112, 135
Koschel, W. H., 100, 113, 128, 129, 137, 138, 139

Kosonocky, W. F., 258(124), 269
Kotz, S., 194(21), 266
Kranz, H., 220(90), 232(69), 241(84), 244(110), 246(109), 250(137), 268, 269
Krebs, J. J., 97, 107, 121, 122, 123, 124, 127, 128, 138, 140, 141
Kressel, H., 81, 99, 138
Kröger, F. A., 101, 103, 138
Kroemer, H., 338, 389
Kronig, R. de L., 4, 77
Krook, M., 358, 387
Krumm, C. F., 336, 389
Ku, S. M., 206(40), 267
Kubach, C., 173, 189
Kubena, R. L., 244(106), 269
Kubo, M., 325, 389
Kubo, R., 351, 389
Kuhn, L., 193(14), 266
Kuijpers, F. P., 133, 140
Kukimoto, H., 94, 138
Kusakawa, H., 336, 376, 390
Kutukova, O. G., 192(12), 266
Kuznetsov, Yu. N., 116, 117, 127, 129, 135
Kwan, C. C. Y., 370, 389

L

Ladonisi, G., 135
Lam, S. K., 147, 151, 152, 153, 178, 187, 188
Lamb, W., 300, 310
Lang, D. V., 84, 101, 102, 138, 139
Lark-Horovitz, K., 5(22), 77
Lashinsky, H., 272, 310
Layman, R. W., 273(22, 23), 310
Leamy, H. J., 225(95), 268
Leblanc, A. R., 315, 325, 387
Lee, S. H., 121, 122, 140
Lee, T. C., 117, 138
Lehovec, K., 268, 314, 335, 336, 342, 343, 389, 390
Lepselter, M., 376, 389
Leslie, T. E., 173, 188
Leutwein, K., 99, 138
Levine, R. D., 171, 188
Leyral, P., 127, 138
Liau, Z. L., 208(87, 88), 268
Liberty, S. R., 386, 390
Liechti, C., 313, 336, 390
Liechti, C. A., 81, 138, 336, 337, 339, 389, 388, 389
Lietoila, A., 227, 268

Lightowlers, E. C., 126, *138*
Lilienfeld, J. E., 314, *389*
Lin, A. L., 119, *138*
Lin, Y. T., 313, 327, 376, *387*
Lindhard, J., 194(17), 230(17), *266*
Lindmayer, J., 40, *78*
Lindner, T., 314, *389*
Lindquist, P. F., 119, *138*
Lineberger, W. C., 144, *187*, *188*
Linhart, J. G., 272(17), *310*
Lipari, N. O., 97, *140*
Littlejohn, M. A., 370, 372, *389*
Litovchenko, V. G., 14(75), *78*
Litton, C. W., 104, 108, *140*
Litty, F., 127, *138*
Litvinov, R. O., 14(75), *78*
Liu, B., 149, 152, 153, 155, 157, 164, 166, 172, 173, 174, *189*
Loescher, D. H., 114, *138*
Loeschner, H., 247(113), *269*
Logan, R. A., 101, *138*
Logan, R. M., 103, *138*
Long, D. L., 353, *389*
Long, S. I., 81, 119, *141*, 312, 313, 340, *388*
Lopantseva, G. B., 158, *188*
Lorentz, D. C., 147, *189*
Lorenz, M., 91, 92, *136*
Los, J., 153, 168, 169, 170, 171, 172, *188*, *189*
Loualiche, S., 107, 127, *136*, *138*
Lowther, J. E., 130, *139*
Lucovsky, G., 111, *139*
Ludwig, G. W., 109, *139*
Luhn, H. E., 252(117), 255(117), *269*
Lukomskii, A. I., 371, *388*
Lum, W. Y., 100, *138*, *139*
Lundgren, R. E., 336, *389*
Luttinger, J. M., 351, *389*

M

McCaughey, M. P., 179, *188*
McCombe, B. D., *138*
McCombe, B. D., 100, *139*
McCoy, G. L., 104, 108, *140*
MacDonald, N. C., 46, *79*
Macdougall, J., 192, *266*
McFarland, M., 177, *189*
McGahan, T. E., 237(73), 238(73), *268*
McGarrity, J. M., 12(49, 50, 51, 52), 26, *77*

McLean, F. B., 12(49, 50), *77*
MacRae, A. U., 237(72), *268*
Madden, P. J., 161, 162, *187*
Madelung, O., 81, *139*
Mader, S., 216(57), 217(57), *267*
Madey, J., 307(25), *310*
Magee, T. J., 227(99), *268*
Mahadevan, P., 147, 178, 180, 183, 184, 185, *187*
Maier, R. J., 23, 32, 53(86), *78*
Mallet, L., 272, *310*
Maloney, T. J., 356, *389*
Manchester, K., 192(8), *266*
Mandel, F., 161, *187*
Mandel, G., 101, *139*
Mandl, A., 177, *189*
Marchi, R. P., 147, *189*
Marciniak, W., 49, 68, *79*
Marenko, V. G., 134, *137*
Marquardt, C. L., 74, *79*
Marsh, J. H., 375, *389*
Marshall, T. C., 273(21), *310*
Martin, G. M., 85, 102, 124, 125, 126, 130, *136*, *139*
Martin, J. D., 166, 176, 184, 185, 186, *189*
Martinez, A., 12(48), *77*
Martinez, G., 119, 124, 126, 127, *136*, *137*, *139*
Martinot, H., 12(48), *77*
Marushchack, V. A., 112, *136*
Mason, E. A., 149, *189*
Massey, H. S. W., 144, *189*
Massoud, H. Z., 12(57), *78*
Masterov, V. F., 110, 130, *139*
Masters, B. J., 232(65), 250(115), *267*, *269*
Masterov, V. F., 134, *137*
Masuda, H., 325, *389*
Masuda, K., 192, *266*
Mathews, J. R., 232(67), *267*
Matić, M., 176, *189*
Matsumura, H., 200, *267*
Matta, R. K., 7(36), *77*
Mayer, J. W., 191, 193, 208(87, 88), 211(46, 47), 213(47, 48), 214(48, 49, 50), *266*, *267*, *268*
Mazey, D. J., 241(81), 248(81), *268*
Mead, C. A., 57(105), *79*, 193(15), *266*, 313, 327, 333, *388*, *389*
Meek, R. L., 232(66), *267*
Mehran, F., 92, 105, *139*
Meignant, D., 120, 124, *139*
Meindl, J. D., 220(91), *268*

Melius, C. F., 153, *189*
Merenda, P., 119, *137*
Merrit, F. R., 94, *138*
Meyer, A., 6(29), *77*
Meyer, F. W., 147, 172, 173, *189*
Michel, A., 216(57), 217(57), *267*
Michel, A. E., 248(133), 264(133), *269*
Middelhoek, J., 49(98), 68(98), *79*
Miller, E. A., 134, *139*
Miller, G. L., 84, *139*
Miller, S. E., 103, *140*
Milnes, A. G., 83, 84, 116, 133, 134, *139*
Mircea, A., 83, 101, *139*
Mircea-Roussel, A., 130, *139*
Mitchell, J. P., 26, 71(113), *78*, *79*
Mitonneau, A., *139*
Mittra, R., 359, *390*
Mizuno, O., 114, *139*
Mizuno, J., 164, *189*
Mizutani, T., 336, 379, *389*
Mohr, T. O., 333, *388*
Moline, R. A., 205, 215(56), 218(58), 237, 260 (127), 261(127), *267*, *268*, *269*
Moll, J. L., 100, *139*, 314, *389*
Monemar, B., 112, *139*
Monk, D. J., 122, *135*
Moon, R. L., 370, 371, *389*, *390*
Moore, B. T., 324, 325, *389*
Morehead, F. F., 209(45), *267*
Morgan, T. N., 91, 92, 94, 105, 106, *136*, *139*
Mori, H., 351, *389*
Morillot, G., 119, *137*
Morimoto, M., 332, *389*
Morino, A., 332, *389*
Morita, H., 243(104), *268*
Morris, S. A., 261(128), 262(128), *269*
Moschwitzer, A., 12(71), 22(71), *78*
Mosei, A., 333, *388*
Moseley, J. T., 147, 176, 183, 185, 186, *187*
Motsch, C., 323, *388*
Mott, N. F., 3, *77*, 83, *139*
Mourier, G., 272(16), *310*
Müller, H., 209(44), 211(47), 213(47), *267*
Müller, K., 194(19), 195(26), 196(26), 220(90), *266*, *268*
Müller, R., 341, 342, 343, *389*
Müller, W., 323, 325, *389*
Murray, R. B., 6(29), *77*
Munoz, E., 100, *139*
Mylroie, S. W., 194(20), 195(23), 197(23), 198 (23), 242(23), *266*

N

Nacci, J. M., 260(127), 261(127), *269*
Nagata, T., 173, *189*
Nagi, K. L., 73(117), *79*
Mahory, R. E., 375, *389*
Nakai, M., 264(129), *269*, 325, *389*
Nakamura, H., 243(104), *268*
Nakashima, H., 131, *139*
Nam, S. B., 104, 108, *140*
Namba, S., 192, *266*
Nashel'skii, A. Ya., 115, *137*
Nassibian, A. G., 232, 236, 237(70), *267*, *268*
Nelson, C. M., 6, *77*
Nelson, D. F., 323, 324, 325, *387*
Nelson, H., 99, *138*
Nelson, R., 191(3), 192(3), 193(3), *266*
Newman, R. C., 95, 96, 119, 121, 129, *135*, *136*, *139*, *140*
Ngu, T. P., 336, 376, *389*
Nicholas, R. J., 127, *136*, 375, *389*, *390*
Nichols, D. K., 47(88), *78*
Nievenctick, W., 339, *390*
Nikolaev, V. S., 162, *188*
Ning, T. H., 56(104), *79*, 322, 327, *389*, *390*
Nishi, H., 206(43), 207(43), 208(43), *267*
Nishi, N., 208(86), *268*
Nishizawa, J., 115, *140*
Nolen, J. J., 252(116), 253(116), *269*
Nomura, K., 237(75), *268*
Nordheim, L., 55(100), *79*
Nordquist, P. E. R., 100, 117, 118, *138*
North, J. C., 205(38), 237, 238(73), *267*, *268*
Notthoff, J. K., 335, 336, *390*
Nouailhat, A., 107, 127, *136*, *138*
Nuzillat, G., 120, 124, *141*, 336, 339, 340, 376, *387*, *389*, *390*

O

Ogura, S., 332, *389*
Ohl, R., 5(23), *77*
Ohmani, M., 336, 379, *389*
Ohta, K., 332, *389*
Ojha, P., 157, 160, 163, *188*
O'Keefe, T. W., 7(36), *77*
Okunev, Yu. A., 116, 117, 127, 129, *135*
Olney, R. D., 244(106), 247(112), *269*
Olson, K. H., 260(127), 261(127), *269*
Olson, R. E., 149, 152, 153, 155, 157, 164, 166, 172, 173, 174, *189*

Omel'yanovskii, E. M., 115, *135, 137, 140*
O'Neill, J. J., Jr., 56(101), *79*
Ono, M., 259(125), *269*
Osafune, K., 336, 379, *389*
Osburn, C. M., 252(117), 255(117), *269, 389*
Ostrovskii, V. N., 161, 162, *189*
Ozawa, O., 262(129), 264(129), *269*

P

Paffen, M. M., 241(78, 79), *268*
Page, D. F., 9, 10(43), *77*
Palmer, R. B., 192(8), *266*
Pandey, K. C., *389*
Panish, M. B., 81, *135*
Pankove, J. I., 134, *139*
Pantelides, S. T., 83, 97, *139, 140*
Pantell, R. H., 273(18, 19), *310*
Parker, D., 124, *135*
Parry, P. D., 237, *268*
Partin, D. L., 133, 134, *139*
Pavlov, N. M., 115, *137*
Payne, R. S., 192, 215(56), 260(9, 127), 261 (127), *266, 267, 269*
Peacher, J. L., *187*
Peaker, A. R., 112, 113, 127, 132, 133, *135, 136, 137*
Pearsall, T., 375, *389, 390*
Pearson, C. E., 358, *387*
Pearson, G. L., 5, 77, 99, 112, 114, 115, 120, 124, 134, *135, 137, 138*, 314, 339, 340, 370, *388, 390*
Peck, D. S., 6, *77*
Peltier, M., 340, *387*
Penchina, C. M., 126, *138*
Peng, J., 227(99), *268*
Penn, D., 362, *389*
Penney, W. G., 4, *77*
Pepper, M., 336, *389*
Perel, V. I., 83, *135*
Perrier, P., 375, *390*
Pervova, L. Y., 124, 125, *137*
Peterson, J. R., 147, 176, 183, 185, 186, *187*
Pettit, G. D., 195(24), *266*
Pfann, W. G., 314, *389*
Pfleiderer, H.-J., 341, 342, 343, *389*
Phillips, A. B., *77*
Phillips, B., 12(63), 53(63), 67(63), 71(63), 72, *78*
Phillips, D. H., 67(107), *79*

Phillips, J. C., 362, 363, 367, *389*
Pickar, K. A., 232(64, 67), *267*
Picoli, G., 127, *136, 139*
Pierson, R. L., 336, *389*
Piestrup, M. A., 273(18, 19), *310*
Pikktin, A. N., 90, *138*
Pilkuhn, M., *139*
Pinard, P., 124, 125, *135*
Pinczuk, A., 375, *389*
Plesiewicz, W., *139*
Plumer, J. D., 220(91), *268*
Poate, J. M., 192(13), 208(85), 225(95), 232 (64), *266, 267, 268*
Poiblaud, G., 124, 125, 126, *138*
Pollack, M. A., 375, *389*
Pone, J.-F., 340, *387*
Pons, D., 101, *138, 139*
Portal, J. C., 375, *389, 390*
Porte, C., 127, *136*
Porteous, P., 84, 113, 119, *135, 140, 141*
Powell, R. Q., 273(18), *310*
Powell, R. J., 11, 12(47), 28(47), 44(47), *77*
Prandtl, L., 2, *76*
Prener, J. S., 101, *135, 139*
Preziosi, B., *135*
Pribylov, N. N., 128, *138*
Prince, J. L., 220(89), *268*
Prinke, G., 194(19), 200(29), 221(29), 225(29), *266, 267*
Protschka, H., 57(106), *79*
Prussin, S., 215(51, 52, 55), *267*
Przewlocki, H. M., 49, 68(97), *79*

Q

Queisser, H. J., 83, 85, 99, 116, 117, 134, *139, 140*

R

Räuber, A., 97, 104, *138*
Ramdane, A., 122, *135*
Ramian, G., 307(25), *310*
Ramin, M., 241, 248(134), *268, 269*
Rampton, V. W., 122, *135*
Ransom, M. N., 385, *390*
Read, W. T., Jr., 5, 58(21), *77*
Reddi, V. G., 205(32), 206(32), *267*
Reddi, V. G. R., 192, *266*
Reddi, V. G. K., 194(22), 198(22), 199(22), *266*

Reekstin, J. P., 330, 375, *388*
Rees, H. D., 383, *390*
Regolini, J. L., 227(99), *268*
Reid, F. J., 103, *140*
Reiser, M., 315, *390*
Rembeza, S. I., 128, *138*
Rensch, D. B., 247(112), *269*
Renucci, M. A., 375, *390*
Reutlinger, G. W., 205(38), *267*
Revesz, A. G., 34(89), *78*
Reynolds, D. C., 92, 104, 108, *140*
Rice, D. W., 237(73), 238(73), *268*
Rideout, V. L., 315, 325, *387*
Rimini, E., 225(96, 97), 227(97), *268*
Risley, J. S., 146, 147, 148, 153, 158, 159, 160, 161, 178, 179, 180, 181, *188*, *189*
Ritson, A., 119, *135*
Robbins, D. J., 113, *135*
Robson, P. N., 375, *389*
Roche, A. E., 147, 176, 183, 184, 185, *189*
Rockstad, H. K., 17(77), 28(77), 46(77), 51, 53(77), 71, *78*
Röschenthaler, D., 225(98), *268*, *269*
Roitsin, A. B., 83, *140*
Romanov, V. P., 68(109), *79*
Roosild, S. A., 215(53), *267*
Rosier, L. L., 84, *140*
Rosler, R. S., 237(71), *268*
Rossel, P., 339, *390*
Rothbart, G. B., 273(18), *310*
Rothemund, W., 241(82), *268*
Rozgonyi, G. A., 225(94), *268*
Ruge, J., 192, 201(4), 225(98), *266*, *268*, *269*
Runge, H., *267*
Russek, A., 160, *189*
Russo, P. M., 312, *390*
Ryan, J. F., 113, 114, *137*, *140*
Ryskin, A. I., *140*
Ryssel, H., 192, 194(19), 195(26), 196(26), 200 (29), 201(4), 209(44), 220(90), 221, 225 (29, 98), 232, 237(76), 241(84), 244(110), 246(109), 248(134), 250(137), *266*, *267*, *268*, *269*

S

Sabnis, A. G., 322, 323, *390*
Sacher, R., 247(113), *269*
Sachs, A., 200(29), 221(29), 225(29), *267*
Saeks, R., 385, 386, *390*
Safarov, V. I., 112, *136*
Sah, C. T., 7(37), 12, 14(53), 73(116), 84, *77*, *78*, *79*, *140*, 322, *390*
Saitoh, K., 332, *389*
Saitoh, M., 332, *389*
Sakurai, J., 255(119), 256(119, 120), *268*, *269*
Sakurai, T., 208(86), *268*
Samorukov, B. E., 110, 130, 134, *137*, *139*
Sandor, J. E., 7, 14(35), *77*
Sangani, S. H., 386, *390*
Sanghera, G. S., 383, *390*
Sankarar, R., 370, *390*
Sarver, K. P., 173, *188*
Sasaki, G., 263(130), *269*
Sasaki, Y., 262(129), 264(129), *269*
Sasso, G., 333, *388*
Scavuzzo, R. J., 192(9), 260(9, 127), 261(127), *266*, *269*
Schairer, W., 91, 92, 117, 118, *136*, *141*
Scharff, M., 194(17), 230(17), *266*
Schatorje, J. J. H., 241(78), *268*
Scheffer, M., 97, *140*
Schiøtt, H. E., 194(17), 230(17), *266*
Schirmer, O., 105, *136*
Schlachter, A. S., 174, *189*
Schlesinger, S. P., 273(21), *310*
Schmeltekopf, A. L., 177, *189*
Schmid, K., 209(44), *267*
Schmidt, M., 117, 118, 126, *140*
Schmiedt, B., 232(69), *268*
Schneider, J., 88, 89, 97, 98, 99, 104, 105, 106, 108, 109, 112, 113, 114, 115, 117, 121, 122, 123, 129, 130, 131, 132, 133, *135*, *136*, *137*, *138*, *140*
Schottky, W., 3, *77*
Schultz, G. J., 180, *189*
Schuster, S. E., *389*
Schwab, C., 124, *139*
Schwettmann, F. N., 220(89), *268*
Schwettman, H., 307(25), *310*
Sedov, V. E., 112, *136*
Segsa, K. H., 117, *140*
Seidel, H. D., 232(67), *267*
Seidel, T. E., 215, 232(66), *267*
Seitz, F., 3, *76*
Seliger, R. L., 244(106), 247(112), *269*
Sequin, C. H., 258(123), *269*
Serrano, C. M., 12(55), *78*
Shah, M. B., 147, 161, 162, *188*
Shand, W. A., 134, *135*

Shannon, J. M., 206(42), 267
Shanurin, Yu. E., 127, 129, 135
Shatzkes, M., 206(40), 267
Shaw, M. P., 337, 338, 390
Shen, Y. D., 257(121), 258(121), 269
Sheng, T. T., 376, 388
Shepard, J. F., 332, 389
Shibatomi, Ohkawa, S., 124, 141
Shimizu, K., 332, 389
Shimizu, S., 259(125), 269
Shinoda, M., 206(43), 207(43), 208(43), 267
Shipsey, E. J., 173, 189
Shockley, W., 4, 5, 58(21), 77, 191, 266, 314, 390
Shottky, W., 314, 332, 390
Shtyrkov, E. I., 192(11), 266
Shui, V. H., 177, 189
Shur, M., 339, 349, 388, 390
Sidis, V., 173, 189
Siebenmann, P. G., 100, 117, 118, 138
Siegel, B. M., 244(107), 269
Siegel, S., 5(22), 77
Sigel, G. H., Jr., 74, 79
Sigmon, T. W., 211(46), 214(50), 227(99), 267, 268
Sigmund, P., 203, 204(31), 205, 267
Simons, J., 169, 172, 188, 189
Simpson, F. R., 178, 188
Sirtl, E., 105, 136
Sivo, L. L., 12(64), 78
Skolnick, M. S., 89, 135
Slack, G. A., 115, 140
Slater, J. C., 6(28), 77
Slobodskoy, A., 268, 314, 389
Smith, B. T., 147, 165, 166, 167, 169, 170, 171, 172, 182, 183, 184, 188, 189
Smith, F. T., 147, 189
Smith, R. A., 184, 185, 188
Smith, T., 307(25), 310
Smits, F. M., 6(32), 77
Snow, E. H., 7, 11, 46, 57(105), 74(44), 77, 79
Snyder, W. L., 100, 139
Solomon, P. R., 337, 338, 390
Sommerfeld, A., 2, 10(3), 76, 271, 310
Spenke, S., 117, 140
Speth, A. J., 46(96), 79
Spirin, A. I., 128, 138
Splinter, M. R., 330, 375, 388
Sprague, J. L., 314, 389
Stauss, G. H., 97, 107, 121, 122, 123, 124, 127, 128, 138, 140, 141

Stebbings, R. F., 147, 161, 162, 188
Stein, H. J., 12, 69(62), 78
Stein, K.-U., 342, 343, 389
Stengl, G., 247(113), 269
Stephen, J., 191(3), 192(3), 193(3), 266
Stephen, J. H., 241(81), 248(81), 268
Stier, P. M., 179, 189
Stirland, D. J., 119, 135
Stivers, A. R., 73(116), 79
Stocker, H. J., 119, 126, 140
Stockton, M., 273(20), 310
Stour, R., 15(76), 20(76), 71(76), 78
Stover, H. L., 247(112), 269
Stradling, R. A., 89, 135, 136, 375, 389
Stratton, R., 357, 390
Streetma, B. G., 100, 135
Strel'sov, L. N., 192(12), 266
Suchkova, N. I., 115, 135
Sullivan, M. J., 206(40), 267
Suto, K., 115, 140
Suyama, K., 336, 376, 390
Suzuki, K., 339, 390
Swiggard, E. M., 121, 122, 140
Szawelska, H. R., 119, 140
Sze, S. M., 26, 78
Szedon, J. R., 5(27), 7, 14(35), 77
Szuszkiewicz, W., 119, 124, 126, 127, 136, 137, 139

T

Taguchi, M., 263(130), 269
Takayanagi, T., 205(36), 207(36), 267
Tallon, R. W., 23, 32, 53(86), 78
Tamm, I., 271, 310
Tamura, M., 216, 218(59), 219(59), 267
Tangena, A. G., 49, 68, 79
Tanimoto, M., 339, 390
Tarui, Y., 332, 389
Tasch, A. F., 259(126), 260(126), 269, 313, 376, 389
Tawara, H., 148, 160, 189
Taylor, C. W., 259(126), 260(126), 269
Taylor, D. M., 69, 79
Taylor, G. W., 328, 390
Taylor, R., 148, 155, 157, 189
Tell, B., 133, 140
Terman, L. M., 314, 390
Tanji, T., 264(129), 269
Teslenko, V. V., 115, 138
Theodorou, D. E., 140

AUTHOR INDEX

Thibault, L. R., 375, 376, *390*
Thomas, D. G., 93, 95, *137*
Thompson, F., 96, *140*
Tihanyi, J., 254(118), *269*
Tisone, T. C., 208(85), *268*
Title, R. J., 209(45), *267*
Title, R. S., 92, 105, 117, *139, 140*, 195(24), *266*
Tokumaru, Y., 262(129), 264(129), *269*
Tokuyama, T., 249(135), *269*
Tooi, A., 241(80), *268*
Träger, L., 244(110), *269*
Tranduc, H., 339, *390*
Tribes, R., 323, *388*
Troeger, G. L., 257(122), 259(122), *269*
Trofimenkoff, F. N., 324, *390*
Troutman, R., 328, *390*
Tsai, J. C. C., 215(56), 220(62), *267*
Tsang, P. J., 332, *389*
Tsaur, B. Y., 208(88), *268*
Tschopp, L. L., 322, *390*
Tsien, P. H., 225(98), *268, 269*
Tsironis, C., 339, *390*
Tsu, R., 99, *135*
Turner, J., 124, *135*
Turner, J. A., 337, *390*

U

Uiklein, C., *136*
Ulrich, R., 272, *310*
Ushakov, V. V., 134, *140*

V

Vaidyanathan, K. V., 100, *135*
van der Meulen, Y. J., 103, *140*
Vanderslice, I. E., 149, *189*
van Gorkom, G. G. P., 117, *140*
Van Tuyl, R., 313, 336, *390*
van Vechten, J. A., 102, 103, 104, 105, 106, 108, *140*, 362, 363, 365, 369, 373, *390*
Vasil'ev, A. V., *140*
Vassamillet, L. F., 133, 134, *139*
Vavilov, P. N., 134, *140*
Verheijke, M. L., 124, 125, 126, *138*
Vink, A. T., 117, *140*
Vink, H. J., 101, 103, *138*
Vogel, P., 83, *137*
Vogl, P., 322, *388*

von Almen, M., 231(102), *268*
von Neida, A. R., 103, *137*

W

Wada, Y., 241(83), *268*
Wadehra, J. M., 161, 162, *187*
Wagner, P., 105, *136*
Wagner, R. J., 97, 107, 121, *140*
Walker, J. C. G., 161, 162, *187*
Walker, W. W., 332, *389*
Walsh, J. E., 273(20, 21, 22, 23), *310*
Walters, H. R. J., 158, *188*
Wang, V., 244(106), *269*
Ward, J. W., 244(106), *269*
Warfield, G., 323, *388*
Warriner, R. A., 383, *390*
Watanabe, H., 114, *139*
Watanabe, M., 241(80), *268*
Watanabe, S. H., 257(122), 259(122), *269*
Watkins, G. D., 85, 97, 98, *140*
Watts, R. K., 112, *137, 140*, 375, 376, *388, 390*
Webb, R., 205(34), *267*
Weber, J., 113, 114, 117, 130, *138, 140*
Weinberg, Z. A., 12(54), *78*
Weiner, M. E., 103, *137*
Weisberg, L. R., 103, 107, 108, *135*
Weiss, J. A., 360, *387*
Welch, B. M., 81, 119, *141*, 257(121), 258(121), *269*, 312, 313, 340, *388*
Wendoloski, J. J., 170, 171, *188*
Wertime, T. A., 1(1), *76*
West, C. L., 113, 114, 117, *137, 140*
Wheeler, H. A., 358, 359, *390*
White, A. M., 84, 97, 107, 112, 113, 119, 121, 123, 124, *136, 140, 141*
White, C. T., 73(117), *79*
Whitehead, C., 119, *135*
Whitehouse, J. E., 121, 129, *136*
Whiting, F. B., 241(81), 248(81), *268*
Widman, D., 254(118), *269*
Widmann, D., 193(16), *266*
Wieder, H. H., 100, *138, 139*
Wight, D. R., 126, 127, 133, *137, 141*
Wijnaendts, van Resandt, R. W., 153, 168, 169, 170, 171, 172, *188, 189*
Willardson, R. K., 81, *141*
Williams, E. W., 99, 102, 103, *141*
Williams, F. E., 101, *139*
Williams, J. F., 179, *189*
Williams, J. M., 134, *135*

Williams, P. J., *136*
Williams, R., 21(81), 44, 55(81), *78*
Willmann, F., 112, *141*
Wilsey, N. D., 97, 98, 105, 106, 108, 130, 131, *138*
Wilson, A. H., 2, *76*
Wilson, B. L. H., 337, *390*
Winokur, P. S., 12(52), *77*
Wiscombe, P., 122, *135*
Wittmer, M., 231(102, 103), *268*
Wolf, P., 247(113), *269*
Wolff, P. A., 357, *390*
Wolford, D. J., 83, *137*
Wolniewicz, L., 162, *188*
Wörner, R., 105, 106, 108, 130, 131, *136*, *137*, *138*
Woodard, D. W., *139*
Woodbury, H. H., 109, 116, 117, *136*, *139*
Woods, M. H., 11(47), 12(47), 21(81), 28(47), 44, 55(81), *77*, *78*
Woolley, J. C., 370, *389*, *390*
Worlock, J. M., 375, *389*
Wright, H. C., *136*
Wu, T. Y., 370, *390*
Wynn, M. J., 166, 176, 184, 185, 186, *189*

Y

Yakobson, S. V., 115, *137*

Yamada, K., 332, *389*
Yanai, H., 339, 340, 360, *388*, *389*, *390*
Yamashita, E., 359, *390*
Yasaitis, J. A., 249(136), *269*
Yu, A. Y. C., 192, 194(22), 198(22), 199(22), *266*
Yu, H. N., 193(14), *266*, 315, 325, *387*, *389*
Yokoyama, N., 124, *141*
Young, D. R., 12(55, 56, 57, 58), 67(58), *78*

Z

Zaininger, K. H., 8, 9, 14(42), 21(41), 28(42), 53(41, 42), 71(42, 112), 73(112), 74(42, 113), *77*, *79*
Zakzouk, A.-K. M., 20(78), *78*
Zandberg, E. I., 144, 167, *188*
Zappert, F., 57(106), *79*
Zaripov, M. M., 192(11), *266*
Ziegler, J. F., 248(132), *269*
Zrelov, Z. P., 272, *310*
Zucca, R., 81, 119, *141*, 257(121), 258(121), *269*, 312, 313, 340, *388*
Zuleeg, R., 335, 336, 342, 343, *389*, *390*
Zwanzig, R., 351, *390*
Zylbersztejn, A., 120, 124, *141*

Subject Index

A

A center, *see* Self-activated luminescent center
Acceptor, semiconductor, 86, 92–96
  3d transition metals, 109–134
  and vacancies, 101–103
Alkali atom, collisional detachment of negative ions, 169–174
Alloy, random, ordering in, 373–375
Alloy scattering effect, 369
Aluminum arsenide, 374
Anion vacancy, 96–97
Annealing process, in ion implantation, 210–231
Antimony, ion implantation, 203–205
Antisite defect, 97, 103–109
Argon
  collisional detachment of negative ions, 151–152, 159–161, 164–170, 174–176, 179, 186
  ion implantation
    annealing, 214
    damage-enhanced etching, 237–240
    ion-beam gettering, 232–236
    MOS devices, 252
Argon laser, annealing, in ion implantation, 229–230
Arsenic
  ion implantation, 192, 195, 201–202, 204–207
    annealing, 216–218, 221–229
    bipolar devices, 260–262, 264
    charge-coupled devices, 259
    damage-enhanced etching, 237–240
    MOS devices, 252–254
    radiation-enhanced diffusion, 250
Atomic mixing, in ion implantation, 208
Atomic negative ion, collisional detachment, 148–184
Attenuation technique, detachment cross section determination, 147

B

Ballistic transport, 349
Binary compound, intrinsic defects, 96–97, 103
Bipolar transistor, 192–193, 210, 225
  diffusion profiles, 224
  ion implantation, 260–264
Boltzmann transport equation (BTE), 347, 350–352
BOMOS, *see* High-density metal–oxide–semiconductor circuit
Boron
  ion implantation, 195–198, 204–205, 209
    annealing, 210, 214–216, 220–224, 230
    bipolar devices, 260–262, 264
    damage-enhanced etching, 237, 240
    MOS devices, 252, 254
Boron difluoride
  ion implantation, 209
    annealing, 210
    ion-beam gettering, 232–235
    MOS devices, 255–256
Bound exciton spectra, semiconductor donors and acceptors, 92–93
Bromine, collisional detachment of negative ions, 165–166, 171, 182–183
BTE, *see* Boltzmann transport equation

C

Cadmium
  luminescence bands in Cd-doped GaAs, 99–100
  in semiconductor donor–acceptor pairs, 93–94
Capture luminescence, in GaP semiconductors, 95
Carbon dioxide laser, annealing, in ion implantation, 226–231
Cation vacancy, 96–97

CCD, *see* Charge-coupled device
Čerenkov gas laser, 275
Čerenkov maser, 273–275
Čerenkov radiation, *see* Stimulated Čerenkov radiation
Cesium, collisional detachment of negative ions, 172–173
Charge-coupled device (CCD), ion implantation, 258–260
Charging–discharging asymmetry, MOS structures, 7
Chlorine, collisional detachment of negative ions, 164–172, 180, 182–183
Chromium, semiconductor impurity defects, 118–130
CMOS, *see* Combination metal–oxide–semiconductor
Cobalt, semiconductor impurity defects, 114
^{60}Cobalt radiation, 3, 6, 20
Collisional detachment, negative ion, 143–189
  atomic reactants, 148–177
  detachment rate constants, 176–177
  molecular reactants, 177–187
  nomenclature and experimental techniques, 146–148
Combination metal–oxide–semiconductor (CMOS), 252, 254
Copper
  luminescence bands in doped GaAs, 99–100
  semiconductor impurity defects, 112
Cylindrical waveguide, 285–289
Czochralski silicon, ion implantation, 234

D

Deep defect (deep center), semiconductors, 82
  assessment, 84
Deep-level transient spectroscopy (DLTS), point defect identification, 84–85
Depletion metal–semiconductor field-effect transistor, 378–379
Deuterium, collisional detachment of negative ions, 148–161, 163, 172–174, 178, 183–184
Dielectric resonator, Čerenkov radiation, 273–275, 285, 305–306
Dielectric theory of electronegativity in covalent systems, 362–364, 369–371, 373

Dielectric waveguide, Čerenkov radiation, 285–289, 291, 307–308
Differential cross section, of collisional detachment, 146–148
DIMOS transistor, *see* Double implanted metal–oxide–semiconductor transistor
Diode
  ion implantation, 257
  ion-beam gettering, 235
Direct band-gap semiconductor, 336
DLTS, *see* Deep-level transient spectroscopy
Donor, semiconductor, 86–96
  nickel–donor pairs, 131–134
  and vacancies, 101–103
Dopant, 82
  annealing dependence, in ion implantation, 214
  donors and acceptors, 86–96
  doping technology, 192–193
Double implanted metal–oxide–semiconductor (DIMOS) transistor, 254–255
DX center, 102

E

EBIC, *see* Electron beam-induced conductivity
Effective mass theory (EMT), 83, 87–89, 92
Elastic differential cross section, of collisional detachment, 146
Electrical conductivity, semiconductors, 83
Electron affinities, 144–145
Electron beam, Čerenkov radiation, 301–304
Electron beam-induced conductivity (EBIC), 69
Electron-beam lithography, 244, 246
Electron detachment cross section, in collisional detachment of negative ions, 146–147
Electron–hole generation and density, irradiated MOS structures, 8–9, 54–55, 67–70
Electron nuclear double resonance (ENDOR) spectroscopy, point defect identification, 86
Electron spin resonance (ESR)
  chromium in semiconductors, 120–124, 127–129
  point defect identification, 85–86, 97–98, 104–109
  vanadium in semiconductors, 130

Electron trap, *see* Trap and trapping mechanism
Electron wave function, collisional detachment of negative ions, 154–155
Electrostatic energy analysis, differential cross section measurement, 147
EMT, *see* Effective mass theory
ENDOR spectroscopy, *see* Electron nuclear double resonance spectroscopy
Energy, collisional, of negative ion, 146
Energy-band structure
 irradiated MOS devices, 55–56
 and materials considerations for very large scale integration, 361–375
Enhancement–depletion inverter circuit, 320–322, 332–333
Enhancement metal–semiconductor field-effect transistor, 379
Etching, damage-enhanced, in semiconductor technology, 236–240

F

Fabry–Perot interferometer, Čerenkov radiation, 306
FEL, *see* Free-electron laser
FET, *see* Field-effect transistor
Field-dependent mobility
 metal–oxide–semiconductor field-effect transistor, 322–325
 metal–semiconductor field-effect transistor, 336–338
Field-effect transistor (FET), 5; *see also* specific types of field-effect transistors
 gallium arsenide, 119
 ion implantation, 208, 257, 259
 radiation-induced behavior, 10
Field-emission diode generator, 303
Flat-band shift
 in irradiated MOS structures, 14, 16, 21, 44, 58, 60–61, 64–66, 68, 70, 72, 74
 dynamic equilibrium models, 46, 48–49, 51
 high-dose range, 15
Flat-band voltage
 in irradiated MOS structures, 14, 18, 22–40, 43, 45, 58–65, 72–74
 dynamic equilibrium models, 46, 48–49, 51–53
Float-zone silicon, ion implantation, 234–235

Flow-drift-tube technique, detachment rate constant measurement, 177
Flowing afterglow technique, detachment rate constant measurement, 177
Fluorine
 collisional detachment of negative ions, 171, 186
 ion implantation, MOS devices, 254–256
Free-electron laser (FEL), 307
Frenkel pair, 3
F-type center, 96
Fused silica, radiation effects, 6

G

Gallium aluminum arsenide, 374
Gallium arsenide, 336
 band structure and dependent properties, 363, 366–367, 370–374
 ion implantation, 257–259
 point defects, 81–141
  donors and acceptors, 87–90, 92–93
  intrinsic defects, 97, 99–103, 107–108
  rare earth metals, 134
  transition metals, 109, 112, 114–127, 130–134
 saturated velocity, 356–357
 short-channel effects, 339
 thermal conductivity, 382
 ultra-large-scale integration devices, 375
  metal–semiconductor field-effect transistor, 378–380
 velocity–field curves, 349–350, 356
 very large scale integration, 313
 junction field-effect transistor, 333–335
Gallium phosphide
 band structure and dependent properties, 363, 366–367, 372, 374
 point defects, 81–141
  donors and acceptors, 87–96
  intrinsic defects, 97–99, 101, 103–106, 108–109
  transition metals, 109, 112–117, 120, 127–134
Gate, floating, in irradiated MOS experiments, 21
Gate bias, irradiated MOS structures, 7–10
 high-dose range, 15

SUBJECT INDEX

Gate-bias voltage, in irradiated MOS structures, 14, 18, 20–40, 43–45, 58–70, 72–75
  dynamic equilibrium models, 46–49, 52–53
Germanium, 373–374
  point defects, 89
Gunn effect, 337

H

Hall effect, semiconductors, 84
Hall–Shockley–Read (HSR) trap site, 5
Halogen ion, collisional detachment of negative ions, 164–172, 177, 182–184
Helium, collisional detachment of negative ions, 148–161, 164–166, 179, 184–186
High-density metal–oxide–semiconductor circuit (BOMOS), 255–256
High-energy electron, irradiation of MOS structures, 8, 20
  oxide heating, 21
High-speed semiconductor logic, switching of, 340–361
  drive currents, 345–347
  effective saturation velocity, 355–357
  interconnection capacitance, 358–361
  transient switching, 341–345
  velocity response, 347–355
Hot-electron effects
  metal–oxide–semiconductor field-effect transistor, 323
  metal–semiconductor field-effect transistor, 336
HSR trap site, see Hall–Shockley–Read trap site
Hydrogen
  collisional detachment of negative ions, 148–164, 172–174, 178–181, 183–184, 186
  ion-beam lithography, 246–247

I

ICFE, see Intracollisional field effect
Indium arsenic phosphide, 374–375
Indium arsenide, band structure and dependent properties, 363, 366–367, 373
Indium gallium arsenic phosphide, 370
Indium gallium arsenide, 336, 367–375
  metal–semiconductor field-effect transistor, 378
  saturated velocity, 357
  thermal conductivity, 382
Indium phosphide, 336
  band structure and dependent properties, 363, 366–367
  metal–oxide–semiconductor devices, 376–377
  metal–semiconductor field-effect transistor, 378
  point defects, 81–141
    donors and acceptors, 88, 92–93
    intrinsic defects, 98, 101
    rare earth metals, 134
    transistion metals, 109, 114–115, 120, 127, 129
  saturated velocity, 356–357
  thermal conductivity, 382
Insulated-gate field-effect transistor, 5, 316
Integrated circuit
  in GaAs, 257–258
  gate-bias voltage, 22
  insulation, 248
  ion implantation, 192
Intracollisional field effect (ICFE), 351–353
Inverter circuit
  high-speed logic, 342–343
  metal–oxide–semiconductor field-effect transistor, 320–322
  metal–semiconductor field-effect transistor, 335–336
Iodine, collisional detachment of negative ions, 166
Ion, negative, collisional detachment, 143–189
  atomic reactants, 148–177
  detachment rate constants, 176–177
  molecular reactants, 177–187
  nomenclature and experimental techniques, 146–148
Ion-beam gettering, in ion implantation, 232–236
Ion-beam lithography, in ion implantation, 244–247
Ion implantation, for very large scale integration, 191–269
  annealing of implanted layers, 210–231
  application to devices, 251–264
  new applications, 231–250
  range distribution of implanted ions, 193–210
Ionization, collisional, of negative ion, 146

SUBJECT INDEX 409

Iron, semiconductor impurity defects, 114–115
Isoelectronic molecule, 94

J

Jahn–Teller distortion, of vacancy centers, 98
Junction field-effect transistor (JFET), 313, 332–336
  drive current, 346–347
  ion implantation, 257, 259
  scaling, 338–339
  short-channel effects, 339–340
Junction transistor, 5

K

Knock-on ion implantation, 204–208
  residual defects after annealing, 216
Kronig–Penney model of electron motion, 4
Krypton, collisional detachment of negative ions, 165

L

Large-scale integration (LSI), 192; *see also* Ultra-large-scale integration; Very large scale integration
Laser annealing, in ion implantation, 192, 225–231
Laser-induced alloying of metallic layers, in ion implantation, 231
LEC, *see* Liquid encapsulation Czochralski technique
Light-emitting diode (LED), 93
  nickel contamination, 113
Liquid encapsulation Czochralski (LEC) technique, 97
LOCOS (local oxidation of silicon) process, in ion implantation, 241–245, 253
Low-energy ion deposition, 249
LSI, *see* Large-scale integration
Luminescence
  gallium arsenide:chromium, 126–127
  point defect identification, 85

M

Magnesium, ion implantation, 257

Manganese
  luminescence bands in doped GaAs, 100
  semiconductor impurity defects, 116–118
Memory effects, in quantum kinetic equation of submicron devices, 351
MESFET, *see* Metal–semiconductor field-effect transistor
Metal–oxide–semiconductor
  buried-nitride fabrication, 249
  integrated circuit, 192
  ion implantation applications to devices, 251–256
  irradiation-induced changes, 1–79
    captured during transit (CDT) models, 42–45, 74
    complete computer simulation, 53–75
    device applications, 17
    dynamic equilibrium models, 41–42, 45–53, 73
    flat-band voltage and gate-bias voltage, 22–40
    high-dose range, 15–17, 19, 28–29, 31, 33–34, 74–75
    low-dose range, 15–16, 19, 28–29, 33, 74–75
    medium-dose range, 15–16, 19, 29–33, 74–75
    modeling considerations, 12–22
    relaxation behavior, 40
    simple descriptive models, 41–53
    transient phenomena, 41
    traps and trapping mechanisms, 18–20
    tunneling into insulator, 21
    ultra-large-scale integration devices, 375–378
Metal–oxide semiconductor capacitor, *c–v* curves before and after irradiation, 13–14
Metal–oxide–semiconductor field-effect transistor (MOSFET), 313–332
  drive current, 345–347
  field-dependent mobility, 322–325
  inverter circuit, 320–322
  scaling, 325–327
  short-channel effects, 327–332
Metal–oxide semiconductor (MOS) transistor, 192–193, 251–252
Metal–semiconductor field-effect transistor (MESFET), 313, 332–340
  depletion devices, 378–379
  drive current, 346–347

SUBJECT INDEX

enhancement devices, 379
field-dependent mobility 336–338
inverter circuit, 335–336
scaling, 338–339
short-channel effects, 339–340
Microstrip transmission line, 358–361
Microwave diode, 306
Millimeter-wavelength stimulated Čerenkov radiation, 304–306
Minority carrier lifetime, 83
MNOS memory field-effect transistor, 208
Mobile-carrier capture, in irradiated MOS structures, 42–45
Mobile-carrier generation, in irradiated MOS structures, 21–22
Molecular ion implantation, 209–210
Molecular reactant, collisional detachment of negative ions, 177–187
MOS, see Metal–oxide semiconductor
MOSFET, see Metal–oxide–semiconductor field-effect transistor
Mössbauer spectroscopic investigation, point defect identification, 86

N

NDC, see Negative differential conductivity
Near-infrared absorption
  gallium arsenide:chromium, 124–126
  gallium phosphide:chromium, 129
Negative differential conductivity (NDC), 337
Neodymium:YAG laser annealing, in ion implantation, 225–228
Neon
  collisional detachment of negative ions, 151–153, 155, 165–166, 168, 175–176, 179
  ion implantation, insulated structures, 249
Nickel, in semiconductor
  impurity defects, 112–113
  nickel–donor pairs, 131–134
Nitrogen
  collisional detachment of negative ions, 178–183
  ion implantation
    annealing, 236
    insulated structures, 248–249
    LOCOS process, 241–245, 253
NMOS, see n-type-channel metal–oxide–semiconductor

NMR, see Nuclear magnetic resonance
n-type capacitor, effect of radiation on, 7
n-type-channel metal–oxide–semiconductor (NMOS), 252–253
Nuclear magnetic resonance (NMR), point defect identification, 86

O

Optical detection of magnetic resonance (ODMR), point defect identification, 86
Optical spectroscopy, point defect identification, 85
Oxidation, LOCOS process, in ion implantation, 241–245
Oxide-thickness dependence, in irradiated MOS structures, 44, 48–49
Oxidizing annealing, in ion implantation, 214–215
Oxygen
  collisional detachment of negative ions, 174–176, 178–180, 183–185
  ion implantation, 205–206
  in semiconductor donor–acceptor pairs, 93–94

P

Parasitic capacitance, 361, 384–387
Phase space theory, for collisional detachment, 177
Phosphorus
  ion implantation, 199, 204–205
  annealing, 214, 216, 218
  bipolar devices, 260–262, 264
  charge-coupled devices, 260
  ion-beam gettering, 232–233
  MOS devices, 252
Phosphorus–silicon–glass (PSG), window tapering, 237, 240
PMMA, ion-beam lithography, 246–247
Point defect, semiconductor, 81–141
  after annealing of ion implants, 216–219
  assessment, 84–86
  donors and acceptors, 86–96
  intrinsic defects, 96–109
  3d transition metals, 109–134
Ponderomotive wave, 307
Position-sensitive channel plate detector, differential cross section measurement, 147–148

SUBJECT INDEX

Potassium chloride, collisional detachment of negative ions, 171–172
Potential barrier, in irradiated MOS structures, 55–57
Power-delay considerations, in logic technologies, 379–382
Pseudobinary alloy, 367–373
PSG, see Phosphorus–silicon–glass

Q

Quartz, radiation effects, 6
Quaternary alloy, 368, 370–373

R

Radiation
  crystalline solids, 6
  metal–oxide–semiconductors, 1–79
    complete computer simulation, 53–75
    flat-band voltage and gate-bias voltage, 22–40
    modeling considerations, 12–22
    simple descriptive models, 41–53
    stimulated Čerenkov radiation, 271–310
Radiation damage, in ion implantation, 231
Radiation-enhanced diffusion (RED), in ion implantation, 250
Radiation-hardened device, 17
Rare earth metals, point defects in semiconductors, 134
Rare gas, collisional detachment of negative ions, 148–161, 164–169, 174–177
Recoil ion implantation, 204–208
Recombination, in semiconductors, 83
RED, see Radiation-enhanced diffusion
Resistivity, semiconductors, 84
Retarded transport equations, 353–355
Ruby laser annealing, in ion implantation, 225–226

S

Saturated velocity
  metal–oxide–semiconductor field-effect transistors, 323–325, 345
  Schottky-gate field-effect transistor, 383
  short-channel devices, 356–357
  silicon devices, 384
Schottky-gate field-effect transistor, 383
  ion implantation, 257

Schwarz–Christoffel conformal transformation, 358
Selenium, ion implantation, 257
Self-activated luminescent center (A center), 101
Self-aligned gate, ion implantation, 251–252
Semiconductor, see also specific types of semiconductors
  ion implantation, 191–269
  point defects, 81–141
  radiation-induced degradation, 12
Sequential gate bias, irradiated MOS structures, 61–67
Shallow defect (shallow center), semiconductors, 82–83
  assessment, 84–85
  donors and acceptors, 87–92
Shock tube technique, detachment rate constant measurement, 177
Short-channel effects, in metal–oxide–semiconductor field-effect transistor, 327–332, 339–340
Short-wavelength Čerenkov devices, 307–309
Silica, fused, see Fused silica
Silicon
  band structure and dependent properties, 361–365, 373–374
  electron transit time, 343
  ion implantation, 194–198, 201, 203–206, 209
    annealing, 210–218, 220, 225–227, 229
    charge-coupled devices, 259
    damage-enhanced etching, 239–240
    insulated structures, 248–249
    ion-beam gettering, 232–236
    LOCOS process, 241–245
    radiation-enhanced diffusion, 250
  oxidized, effects of ionizing radiation on, 6
  point defects, donors, 89
  saturated velocity, 356–357, 384
  thermal conductivity, 381–382
  transient response, 354–355
  ultra-large-scale integration devices, 375–380
  vacancy centers, 97–98
  velocity response of electrons, 349, 352–353
  very large scale integration, 312–314
Silicon dioxide
  electron mobilities in, 69, 71
  ion implantation, 195–198, 205–207
    annealing, 216–218, 220, 225

charge-coupled devices, 259
damage-enhanced etching, 237–240
LOCOS process, 241
MOS devices, 253–254
intracollisional field effect, 352
radiation effects on, 12, 20
  hole trapping by absorbed radiation, 11
  ion transport phenomena, 7
Silicon nitride, ion implantation, 195, 205–207
  damage-enhanced etching, 237–238, 240
  insulated structures, 248–249
  LOCOS process, 241
  Schottky FET, 257
Slab waveguide, 275, 288–289, 291–294, 308
Slow-wave structure, Čerenkov maser, 273
Sodium chloride, collisional detachment of negative ions, 169–172
Sodium hydride, collisional detachment of negative ions, 172–174
Sputtering, in ion implantation, 200–204
Stimulated Čerenkov radiation, 271–310
  beam velocity spread, 294–300
  bounded structures, 285–294
  experiments, 301–309
  gain from unmagnetized beam, 283–285
  gain on strongly magnetized beam, 276–283
  nonlinear behavior, 300–301
  theory, 275–301
Submicron very large-scale integration, 311–390
  band structure considerations, 361–375
  metal–oxide–semiconductor field-effect transistors, 314–332
  metal–semiconductor field-effect transistor, 332–340
  switching of high-speed logic, 340–361
Sulfur
  ion implantation, 257
  single-crystal, electron mobilities in, 71
Surface properties
  oxidized silicon, 6
  semiconductors, 7

T

Ternary alloy, 367–368
Thermal considerations, in high-speed circuits, 380–383
Time-of-flight method, differential cross section measurement, 147–148

Transformer, millimeter-wavelength stimulated Čerenkov radiation, 304
Transistor, 4; *see also* specific transistors
  effect of low ionizing irradiation on, 9–11
  electron-beam irradiation, 7
Transition metals, point defects in semiconductors, 109–134
Trap and trapping methanisms
  chromium-doped gallium arsenide, 119–120
  detachment cross section determination, 147
  Hall–Shockley–Read trap sites, 5
  intrinsic defects of binary compounds, 96
  irradiated MOS structures, 11, 18–20, 42–43, 54, 65, 67–71, 73–74
    dynamic equilibrium models, 46–47, 52
Traveling-wave tube, Čerenkov maser, 273
Tungsten, ion implantation, 206–207

U

Ultra–large-scale integration (ULSI), 313, 375–387
  cooled logic, 383–384
  device–device interactions, 384–386
  device parameters, 376–380
  thermal considerations, 380–383
Ultraviolet radiation, MOS structures, 11
Uranium hexafluoride, collisional detachment of negative ions, 186

V

Vacancy-related defect, binary compounds, 96–103
Vanadium, semiconductor impurity defects, 130
Very large-scale integration (VLSI)
  ion implantation, 191–269
    annealing of implanted layers, 210–231
    application to devices, 251–264
    new applications, 231–250
    range distribution of implanted ions, 193–210
  materials considerations for submicron region, 311–390
    band structure considerations, 361–375
    logic, comparisons and limitations of, for ultra large-scale integration, 375–386

metal–oxide–semiconductor field-effect transistors, 314–332
metal–semiconductor field-effect transistor, 332–340
switching of high-speed logic, 340–361
V-type center, 96–97

X

Xenon, collisional detachment of negative ions, 168–169

Z

Zero-radius model, collisional detachment of negative ions, 153–155, 157
Zinc
  luminescence bands in Zn-doped GaAs, 99–100
  in semiconductor donor–acceptor pairs, 93–94
  vacancies, 97–98, 101
Zincblende structure, 361–362, 367

RETURN TO ➡ ENGINEERING LIBRARY

642-3366

LOAN PERIOD 1	2	3
4	5	6

ALL BOOKS MAY BE RECALLED AFTER 7 DAYS
Overdues subject to replacement charges

DUE AS STAMPED BELOW

JUL 15 1982

SEP 24 1983

DEC 3 1983

AUG 10 1984

JUL 26 1985

JUN 6 1986

FORM NO. DD11, 80m, 8/80 UNIVERSITY OF CALIFORNIA, BERKELEY
BERKELEY, CA 94720